Nanostructure Science and Technology

Series Editor:
David J. Lockwood, FRSC
National Research Council of Canada
Ottawa, Ontario, Canada

For further volumes:
http://www.springer.com/series/6331

Yaser Abu-Lebdeh • Isobel Davidson
Editors

Nanotechnology for Lithium-Ion Batteries

Springer

Editors
Yaser Abu-Lebdeh
NRC Energy, Mining and Environment Portfolio
National Research Council of Canada
Ottawa, Ontario
Canada

Isobel Davidson
NRC Energy, Mining and Environment Portfolio
National Research Council of Canada
Ottawa, Ontario
Canada

ISSN 1571-5744
ISBN 978-1-4614-4604-0 ISBN 978-1-4614-4605-7 (eBook)
DOI 10.1007/978-1-4614-4605-7
Springer New York Heidelberg Dordrecht London

Library of Congress Control Number: 2012946720

Printed on acid-free paper

Springer is part of Springer Science+Business Media (www.springer.com)

Preface

The rapid progress in nanotechnology and its tremendous achievements in many fields of science have motivated researchers in materials science, especially those working towards applications in energy, to adopt some of these technologies in their research. Li-ion battery researchers, who are highly motivated by the successful introduction of Li-ion batteries in many consumer electronic devices and their projected use in large-scale storage for the electrical grid and electric vehicles, are no exception, and this adoption has led to profound enhancement of battery performance.

The motivation for putting this book together is that in the last few years there has been a large number of publications and reviews on the subject scattered all over the literature but there is simply no single book that consolidates the developments in all areas of research related to nanotechnology and lithium ion batteries. The combination of the two fields in this book will not only give an insight into how nanotechnology can be used to improve Li-ion battery technology but also serve as a reference textbook for researchers, students, and professors or any person with a desire to understand and explore the relationship between these two interesting scientific fields.

The book starts with a chapter that covers the basic concepts of Li-ion battery technology and the subject of nanotechnology and also the fundamental aspects of ionic materials at the nanoscale and highlights the key differences between them and conventional-scale materials. Chapter 2 discusses the role of nanoarchitectured electrode materials in the performance of Li-ion batteries, while Chapters 3, 4, 5, and 6 demonstrate in detail the role of anode materials that are based on metals, metalloids, or metal oxides and capable of alloying with lithium as alternatives to graphitic conventional anodes when made as nanoparticles or nanostructures thus enhancing the rate capability of the batteries. In some cases, as shown in Chapters 5 and 6, it was demonstrated that certain metal oxides can become active towards lithium only as a result of the "nano" effect. Chapters 7 and 8 describe the benefits of nano-sized cathode materials for example in overcoming some limitations of Li-ions intercalation in lithium metal oxide and phosphate materials and improving their power performance. Chapter 9 is concerned with the work performed on the effect of incorporating nanoparticles on the properties of electrolytes and to a lesser extent the electrode-electrolyte interface. Finally, Chapter 10 gives an outline of the work on interdigitated electrodes and 3D micro-batteries.

The editors would like to thank everyone who helped during the course of the preparation of this work, particularly all the contributors to the book.

Ottawa, Canada — Yaser Abu-Lebdeh
Ottawa, Canada — Isobel Davidson

Contents

Contributors

Ali Abouimrane Argonne National Laboratory, Lemont, CA, USA

Yaser Abu-Lebdeh NRC Energy, Mining and Environment Portfolio, National Research Council of Canada, Ottawa, ON, Canada

Fabrice M. Courtel Atomic Energy of Canada Limited, Chalk River, ON, Canada

Isobel Davidson National Research Council of Canada, Ottawa, ON, Canada

Hugues Duncan Lawrence Berkeley National Laboratory, Berkeley, CA, USA

J.B. Goodenough The University of Texas at Austin, Austin, TX, USA

Phil Johns School of Chemistry, University of Southampton, Southampton, Hampshire, United Kingdom

C.M. Julien Physicochimie des Electrolytes, Colloïdes et Sciences Analytiques (PECSA), Université Pierre et Marie-Curie, Paris, France

Nikhil Koratkar Department of Materials Science and Engineering, Department of Mechanical, Aerospace and Nuclear Engineering, Rensselaer Polytechnic Institute, Troy, NY, USA

Rahul Krishnan Department of Materials Science and Engineering, Rensselaer Polytechnic Institute, Troy, NY, USA

Nathalie Lavoie National Research Council of Canada, Ottawa, ON, Canada

Toh-Ming Lu Department of Physics, Applied Physics, and Astronomy, Rensselaer Polytechnic Institute, Troy, NY, USA

Justin C. Lytle Chemistry Department, Pacific Lutheran University, Tacoma, WA, USA

Patrick R.L. Malenfant National Research Council of Canada, Ottawa, ON, Canada

A. Mauger Institut de Minéralogie et Physique de la Matière Condensée (IMPMC), Université Pierre et Marie Curie-Paris-6, Paris, France

Rahul Mukherjee Department of Mechanical, Aerospace and Nuclear Engineering, Rensselaer Polytechnic Institute, Troy, NY, USA

John Owen School of Chemistry, University of Southampton, Southampton, Hampshire, United Kingdom

Matthew Roberts School of Chemistry, University of Southampton, Southampton, Hampshire, United Kingdom

Nuha Salem National Research Council of Canada, Ottawa, ON, Canada

K. Zaghib Institut de Recherche d'Hydro-Québec (IREQ), Varennes, QC, Canada

Chapter 1
Introduction

Yaser Abu-Lebdeh

Abstract This chapter gives brief introductions to the basic concepts of lithium-ion battery technology, the subject of nanotechnology, and goes further to give insight on the relationship between the two. It also attempts to demonstrate how nanotechnology can improve lithium-ion battery technology. It is emphasized that despite the short life span of lithium-ion battery technology (~20 years) and nanotechnology (~30 years) and the limited knowledge of behavior at the nanoscale, great advances have been made in understanding the relationship between performance and the use of nanofabricated lithium-ion materials and batteries.

1.1 Energy Conversion and Storage: A Global Challenge

The world is witnessing an ever increasing demand for energy in order to meet the needs of modern societies. There are different sources for energy, and they vary from one country to another, but globally it is based on fossil fuel. There are many problems associated with the use of fossil fuels (oil, coal, and gas), most important of which is the release of greenhouse gasses (GHGs), mainly CO_2, that are responsible for not only air pollution and associated health problems but also the devastating effects of raising the Earth's temperature in what has become known as global warming. Also, fossil fuels are finite and nonrenewable source of energy. It is therefore imperative to find alternative energy sources to replace, or at least complement, the use of fossil fuels so that the effects of GHGs can be mitigated by lowering their levels in the atmosphere. This can be achieved by increasing energy efficiency and conservation of processes while utilizing current energy sources. However, it could be more effective by adopting less-polluting, less-harmful, and abundant alternative energy sources such as solar, wind, ocean,

Y. Abu-Lebdeh (✉)
National Research Council of Canada, Ottawa, ON K1A 0R6, Canada
e-mail: Yaser.Abu-Lebdeh@nrc-cnrc.gc.ca

Y. Abu-Lebdeh and I. Davidson (eds.), *Nanotechnology for Lithium-Ion Batteries*, Nanostructure Science and Technology, DOI 10.1007/978-1-4614-4605-7_1,

biomass, geothermal, hydroelectric, and to some extent nuclear energy. Solar and wind are attracting most of the attention due to their ability to produce huge amounts of electrical energy around 10^{12} Wh, but they are intermittent in nature and therefore require a storage system for the converted energy so it can be used when needed [1].

Energy can be converted from one form to another and can also be stored in different forms [2]. Electrical energy, the most widely used in modern times, is mainly produced by power plants that run on fossil fuels, nuclear reactors, or hydroelectric power. Electrochemical devices such as batteries are one popular form of storing electricity in the form of chemical energy and reverse the process by converting the chemical energy to electricity. This allowed for the concept of portable electricity which enabled not only the current revolution in electronic devices but also to a revolution in the automotive industry and the realization of the electric car which is powered fully or partly by a battery. Also, with the invention of high energy density batteries, very large systems can be used to facilitate a more efficient transmission and delivery of electricity by the grid and its integration with solar and wind energy sources [3].

1.2 Batteries: The Ultimate Electrical Energy Storage Device

Batteries have been known for more than two centuries since the pioneering work of Galvani and Volta. They only gained great importance with the discovery of the telegraph, radio, and cars and became very popular in modern days because of their use in most consumer electronic devices, electric cars, and energy storage for the electrical grid. A battery is made of two electrodes (a positive and a negative) separated by an electrolyte with the battery voltage being the difference in the potential buildup between the two electrodes. They can be primary (nonrechargeable) or secondary (rechargeable). The latter can be discharged and recharged over a number of times (cycle life). There are various types of batteries, and they differ by the chemistry of each electrode material and electrolyte. However, only a very few batteries have found great success especially those that are rechargeable due to the ability to reverse the electrochemical reactions at both sides of the electrodes. The four most-used types of commercial rechargeable batteries are lead-acid introduced by G. Planté in 1850, nickel-cadmium introduced by W. Jungner in 1899, nickel-metal hydride commercialized in 1989, and lithium-ion commercialized in 1991 [4]. Table 1.1 summarizes the physical and chemical properties of each battery, while Fig. 1.1 compares their energy densities. Both clearly demonstrate the unequivocal dominance of lithium-ion battery technology over the other batteries particularly for having the highest energy density and cycle number [5].

Table 1.1 Summary of the physical and chemical properties of the four most-used commercial rechargeable batteries

Battery type	Negative electrode	Positive electrode	Electrolyte	Voltage (V)	Cycle life (#)
Lithium-ion	Graphite	$LiCoO_2$	$LiPF_{6(nonaqueous)}$	3.7	>1,000
Lead-acid	Pb	PbO_2	$H_2SO_{4(aqueous)}$	2.1	<500
Ni-Cad	Cd	NiOOH	$KOH_{(aqueous)}$	1.2	2,000
NMH	Intermetallic	NiOOH	$KOH_{(aqueous)}$	1.2	500–1,000

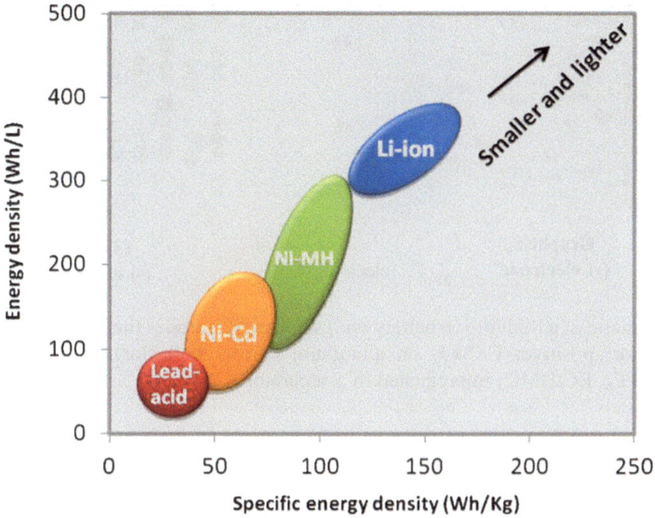

Fig. 1.1 Comparison of the different rechargeable battery technologies in terms of volumetric and gravimetric energy density

1.3 Lithium-Ion Batteries

Lithium-Ion batteries are based on the principle of "electrochemical intercalation" which has been known for two decades and were under research and development in many industrial and academic institutions around the world before their successful commercialization by Sony in 1991 and Asahi Kasei and Toshiba in 1992 [6–8]. Earlier efforts were focused on lithium metal as a negative electrode due to its high electrochemical capacity (3.86 Ah/g), very negative standard redox potential (−3.04 V), and low density (0.53 g/cm^3), but it was soon realized that it lowered the performance of the batteries (lower capacity and cycle life) and made them unsafe due to the formation of dendrites upon recharging. It was later replaced by graphitic carbon. It had already been demonstrated that graphitic carbon can intercalate lithium ions electrochemically and reversibly at low potentials that can reach values very close to that of lithium metal but with ten times lower capacities

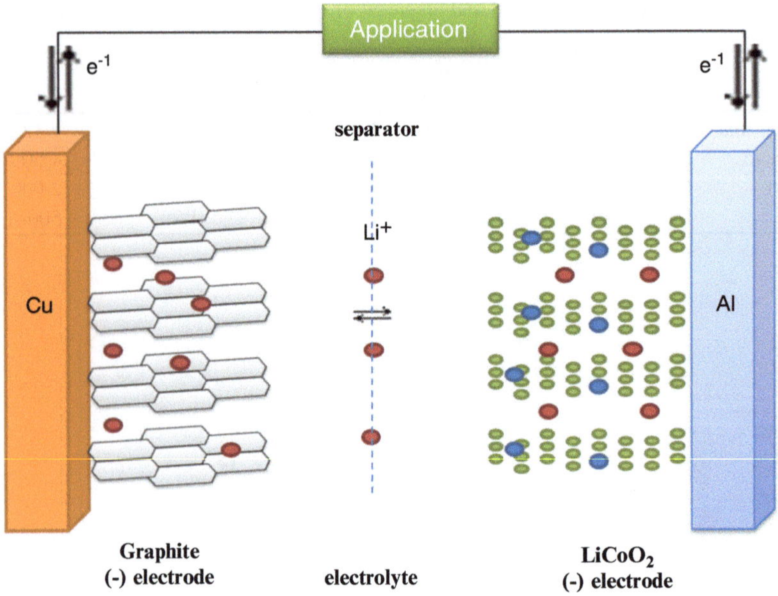

Fig. 1.2 Schematic of a lithium-ion battery with standard electrodes (negative: graphite on copper current collector; positive: $LiCoO_2$ on aluminum current collector) in a nonaqueous liquid electrolyte ($LiPF_6$, EC:DMC) impregnated in a separator

(0.372 Ah/g). Similarly, electrochemical intercalation of Li ions into crystalline layered structures of materials that work as the positive electrode was demonstrated for transition metal sulfides (TiS_2) by M. Whittingham and later by J. Goodenough in oxides ($LiCoO_2$). Herein, a redox reaction of the transition metal (M^n/M^{n+1}, where M is mostly Fe, Mn, Ni, or Co) takes place that along with other factors control the potential of the material that has to be practically high (>3.5 V) for it to act as a positive electrode. The story for the electrolyte was somehow more straightforward as aqueous solutions were not considered due to the high reactivity of water towards lithium metal and also its limited liquid range (100°C) and electrochemical stability (1.23 V) range. Electrolyte solutions of a lithium salt in nonaqueous, aprotic organic solvents with excellent chemical and physical properties were found to successfully operate in a lithium-ion battery. The solvent was commonly a cyclic carbonate like propylene carbonate that was initially used but sooner replaced by ethylene carbonate mixed, due to its high melting point, with a low viscosity linear carbonate solvent of the type ROCOOR where R is either a methyl, an ethyl or both.

Figure 1.2 shows a generic schematic of a lithium-ion battery illustrating its various components and the movement of lithium ions. The negative electrode "anode" in most commercial batteries is a graphitic carbon with theoretical and experimental capacities of 372 and 330 mAh/g, respectively. The positive electrode "cathode" is a layered oxide such as $LiCoO_2$ with theoretical and experimental capacities of 274 and 140 mAh/g, respectively. The electrolyte is a lithium salt ($LiPF_6$) dissolved in a mixture of nonaqueous, aprotic carbonate solvents.

During initial battery cycling (charge/discharge), two multicomponent (organic and inorganic), multilayer passivation films form at each electrode known as the solid electrolyte interface (SEI) [1] or the cathode-electrolyte interface (CEI). These are important in protecting the electrolyte from further reactions with the electrodes. During cycling, lithium ions move back and forth through the ionically conducting electrolyte between the layered crystalline structures of $LiCoO_2$ and graphite, passing through the passivation surface films, in a reversible electrochemical reaction that is accompanied by electron transfer inside and outside the battery as illustrated in the following chemical reactions:

$$LiCo^{III}O_2 \rightleftarrows Li_{1-x}Co^{IV}O_2 + xLi^{+1} + x\ e^{-1} \qquad 0<x<0.5 \qquad E^{\circ} = 3.9V$$

$$6C + xLi^{+1} + x\ e^{-1} \rightleftarrows Li_xC_6 \qquad x = 1 \qquad E^{\circ} = 0.2V$$

$$6C + LiCo^{III}O_2 \rightleftarrows Li_xC_6 + Li_{1-x}Co^{IV}O_2 \qquad E^{\circ} = 3.7V$$

The potential, E, changes as a function of the activity of Li ions, a_{Li+}, according to Nernst equation:

$$E = E^{\circ} - RT/nF\ ln\ a_{Li+}$$

A lot of materials development has taken place since the commercialization of lithium-ion batteries in 1991. Positive electrodes based on structures besides the layered oxides were introduced commercially such as the spinel $LiMn_2O_4$ and the polyanion olivine $LiFePO_4$ and their derivatives with cation/anion substituent or modified surface. As for the negative electrode, alloy composites such as (Sn/Co/C) were introduced in commercial batteries with 30% higher capacities than graphite while keeping low voltages.

The amount of total charge "capacity" that a battery can deliver and its retention over frequent cycling (cycle life) depend on the reversibility of the above reactions and the occurrence of side reactions. The reversibility is usually represented in the form of coulombic efficiency (= discharge capacity/charge capacity). Like all other batteries, lithium-ion ones can be made in different sizes and shapes depending on the application. The voltage for each cell is the same, while the capacity is different. If large batteries are required, cells are stacked into modules that are in turn stacked into packs in series (to maximize voltage) and parallel (to maximize capacity or current) combinations to reach the required energy density (Wh) [9]. Some of the battery components (casing, current collector, binder, and separator) are not active and do not contribute to the electrochemical reaction and obtained capacity. This, also known as "dead weight," can reach up to 30–50% of the battery total weight and lowers the energy density. It is therefore important to design the battery so that the active material content in the electrodes is maximized while the inactive is minimized. Power density is maximized by packing the materials as efficiently as possible [10]. Figure 1.3 shows the relationship between the energy density requirements for an application to the format of the lithium-ion battery used.

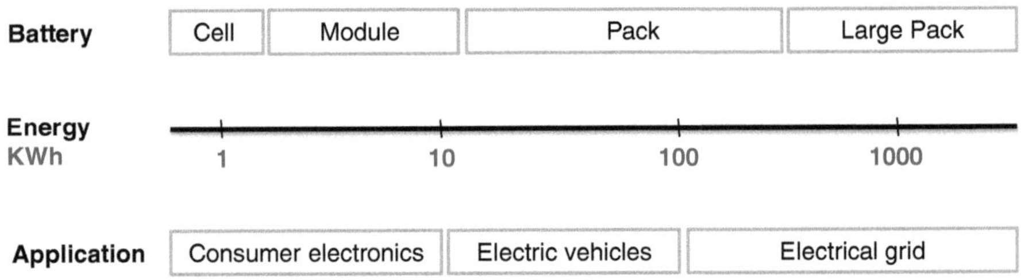

Fig. 1.3 Schematic that illustrates the relationship between the battery format and energy density requirement for three main applications of lithium-ion batteries

In small single cells, the energy density is less than 1 KWh, which is enough to power small communication/electronic devices such as cell phones. Modules of typically six cells connected in series/parallel to increase energy density are used to power laptops. Making battery packs follows a similar fashion, but usually flat prismatic or pouch cells are preferred because of better heat transfer and packing efficiency. These packs which typically have an energy density of 10–30 KWh are used to power commercialized electric vehicles such as Nissan Leaf and Chevy Volt. Even bigger battery packs with energy densities greater than 1 MWh are used as electrical energy storage for the electrical grid to level the load and to help in integrating the renewable solar and wind energy sources into the electrical grid.

1.4 Nanotechnology

The pioneering and visionary work of R. Feynman on the possibility of manipulating and controlling "things" on a small scale led to the emergence of the field of nanotechnology [11]. This was later elaborated upon by K. Drexler [12] with focus on the concept of molecular assemblers. The term nanotechnology was first coined by N. Taniguchi in 1974 to describe "extra high accuracy and ultra fine dimensions" to control semiconductor processes, but since then, more precise and descriptive definitions of the term have emerged [13]. B. Fahlman stated that the definition given by NASA seems to be the most acceptable [14]; it defines nanotechnology as "the creation of functional materials, devices and systems through control of matter on the nanometer length scale (1–100 nanometers), and exploitation of novel phenomena and properties (physical, chemical, biological, mechanical, electrical, etc) at that length scale." So, in essence, one deals with a phenomena at a one billionth (10^{-9}) of a meter. Ancient and medieval civilizations had made use of nanotechnology unintentionally in arts to color glass or in war to make sharp swords. Chemists have worked with what are now known as nanomaterials such as monolayers and metal nanoparticle colloids for a long time, but it was not until the mid-1980s with the discovery of sophisticated characterization techniques such as scanning tunneling microscopy (STM) and atomic force microscopy (AFM) and

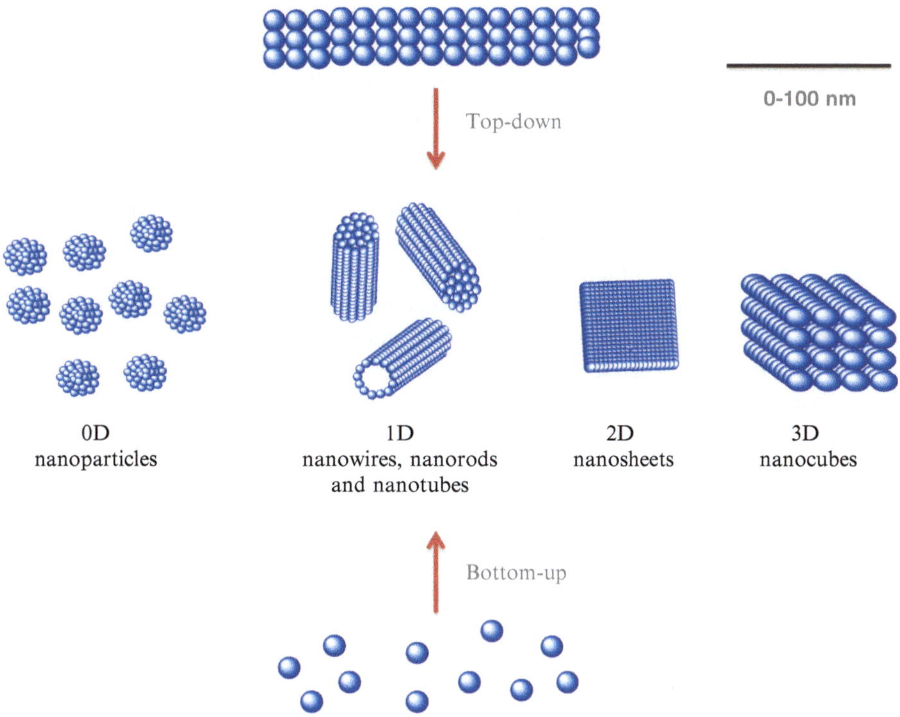

Fig. 1.4 Schematic of various types of nano-sized materials synthesized by top-down or bottom-up chemical or physical methods

later high-resolution transmission electron microscopy (HRTEM) that it was possible to identify materials and later processes and devices at the nanoscale.

The last two decades have witnessed a great interest from various sections of the societies (academia, industry, government, and public) around the world to understand and explore the benefits of nanotechnology. The two most important aspects of nanotechnology are the ones that deal with the development of new materials and devices with reduced "nano" dimensions. In the last two decades, a wide variety of nanomaterials (sized or structured) were prepared in different forms by either physical or chemical "bottom-up" or "top-down" methods as shown in Fig. 1.4 [15].

The most interesting part about these structures at the nanoscale is that size-related phenomena and processes can occur, and hence changes in properties or development of new ones arise only because of size, that cannot be seen in their large-scale counterpart. Some of the observed changes are in the chemical or physical (optical, mechanical, electrical, or magnetic) properties that are brought about by the fact that the surface to volume ratio is high. The surface properties dominate over the bulk, and also quantum effects due to confinement of the electron occur. For example, the large increase in the surface area, accompanied with formation of different crystalline facets, leads to an increase in chemical reactivity, especially enhancement of catalyzed chemical reactions. Another example is the change in the melting point and color of some metals when they become nanoscale in size such as gold whose melting point drops from 1,336 K (bulk) to 310 K for 6-nm particles, and its color changes from bright yellow to red, passing through a wide variety of colors in between [16]. Another is the exceptional mechanical properties of some carbon nanostructures like fullerenes, nanotubes, and graphene that are much stronger and lighter than common structural materials like steel. Finally, changes in electronic and electrochemical properties of some metals, carbons, and ionic compounds were found to change greatly at the nanoscale in terms of magnitude or function.

The benefits of nanomaterials and nanotechnology in general are yet to be fully realized, but some products have already made it to the markets such as in cosmetics. If one sets aside the hype on harder-to-achieve applications related to some aspects of nanotechnology such as Drexler's molecular engineering [12], efforts are focused now on more realistic applications. For example, applications in medicine for drug delivery, diagnostic, and therapeutic effects such as the use of gold nanoparticles for tumor treatment. Another important application is in energy conversion and storage such as the use of metal chalcogenides in photovoltaic cells or nanotubes, graphene, and metal/metal oxide nanoparticles in batteries, capacitors, and fuel cells. Also, applications in the fabrication of highly performing bulk materials such as carbon nanotubes and graphene composite materials, or in electronics such as the use of graphene and nanotubes in transistors.

1.5 Nanotechnology for Lithium-Ion Batteries

The application of nanomaterials and nanotechnology principles in electrochemical energy conversion and storage devices in general caught on at a very early stage due to the obvious benefits in catalysis for fuel cells or enhancing charge and discharge in batteries and capacitors. Lithium-Ion battery technology was no exception, and as discussed in the previous section, one or two of the electrodes are mixed ionic or electronic solid conductors and the electrolyte is an ion conductor. Therefore, it is very important to understand the behavior of ions and electrons when confined in solid materials at the nanoscale and its effect on battery performance.

In the last two decades, there have been rapid advances in the field of nano-electronics to understand the behavior of confined electrons in solids and to a lesser extent in nano-ionics for the behavior of confined ions in solids. The term nano-ionics was first introduced in 1992 by Despotuli and Nikolaichik [17], and since then it was developed to define the study and application of phenomena, properties, effects, and mechanisms of processes connected with fast ion transport in all solid-state nanoscale systems [18, 19]. At the nanoscale, the behavior of ionic solids is dominated by the interface, giving rise to two types of nano-sized effects: trivial and true that alter the kinetic and thermodynamics of the solid material [20]. In the former, any observed change in the behavior of the material is due to increased surface to bulk ratio and the high proportion of the interface. This is evidenced in the observed enhanced conductivity in nanocrystalline CaF_2 compared to its bulk counterpart or the increase in energy due to excess surface energy such as in the excess electromotive force of 62 mV for the 25-nm nanoparticles of rutile (TiO_2) compared to the bulk material. However, the true nano-effects arise only when the dimension is much smaller than the interspace between nanomaterials by at least a factor of four leading to an overlap of space charge layers [21]. This results in various structural changes and redistribution (accumulation or depletion) of charge carriers (ions, electrons, defects) at the boundary. These result in unusually high ionic conductivity such as in $SrTiO_3$ particles, CaF_2/BaF_2 hetero-structure or AgI when mixed with insulating Al_2O_3, or a change in the nature of conductivity from ionic to electronic in the case of CeO_2 [21].

There is little known about the behavior of most ionic solids at the nanoscale in general and those used in lithium-ion batteries in particular. It is, however, expected that similar changes to the energy, kinetics, and transport properties of battery materials, mostly nanoparticles and nanofilms, will take place at nanoscale. Moreover, the most important property of a lithium-ion battery electrode material is its ability to allow lithium ions to pass quickly and reversibly by intercalation, alloying, or conversion reactions in and out of the negative electrode and by intercalation in and out of the positive electrode. In general, the benefits of using nanomaterials in lithium-ion batteries or the fabrication of small "micro or nano" batteries will have the following benefits:

1. Kinetics: Small dimensions will provide a short path for the ion and electron movement in the ionic solid electrodes. Therefore, reactions become faster and batteries can be charged and discharged very quickly. Moreover, the high surface area of the electrode allows for improved contact with the electrolyte that has the advantage of allowing faster movement of Li ions in and out of the electrode from and to the electrolyte. However, it has the disadvantage of having more side decomposition reactions of the electrolyte at the electrode surface. In some cases, small dimensions overcome intrinsic slow diffusion of ions such as in the case of $LiFePO_4$ negative electrode which also becomes electro-active with the application of a thin, porous, conducting surface layer by nano-painting.
2. Thermodynamics: Changes in the potential of certain materials that are inactive to lithium can occur when synthesized at the nanoscale or become nano-sized during battery cycling. When the correct potential is reached, the nanomaterials become active unlike the bulk material. For example, Li_2O and LiF become reactive at the nanoscale but are inactive in the bulk as they are known to be poor ionic and electronic conductors [22]. When certain transition metal oxides (TMO) or fluorides (TMF) react electrochemically with lithium ions, M/Li_2O and M/LiF nano-composites form, and hence Li_2O and LiF can be decomposed and form reversibly.
3. New mechanism for lithium-ion storage: It was shown that excess lithium-ion storage at the interface of boundary regions takes place by a "capacitive-like" charge separation between nano-sized metal and Li_2O grains until the lithium potential approaches the value of pure lithium [23]. This was observed in batteries utilizing RuO_2 nanoparticles as anode material.

4. Enhanced mechanical properties: Lithium-ion storage inside and outside the electrode involves large changes in volume of the electrode material. Nanomaterials are able to accommodate these changes due to faster stress relaxation times and therefore retain the battery high capacity and improve its cycle life. A very good example is silicon whose volume increases by 300% when its atoms form an alloy with lithium ions. When bulk silicon is cycled, the high capacity of 4,200 mAh/g cannot be sustained for more than 30 cycles, but when nanoarchitectured silicon (e.g., nanowires) was used, the high capacity was retained for hundreds of cycles [24].

Scrosati et al. [25] have summarized the advantages and disadvantages of nanomaterials in lithium-ion batteries and pointed out that the synthesis of nanoparticles can be difficult and hard to control and also that their density is low which might lead to lower energy density (Wh/L). They also pointed out that like any new invention that is not fully understood, there is potential hazard to humans and environment, and more studies of their impact are required as more nanomaterials are being developed.

1.6 Summary and Outlook

The successful adoption of lithium-ion batteries in most consumer electronic devices and the huge interest in integrating them in electric vehicles or large energy storage for the electrical grid have made the demand and the research and development in this type of battery technology like never before. Nanotechnology took the world by storm due to the huge potential for applications in various aspects of our lives, but the real benefits have only started to emerge. It has been demonstrated in this chapter that there is still a great deal of research and development to be done in order to understand the behavior of materials at the nanoscale. This is especially true for ionic materials in order to understand nano-ionic effects and how they affect physical and chemical properties of electrode and electrolyte materials. However, certain advances have been made in elucidating the relationship between nanomaterials and nanotechnology and enhanced performance of lithium-ion batteries. Clearly, no one knows what the future holds, but it is certain that nanotechnology has a lot to offer to lithium-ion battery technology, and the best is yet to come.

References

1. Cook T, Dogutan D, Reece S, Surendranath Y, Teets T, Nocera D (2010) Solar energy supply and storage for the legacy and nonlegacy worlds. Chem Rev 110(11):6474–6502
2. Huggins RA (2008) Advanced batteries: materials science aspects. Springer, New York, 474 p
3. Yang Z, Zhang J, Kintner-Meyer M, Lu X, Choi D, Lemmon J, Liu J (2011) Electrochemical energy storage for green grid. Chem Rev 111(5):3577–3613

4. Vincent CA, Scrosati B (1997) Modern batteries: an introduction to electrochemical power sources. Arnold, London/New York, 351 p
5. Linden D, Reddy TB (2002) Handbook of batteries, 3rd ed. McGraw-Hill handbooks. McGraw-Hill, New York, 1200 p
6. Nazri G, Pistoia G (2004) Lithium batteries: science and technology. Kluwer, Boston, 708 p
7. Schalkwijk WAV, Scrosati B (2002) Advances in lithium-ion batteries. Kluwer Academic/Plenum, New York, 513 p
8. Yoshio M, Brodd RJ, Kozawa A (2009) Lithium-ion batteries: science and technologies. Springer, New York, 452 p
9. Garche J (2009) Encyclopedia of electrochemical power sources. Elsevier, Boston
10. Ozawa K (2009) Lithium ion rechargeable batteries. Wiley-VCH, Weinheim, 336 p
11. Feynman R (1959) There's plenty of room at the bottom. In: American Physical Society meeting, California
12. Drexler KE (1986) Engines of creation, 1st edn. Anchor Press/Doubleday, Garden City, 298 p
13. Taniguchi N (1974) On the basic concept of 'Nano-Technology'. In: Proceedings of the International Conference on Production Engineering. Part II. Japan Society of Precision Engineering, Tokyo
14. Fahlman BD (2007) Materials chemistry, 2nd edn. Springer, Dordrecht/New York, 736 p
15. Ozin GA, Arsenault AC, Royal Society of Chemistry (Great Britain) (2005) Nanochemistry: a chemical approach to nanomaterials, RSC nanoscience & nanotechnology series. RSC, Cambridge, 628 p
16. Roduner E, Royal Society of Chemistry (Great Britain) (2006) Nanoscopic materials: size-dependent phenomena, RSC nanoscience & nanotechnology. RSC, Cambridge, xii, 285 p
17. Despotuli AL, Nikolaichik VI (1993) A step towards nanoionics. Solid State Ionics 60(4):275–278
18. http://en.wikipedia.org/wiki/Nanoionics
19. Kharton VV (2009) Solid state electrochemistry, vol 1. Wiley-VCH, 527 p
20. Maier J (2004) Physical chemistry of ionic materials: ions and electrons in solids. Wiley, Chichester/Hoboken, 537 p
21. Maier J (2005) Nanoionics: ion transport and electrochemical storage in confined systems. Nat Mater 4(11):805–815
22. Li H, Richter G, Maier J (2003) Reversible formation and decomposition of LiF clusters using transition metal fluorides as precursors and their application in rechargeable Li batteries. Adv Mater 15(9):736–739
23. Balaya P, Li H, Kienle L, Maier J (2003) Fully reversible homogeneous and heterogeneous Li storage in RuO2 with high capacity. Adv Funct Mater 13(8):621–625
24. Chan CK, Peng H, Liu G, McIlwrath K, Zhang XF, Huggins RA, Cui Y (2008) High-performance lithium battery anodes using silicon nanowires. Nat Nano 3(1):31–35
25. Bruce PG, Scrosati B, Tarascon J-M (2008) Nanomaterials for rechargeable lithium batteries. Angew Chem Int Ed 47(16):2930–2946

Chapter 2
Inverse Opal Nanoarchitectures as Lithium-Ion Battery Materials

Justin C. Lytle

Abstract Lithium-ion battery technologies enable the proliferation of portable electronic devices and are critical to developing energy infrastructures that will supplant fossil fuels. Inverse opal materials have been evaluated as advanced lithium-ion battery components because an inverse opal's three-dimensionally ordered, bicontinuous macropore-solid nanostructure supports high rates of electrochemical charging and discharging and can be synthesized in a wide range of compositions that store and transport Li ions. This chapter appraises the state of research about inverse opal materials for lithium-ion anodes, cathodes, and solid-state electrolytes and examines their role in three-dimensionally interpenetrated lithium-ion microbatteries.

2.1 Introduction

An ongoing theme in battery research has been to maximize the energy densities and power densities of batteries in order to feed our growing need for portable electricity—particularly for transportation—and to efficiently store solar, wind, and other alternative energies. Lithium-ion battery technology has been part of the solution in this regard because it combines greater specific energy density and specific power density than other types of batteries. Lithium-ion electrodes store electrochemical charge by intercalating lithium ions into the lattices of crystalline solids and into disordered vacancies within amorphous hosts [1]. Solid-state ion diffusion into bulk electrode materials is slow and therefore limits the rate at which batteries discharge and recharge. These limits on ionic mobility are at least partially overcome by fabricating battery components that have solid-state ion diffusion path lengths on a nanometer-length scale. Electrode nanoarchitectures are therefore an

J.C. Lytle (✉)
Chemistry Department, Pacific Lutheran University, Tacoma, WA 98447, USA
e-mail: lytlejc@plu.edu

Y. Abu-Lebdeh and I. Davidson (eds.), *Nanotechnology for Lithium-Ion Batteries*, Nanostructure Science and Technology, DOI 10.1007/978-1-4614-4605-7_2,

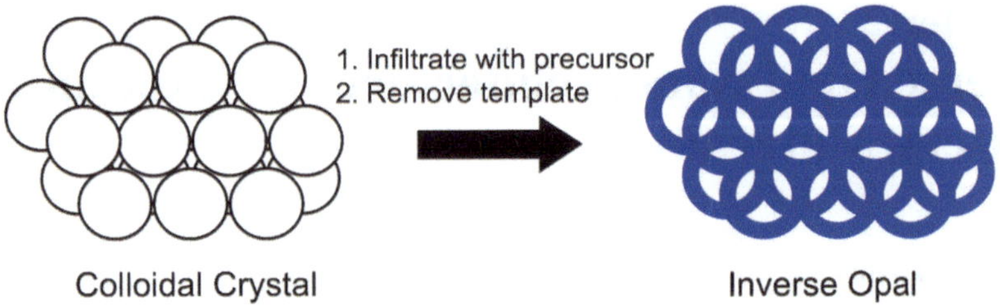

Fig. 2.1 A diagram of colloidal crystal templating and the resulting inverse opal structure

important innovation in battery research because they can have vast and accessible surface areas, nanoscale solid-state ion transport distances, and can be prepared in many lithium-ion battery material compositions.

This chapter focuses specifically on lithium-ion battery nanomaterials that have been designed with an inverse opal geometry [2]. Although inverse opals are just one of the many types of nanostructured solids that have been studied as lithium-ion battery materials during the past decade [3–10], the inverse opal structure is distinguished by close-packed spherical macropores (pore diameters >50 nm) that are interconnected in three dimensions via pore windows (see Fig. 2.1).

Inverse opal materials typically have pore wall thicknesses of 10s–100s nm and can be synthesized in a variety of compositions via sol–gel [11], precipitation [12], polymerization [13, 14], and vapor deposition chemical routes [15, 16]. This three-dimensionally ordered macroporous structure was first reported in the late 1990s [17, 18] and has since been prepared and characterized in multiple compositions of lithium-ion anodes, cathodes, and solid-state electrolyte materials.

2.2 Lithium-Ion Intercalation and the Role of Electrode Structure on Electrochemical Performance

The advent of lithium-ion electrochemistry research is attributed to the graduate thesis of Harris, who demonstrated that lithium metal electrodeposits from Li salts under controlled, nonaqueous conditions [19]. This concept converged with the early sodium-ion transport studies in beta alumina [20, 21] and tungsten bronzes [22], in which ions "intercalate" or conduct through interstitial vacancies, porous one-dimensional channels, and between two-dimensional layers in crystalline solids. Sodium-ion transport research quickly evolved into the seminal studies of reversible lithium-ion intercalation within layered metal dichalcogenides (TiS_2) by Whittingham at Exxon in the 1970s [23–25]. Ions can reversibly intercalate (i.e., in secondary cells) if the electrode material remains electrically conductive as ions enter the structure, although the degree of reversibility can fade if the electrode material undergoes phase changes that alter the dimensions of the unit cell to the extent that the electrode crumbles over multiple charge and discharge cycles [26, 27].

The rate that ions intercalate into electrode materials, and therefore the rate that batteries charge and discharge, is inherently limited by the charge transport characteristics of each electrode composition and the diffusion path lengths that charges must travel in an electrode's architectural design. Many conventional lithium-ion electrode compositions conduct ionic charge with several orders of magnitude less ionic conductivity than liquid electrolytes. As a case in point, graphite [28] and $LiFePO_4$ [29] are common lithium-ion anode and cathode compositions, respectively, that exhibit lithium-ion diffusion coefficients on the order of 10^{-10} cm^2s^{-1}. In contrast, the diffusion coefficient of a typical lithium-ion liquid electrolyte solution is approximately four orders greater in magnitude than in the solid state (10^{-6} cm^2s^{-1}) [30–33].

Of critical importance to this chapter, though, is the impact of an electrode's physical structure on its ability to rapidly transport charge. Most commercially available lithium-ion battery electrodes are themselves dense laminates, on the order of hundreds of micrometers thick, which often comprise micrometer-scale active lithium-ion insertion powders, polymeric binders, and conductive additives. In order to reach and then diffuse into active electrode particles, Li ions must first diffuse into and through these dense composites. Many recent studies have prepared such composite electrodes in order to electrochemically evaluate porous electrode materials such as aerogels [34] and templated nanomaterials [16, 35, 36], but the macromolecular adhesives in the composite can obstruct and pave over the desired nanoscopic features and access to the large surface areas in the nanomaterials being considered. Furthermore, binders and conductive additives add "dead weight" to the electrode because those materials store a negligible amount of ionic charge as compared to the active electrode material. Although the ratio of active material to additives varies in different composite blends, it is common to encounter composite electrodes that contain approximately 30 wt.% of binders and conductive powders.

Once at the interface of an active particle, Li ions enter from the outside inward, filling the atomic-scale vacancies in the active material's exterior microstructure and therefore creating diffusion bottlenecks that impede additional ion transport. Acting in concert, the dense nature of electrode composites, electrode thicknesses of ~100s μm, and micrometer-scale diffusion path lengths in active electrode particles obstruct ion mobility and limit lithium-ion batteries to slower discharging and recharging rates than are otherwise possible.

For these reasons, inverse opals and other nanomaterials have successfully improved the power density of lithium-ion electrodes. With respect to inverse opals, three-dimensionally interconnected macropores facilitate the rapid flux of liquid electrolyte solutions, such that 100s m^2 g^{-1} of electrode interfaces are simultaneously accessible to mobile ionic charges. This degree of macropore interconnectivity is intentionally tailored into inverse opals as spherical colloids self-assemble into close-packed templates. For example, the ionic conductivity of a 1 M $LiPF_6$ liquid electrolyte solution is only impeded by a factor of two when ion transport occurs through the interconnected macropores of monolithic carbon inverse opals (electrode thicknesses: 300 μm–3 mm). Random pore interconnectivities do not preclude ionic mobility through aerogels and other nanomaterials with disordered pore geometries, though. Manganese oxide aerogel monoliths also conduct ionic charge across macroscopic diffusion path lengths (>0.3 mm) [37].

Inverse opal electrodes generally exhibit higher power densities than bulk materials, in part because macropores accommodate a greater flux of liquid electrolyte than mesopores (2 nm < pore diameters < 50 nm) or micropores (<2 nm). At rapid rates of charge and discharge, micropores and mesopores hinder ionic transport in otherwise ultraporous electrodes because these pore sizes too closely approximate the size of solvated ions. This polarization phenomenon is well known for carbon aerogels with hierarchical porosities [38, 39]. In contrast, inverse opal macropore sizes are much larger relative to mobile ionic species, such that increasing the size of inverse opal macropores by even 20% has no measurable effect on the specific lithium-ion capacity of carbon inverse opal monoliths [40, 41].

On a related note, accessible surface area is a well-known factor in the increased power density of inverse opals and other electrode nanoarchitectures. Many nanomaterials have specific surface areas in excess of 100s $m^2\ g^{-1}$, which may or may not aid the electrochemical performance of a lithium-ion battery depending on whether the pore size regimes that contribute the most surface area in the material are electrochemically accessible on a timescale that is meaningful to lithium-ion insertion processes.

For instance, spherical macroporous voids are the predominant source of surface area (< 21 $m^2\ g^{-1}$) in inverse opal $LiCoO_2$ powders that are prepared by a sol–gel chemical route [36]. The macropore surfaces in an inverse opal are simultaneously accessible throughout the 3D network of macropores, which enables Li ions to enter the rock-salt crystal structure of $LiCoO_2$ inverse opals with less polarization than is observed in bulk $LiCoO_2$ (0.21 $m^2\ g^{-1}$). As a result, $LiCoO_2$ inverse opal powders discharge greater specific capacity than bulk $LiCoO_2$ powders at current densities between 100 and 700 mA g^{-1}. It is true that micropores and mesopores contribute large amounts of specific surface area (100s–1,000s $m^2\ g^{-1}$) to many carbonaceous electrode nanoarchitectures. However, small mesopores and micropores experience severe polarization with increasing current density, such that macroporous electrodes deliver greater amounts of electrochemical energy at high rates of discharge [39].

The following section will review the general approaches that have been used to prepare and characterize inverse opals, and subsequent sections will detail the synthesis and electrochemical properties of inverse opals as lithium-ion anode, cathode, and solid electrolyte materials.

2.3 General Fabrication Considerations

A variety of fabrication routes exist to prepare inverse opal materials, yet all fabrication schemes begin by forming colloidal crystal templates (CCTs) from monodisperse polymer [42–46] or silica spheres [47, 48]. Poly(methyl methacrylate) (PMMA) and polystyrene (PS) spherical colloids are two polymer compositions that are compatible with the chemicals and processing conditions in many inverse opal syntheses [2]. Both PMMA and PS are formed as spherical colloids via aqueous free-radical emulsion polymerizations, in which growing free-radical-macromolecules contort themselves into spheres in order to minimize hydrophobic interactions with water in the aqueous monomer solution [49]. Colloidal PMMA and PS spheres generated in this way are typically on the order of 10s–1000s nm in diameter [50]. The surfaces of these polymer spheres are decorated with molecular moieties that are the residual fragments of free-radical initiator molecules. The inherent polarity of some polymer chain end groups that originate from initiator molecules [43], when combined with a polar polymer composition like PMMA, greatly improves the wettability of CCTs by polar precursor solutions in subsequent steps.

Whereas PMMA and PS spheres are compatible with many inverse opal syntheses, colloidal silica spheres are an alternative composition that offer greater degrees of chemical and thermal inertness than organic polymer sphere compositions. Silica spheres are prepared by the Stöber method, in which silicon alkoxides are controllably hydrolyzed and condensed in a nonaqueous solution to form spherical colloids of amorphous silicon oxides [47]. Monodisperse silica spheres are typically prepared in this way with diameters that range from 50 nm to several μm [13]. Silica is considerably less reactive than PMMA and PS and therefore can withstand all but strong bases and HF before dissolving. Likewise, silica spheres maintain their shapes at high temperature and therefore limit excessive crystallite growth that would otherwise densify the inverse opal pore network if silica templates were not present.

The end product of free-radical emulsion polymerizations of PMMA and PS spheres and Stöber silica syntheses are monodisperse spheres with diameters that deviate from each other by $<5–8\%$ [13]. Monodisperse spherical colloids can be crystallized by several techniques, and a non-exhaustive list includes gravitational sedimentation [51–54], centrifugation [18, 55–58], and convective self-assembly into thin films [59–61]. Monodisperse spherical colloids close-pack into crystalline arrays, with a slight entropic preference toward face-centered cubic close-packing over hexagonal close-packing [62–64]. In either case, each sphere is surrounded by six close-packed spheres and a total of six spheres above and below each close-packed plane of spheres. This packing geometry is identical to that of the natural gemstone opal, and therefore colloidal crystals are often referred to as opals (see Fig. 2.2) [65, 66].

Other crystal lattice types are possible (bcc [67–70], diamond [71]) but are not energetically favorable under the conditions normally involved with concentrated suspensions of monodisperse spheres. Colloidal crystals are commonly polycrystalline, unless rigorous controls are applied, and exhibit point, line, and planar defects. Electrochemical applications for inverse opals do not require monocrystalline and defect-free colloidal crystalline lattices as do inverse opals for photonic research. As long as macropore networks are three-dimensionally continuous, a degree of imperfection still enables electrodes to cycle with relatively high power densities [36].

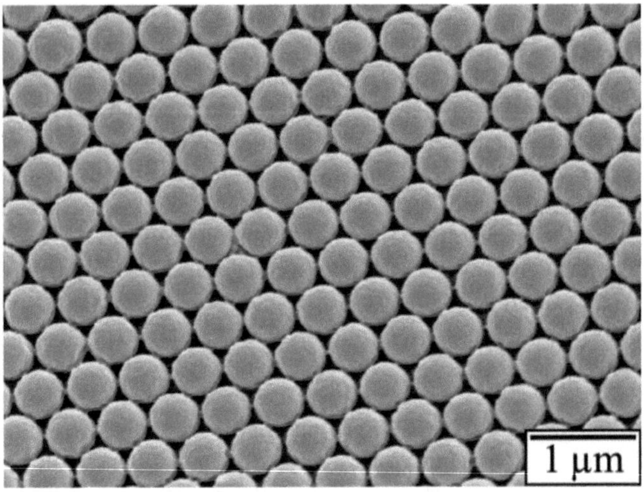

Fig. 2.2 A scanning electron micrograph of a PMMA colloidal crystal

Close-packed colloidal spheres with diameters of 100s nm are surrounded by interstitial voids approximately 50 nm in size. The interstitial voids between close-packed spheres occupy ~26 vol.% of the close-packed unit cell, and these interstices can be infiltrated with either liquid or vapor precursors to deposit the solids that will become the skeletal framework of the resulting inverse opals. While the focus of this chapter is on inverse opal lithium-ion battery materials, in general, inverse opals are synthesized by precipitating metal salts [72, 73] or gelling sols within colloidal crystal interstices [74, 75]. Several electrochemical approaches have been developed that electrodeposit inverse opal metals, alloys, and metal oxides under constant current [2, 76] or constant potential conditions [77]; electropolymerize conductive and semiconducting polymers within colloidal crystal films that are supported on planar current collectors [78, 79]; and electrolessly deposit metals [80, 81]. Rather than deposit inverse opals from liquid precursors, some fabrication schemes infiltrate nanoparticles into colloidal crystals [82, 83], co-sediment nanoparticles with monodisperse colloidal spheres [84, 85], and assemble inverse opal materials via layer-by-layer coatings of polyelectrolytes that are capped with oppositely charged nanoparticles [86, 87]. Organic polymers may be synthesized *in situ* [40, 88] and even melt infiltrated into CCTs [89–91]. Chemical vapors can also deposit solid framework materials within colloidal crystal interstices to form carbon inverse opals from volatile hydrocarbon feedstocks [15, 16] and metal oxides and sulfides from organometallic precursors [60, 92–95].

Once solids are deposited within the interstices of colloidal arrays, the spherical particles have served out their usefulness as three-dimensionally ordered space holders and must be removed to form the inverse opal geometry. Depending on the composition, polymer sphere templates are commonly dispatched when samples are calcined in air or oxygen at temperatures ranging from 300 to 400°C or are thermally decomposed at 400–500°C in inert atmospheres[73]. Pyrolysis conditions in inert gas avoid combusting the infiltrated hydrocarbon precursors in the case of inverse opal carbons [40] and limit crystallite growth as samples oxygenate at high temperatures [36, 61]. It must be noted that pyrolyzed inverse opals contain up to 40 wt.% carbon in the form of molecular remnants of PS and 15 wt.% carbon from PMMA, and these residuals must be considered if the surface chemistry of an inverse opal is decisive to its operation.

2.4 Inverse Opals as lithium-ion Battery Materials

Inverse opals have been prepared and tested as lithium-ion anode, cathode, and solid electrolytes since 2002 [74]. This section will be divided into these three corresponding parts to focus on the required synthetic methods, physical properties, and electrochemical behaviors of each component of a lithium-ion battery.

2.4.1 Lithium-Ion Anodes

In a fully charged lithium-ion battery, the anode contains an excess of Li ions that have chemical potential to diffuse across the electrolyte and into atomic-scale vacancies in a Li-deficient cathode. In early lithium-ion batteries, Li metal was used as an anode because Li has one of the lowest standard reduction potentials (−3.04 V vs. SHE) and a large specific capacity (3,861 mAh g^{-1}) [96]. However, Li forms dendrites on anode surfaces after repeated cycling, and these metallic offshoots can cause a hard short circuit, in which electrochemical energy is violently released from the battery.

2.4.1.1 Carbonaceous Anodes

Carbon is the most common lithium-ion anode composition in commercial batteries. Carbon electrodes are lightweight, reversibly discharge 100s mAh g^{-1} depending on the chemical and microstructural properties of each particular form of carbon (graphite has a reversible specific capacity of 372 mAh g^{-1}), and can have metallic electrical conductivities when prepared from aromatic or conjugated macromolecular hydrocarbons [97]. For these reasons, it is little wonder why carbon continues to be one of the most investigated compositions of inverse opal materials. Carbon inverse opal films were first reported by Zakhidov et al. in 1998 via a vapor deposition technique that grows volume-templated diamond and amorphous carbon inverse opals and also directs the growth of graphite coatings at the surface of silica CCTs [15]. Solution-phase synthetic schemes were subsequently reported using sucrose [98] and phenolic resin [88, 99–102] precursors beginning in 2001.

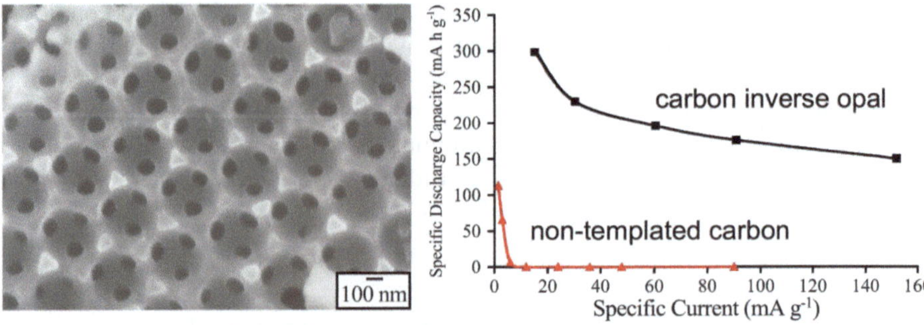

Fig. 2.3 A scanning electron micrograph of the ordered macropores in a carbon inverse opal; carbon inverse opals polarize to a lesser extent than bulk carbon of the same composition [40]

Two separate papers in 2005 investigated the lithium-ion insertion properties of inverse opal carbons that were prepared from solution and vapor precursors, respectively. In one approach, carbon inverse opal monoliths were fabricated by first infiltrating and curing resorcinol-formaldehyde (RF) precursors within PMMA CCTs, and then pyrolyzing these composites to carbonize the RF and thermally decompose the PMMA spheres (see Fig. 2.3) [40].

Smooth macropore walls without mesoporous textures are formed in this synthesis, such that 300–400 m^2 g^{-1} of specific surface area result only from the macroporous pore network and microporous vacancies that exist between turbostratically disordered graphene sheets in RF carbon [97]. The monoliths are electronically (0.2 S cm^{-1}) and ionically conductive, and therefore do not greatly impede the flow of lithium-ion liquid electrolyte solution according to impedance measurements. These conductivities, the degree of plumbing throughout the carbon inverse opal, and the nanometer-scale lithium-ion diffusion path lengths of the macropore walls allow carbon inverse opals to experience far less polarization than non-templated RF carbon monoliths at specific current densities up to 152 mA g^{-1}. Incorporating SnO_2 nanoparticles onto the pore walls of inverse opal carbons further increases the energy density of electrodes. This point will be discussed in more detail in the following section.

Carbon inverse opal powders for lithium-ion batteries have also been fabricated by a vapor deposition route that is significantly different from the one employed by Zakhidov et al. [15]. Silica inverse opals, themselves prepared from PS sphere CCTs, serve as a mold that supports the growth of nanoscopic graphitic coatings from benzene vapor at elevated temperatures [16]. This technique replicates the structure of the silica inverse opal with <10-nm resolution, depositing 30-nm-thick walls of graphitic carbon. Graphite is a metallic electronic conductor and is well known as a lithium-ion anode in commercial batteries. This combination endows carbon inverse opals, when formed as a composite with carbon black conductive additive and a polyvinylidene difluoride (PVdF) binder, with a lithium-ion specific capacity of up to 260 mAh g^{-1} at 1,000 mA g^{-1}. Like the previous carbon material, SnO_2 can be incorporated into the macroporous structure to increase the specific capacity. In this case, the capacity of SnO_2 inverse opal carbon composites (18.6 wt.% 5-nm SnO_2 nanoparticles) fades more slowly than in the previous carbon inverse opal.

A variation on the previous nanocasting methodology adds hierarchical porosity in the form of mesopores that are directed by Brij 56 surfactant into silica inverse opals [103]. Acidic Al^{3+} sites are incorporated into the silica prior to introducing phenol and paraformaldehyde precursors in the vapor phase. After pyrolysis, the resulting carbon contains wormlike mesopores (4.1 nm) that can themselves collect vapor-deposited graphitic carbons. Even nanoscopic domains of graphite substantially increase the electronic conductivity of the phenol-derived carbon by 67% to 0.25 S cm^{-1}. In a comparison between the hard carbons that are generated from RF precursors and the soft, graphitizable carbons that are vapor deposited from acetonitrile, the carbon inverse opal nanocast with a graphitic carbon layer delivers the lowest specific discharge capacity (272 mAh g^{-1})—which is to be expected because graphite's specific capacity (372 mAh g^{-1}) is less than that of amorphous hard carbons [97]. The authors note that a graphitic coating reduces the formation of a solid electrolyte interface (SEI), the absence of which is advantageous for preserving coulombic efficiency during cycling [103]. Other nanocasting approaches that graphitize carbon inverse opals have since been reported [104].

Further enhancing the specific discharge capacity of carbon inverse opals, Si is nanocast from diiodosilane onto mesoporous carbon inverse opals that are derived from a phenol-formaldehyde precursor [105]. Silicon is an attractive lithium-ion anode composition because Si reversibly forms Li alloys and has a theoretical specific capacity of 4,200 mAh g^{-1} [106]. Carbon inverse opal monoliths that are nanocast with 22–40 wt.% Si lose all trace of mesoporosity, according to nitrogen porosimetry data, and behave as type I microporous adsorbents. Interestingly, the reversible specific charge capacities of Si-nanocast carbon inverse opals is 332 mAh g^{-1}, which is smaller than the 587 mAh g^{-1} that is measured during the reversible charge of the mesoporous carbon inverse opal framework without Si. The authors attribute the unexpectedly low charge capacity of their Si-nanocast electrodes to silicon oxidation and to residual, unreacted diiodosilane.

Hard-carbon inverse opals are also nanocast from sucrose on titania inverse opals [107]. The reversible capacity of these electrodes is quite modest ($<$100 mAh g^{-1} at cycling rates of 0.5–2.0 C) because the underlying semiconducting TiO_2 framework does not transport electrons with the relative ease of the inverse opal carbons hereto mentioned.

2.4.1.2 Li-Alloy Anode Materials

One of the early commercial alternatives to metallic Li anodes were tin-based amorphous solids [108]. Tin and tin oxides (SnO and SnO_2) semi-reversibly form a series of binary Li–Sn alloys up to a stoichiometric maximum of $Li_{4.4}Sn$ (993 mAh g^{-1}) but undergo tremendous volumetric expansions and contractions in the tin unit cell during alloying and dealloying. Repeated cycling causes electrodes to gradually disintegrate and increasingly lose capacity [26, 27, 109]. Inverse opal SnO_2 films were first prepared as gas sensors by melt-infiltrating tin *tert*-butoxide into PS CCTs [110], and other tin oxide gas sensors have since been reported [111]. A study released in 2003 describes the synthesis and physical characterization of inverse opal SnO_2 powders that were prepared by infiltrating PMMA CCTs with $SnCl_2$ in ethanol and subsequently adding ammonium hydroxide to precipitate $Sn(OH)_2$ within the interstices of the template [75].

The lithium-ion electrochemistry of inverse opal SnO_2 thin-film electrodes was first described in 2004 [61]. These electrodes are sequentially constructed by first assembling thin films of PMMA colloidal crystals on Pt-sputtered glass microscope slide current collectors. Droplets of aqueous Sn^{4+} precursor solution are then wicked into colloidal crystal thin films and rely on a surfactant additive to overcome surface tension at the film interface. Infiltrated films are dried and then calcined in air at 450–800°C to form cassiterite SnO_2 and combust the polymer spheres. The resulting inverse opal films comprise ~10-μm-wide islands of SnO_2 inverse opals, and the texture of the pore walls becomes more granular with increasing calcination temperatures. These materials were electrochemically evaluated using cyclic voltammetry and galvanostatic cycling in a three-electrode cell configuration with a 1 M $LiClO_4$ liquid electrolyte in propylene carbonate. Morphological changes are observed during the first lithiation of this material, and the interconnected macropore structure becomes occluded by volumetric expansions no later than the fifth cycle. These physical changes correspond to a fourfold decrease in the charge capacity. Inverse opal SnO_2 films severely polarize at a 10 C cycling rate, which indicates that the electrical resistance of nanostructured SnO_2 grains dominates the electrochemical response of the three-dimensionally interconnected macropore structure.

Three newer synthetic approaches have since been reported that aim to take advantage of the large specific capacity that tin-based inverse opal electrodes offer while attempting to minimize the irreversible expansion of the unit cell over repeated electrochemical cycles. In one approach, tin dioxide is decorated as 10–40-nm nanoparticles on the electrically conductive macropore walls of inverse opal carbon monoliths by first wetting the supporting inverse opal framework with aqueous tin(II) sulfate solution [40]. The sulfate salt precipitates during drying and disproportionates into SnO_2 at 400°C in nitrogen gas. While the presence of SnO_2 nanoparticles initially boosts the specific capacity of inverse opal carbon by 55 mAh g^{-1}, the particles lose electrical contact with the inverse opal carbon electrode over 30 cycles.

In a second approach, tin oxide nanoparticles are deposited from $SnCl_4$ vapor within 5-nm mesopores on the solid skeleton of inverse opal carbon monoliths [112]. The mesopores form from a triblock copolymer that is added to a resorcinol-formaldehyde precursor and are designed to confine tin oxide nanoparticles within mesoporous voids and thereby prevent Li_xSn_y alloys from readily disconnecting from the carbon inverse opal during cycling. This composite contains up to 38 wt.% SnO_2 and has 68% more specific capacity over 100 cycles than the tin–carbon inverse opals in the previously mentioned study after 30 cycles [40]. The confinement effect employed here is very successful and bears repeating with other compositions that experience lattice changes during lithiation and delithiation.

Unlike the previous examples, a third approach incorporates nickel as a reasonably inert atomic-scale matrix that is hypothesized to resist the volumetric expansions that plague tin-based materials during cycling [76]. Nickel–tin alloy inverse opal films are fabricated by first electrophoretically depositing PS spheres that are capped with amidine cationic groups. The CCTs briefly anneal and then are pre-infiltrated with ethanol before electrodepositing the alloy from a bath containing $NiCl_2·6H_2O$ and $SnCl_2·2H_2O$. After the spheres dissolve in room-temperature toluene overnight, a 16-μm-thick inverse opal alloy is formed that contains predominantly crystalline Ni_3Sn_4. The pore walls generated in this way are relatively smooth and as thick as 450 nm. After 30 cycles of electrochemical cycling, the film detaches from the current collector, presumably as a result of an epitaxial mismatch between the lithiated alloy and the current collector. It is not clear from the article when the inverse opal structure is lost to volumetric expansion, only that the material undergoes relatively rapid morphological changes.

Several other metallic inverse opal compositions have been prepared, but none have been studied as lithium-ion battery materials, to the best of my knowledge [73, 80, 81, 113–116]. One metalloid composition, silicon, has drawn the attention of two research groups as a way to couple the theoretical specific capacity that Si offers (4,200 mAh g^{-1}) with the rapid electrochemical kinetics that are inherent to the nanoarchitectured inverse opal geometry [105, 117]. One technique, which was discussed in Sect. 2.4.1.1, nanocasts amorphous Si nanoparticles into surfactant-templated mesopores that reside in carbon inverse opal monoliths [105]. A second method deposits amorphous Si from silane vapors within 5–15-μm-thick silica CCT films [117].

Hydrogenated amorphous Si inverse opals undergo volumetric changes during lithium-ion insertion–de-insertion, and as expected, macropores templated from 850-nm spheres accommodate alloy expansion better than those templated from 250-nm spheres [117]. Overall, the structure of delithiated amorphous Si inverse opals appears similar to that of uncycled electrodes, but macropore windows swell closed upon full lithiation. The amorphous material has an initial delithiation specific capacity of 2,569 mAh g^{-1}, but contributes a negligible capacity at a rate of 10 C. To achieve greater rate capabilities, amorphous silicon is converted into a more conductive (10^{-4} to 10^{-3} S cm^{-1}) nanocrystalline microstructure, although the discharge capacity of this form of Si quickly fades within 15 cycles as the material rapidly and irreversibly swells. The authors sought to restrict capacity fade by incorporating a hard-carbon coating as a conductor on amorphous Si inverse opals and subsequently crystallizing the Si inverse opal framework at 800°C in $N_2(g)$. Still, the capacity fades in less than 20 cycles as the structure irreversibly expands. The optimized electrode configuration in the referenced study consists of a sucrose-derived carbon inverse opal film that is decorated with amorphous silicon. This material has an electronic conductivity up to 0.6 S/cm, an initial capacity over 1,700 mAh g^{-1}, maintains >80% of its initial specific capacity after 145 cycles, and retains capacity at a cycling rate of 1 C. Based on these observations and on the work by Lee et al. [40] and Wang et al. [105], the ideal configuration of an alloying electrode phase is as nanoparticles that are confined within mesopores on the surface of carbon inverse opals.

2.4.1.3 Metal Oxide Anodes

Lithium ions can occupy the interstitial sites in many metal oxide compositions. As an example, electrochemically deposited tungsten trioxide (WO_3) inverse opals exhibit an electrochromic transition when Li ions insert into the amorphous oxide microstructure [118]. While WO_3 is not a common lithium-ion anode material, it hosts Li ions without experiencing the destructive volumetric swelling that is so common to the alloys in the previous section. In this regard, lithium titanate ($Li_4Ti_5O_{12}$) has received considerable attention as a "zero-strain" Li-anode material. The spinel-phase $Li_4Ti_5O_{12}$ undergoes negligible changes in lattice dimensions ($\Delta a = 0.006$ Å) as it hosts up to three Li ions in vacant octahedral sites to form the rock-salt $Li_7Ti_5O_{12}$ [119–126]. A zero-strain insertion material seems ideal for the inverse opal geometry, given that the large volumetric expansions during Li alloying reveal the delicate nature of some nanoarchitectured compositions. Inverse opal $Li_4Ti_5O_{12}$ has been prepared by two approaches, each using lithium acetate and either titanyl oxalate [127] or tetraethyl orthotitanate [128] in a 4:5 molar ratio. Anatase- and rutile-phase impurities persist in the inverse opals as fabricated by both methods at temperatures <800°C. Ordered macropores are present in both materials to begin with, and Sorensen et al. provide micrographic evidence that the inverse opal structure is clearly maintained after 70 cycles [127].

The electrochemical performance of $Li_4Ti_5O_{12}$ inverse opals again confirms that through-connected porous nanoarchitectures are capable of faster electrochemical cycling performance than bulk materials of the same composition. When the total volume of $Li_4Ti_5O_{12}$ precursors is carefully controlled in order to underfill, exactly fill, or overfill the interstices in CCTs, the capacity of the overfilled inverse opal material fades fastest because it contains non-templated $Li_4Ti_5O_{12}$ powders that polarize with increasing current densities [127]. The lithium titanate inverse opals made from either precursor discharge specific capacities of approximately 160 mAh g^{-1} (theoretical = 167 mAh g^{-1}) and discharge measurable capacities after $\geq$70 cycles at a cycling rate of 10 C [128] or a current density of 0.625 mA cm^{-2}, respectively [127].

As an interesting comparison to $Li_4Ti_5O_{12}$, anatase titanium dioxide inverse opals reversibly insert up to 0.5 Li per TiO_2 (*ca.* 155 mAh g^{-1}), which suggests that $Li_4Ti_5O_{12}$ cannot be synthesized by electrochemically lithiating anatase TiO_2 [129]. The anatase microstructure transforms into an amorphous matrix upon first lithiation according to X-ray diffraction patterns of pristine and lithiated inverse opal anatase titania. This data is consistent with other known metal oxide disproportionation reactions [5] and with the observation that the inverse opal structure morphs into a dense film after lithiation.

While $Li_4Ti_5O_{12}$ is dimensionally stable during reversible lithiation, its theoretical capacity is meager in comparison to that of many carbons and Li alloys. Metal oxide anodes are increasingly well known for their ability to disproportionate into other valence states in order to store relatively large specific capacities [5]. Mixed metal oxides, like $CoFe_2O_4$, store large amounts of electrochemical charge but suffer from electrical resistance problems in the bulk [130]. To counter this effect, a continuous macropore network and pore walls that are tens of nanometers in thickness enhance the power density of $CoFe_2O_4$, which is synthesized from a precursor solution of cobalt and iron nitrates. This composition delivers over 700 mAh g^{-1} of capacity even after 30 cycles and retains *ca.* 68% of maximum specific capacity at a current density of 5 mA cm^{-2}. The capacity fades as the inverse opal structure deteriorates during repeated cycle. As an interesting note, the cobalt metal that is formed during anode reduction catalyzes the formation of a SEI layer, and this passivating film may behave as a diffusion bottleneck to lithium-ion transport.

2.4.2 Lithium-Ion Cathodes

lithium-ion cathodes are in their charged state when they are deficient in Li ions and, vice versa, are discharged when their thermodynamically accessible vacancies are occupied by Li ions. Like many lithium-ion anode compositions [5], the power density of lithium-ion cathodes is increased by electrode structures that are well plumbed, have large amounts of electrochemically accessible surface area, and have nanometric diffusion path lengths [131]. Metal oxides, phosphates, and lithium metal oxides play an important role as cathode materials in lithium-ion batteries.

2.4.2.1 Vanadium Oxide Cathodes

Vanadium oxide was the first inverse opal composition to be electrochemically characterized as a lithium-ion battery material [11]. In that seminal study, a thin film of PS CCTs was assembled onto ITO current collectors and then infiltrated with a vanadyl alkoxide precursor. Once the spheres dissolved in toluene, the pore fluids of the amorphous vanadia ambigel are displaced by cyclohexane, which has a sufficiently small surface tension, to partially avoid the collapse of mesopores in the vanadia inverse opal skeleton. Interconnected macropores reduce the tortuosity of the inverse opal structure (see Fig. 2.4.), the square of which is directly proportional to the electrochemical polarization experienced by ions diffusing into an electrode.

As a result, the specific discharge capacities of hierarchical vanadia inverse opal thin films decrease by a factor of 2.5 as the current density increases by a factor of 100. In contrast, comparable ambigel vanadia electrodes suffer a fourfold decrease in specific discharge capacity when current density increases by just a factor of 25. This data cements the importance of maximizing electrolyte transport with an open macropore network and fabricating nanostructured electrodes to have solid-state

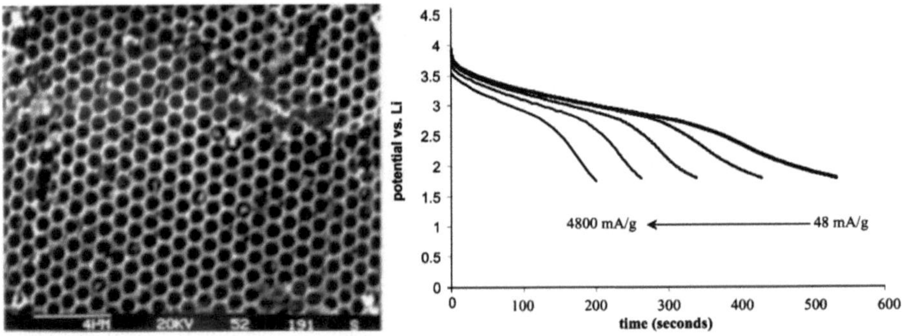

Fig. 2.4 Scanning electron micrograph of an amorphous vanadium oxide monolith and its electrochemical discharge behavior at increasingly greater current densities ([11]—reproduced by permission of the Royal Society of Chemistry)

ion diffusion path lengths on the order of 10s nm. Both structural features are key in order to increase the power densities of vanadia electrodes beyond those of conventional and mesoporous electrodes of the same composition.

Crystalline vanadia inverse opals are not as easily prepared from vanadyl alkoxide precursors and polymer CCTs because vanadium pentoxide nanocrystallites grow to dimensions that obscure the inverse opal structure during calcination at 450°C [75, 132]. Proton-exchanged sodium metavanadate is instead a vanadium precursor that better resists grain growth, yet the dilute concentration of the proton-exchanged solution fills only 2.7 vol.% of CCT interstices, and more concentrated sols of decavanadic acid rapidly precipitate. Multiple precursor loadings into a previously infiltrated sample deposit surface crusts and not additional volume filling in CCT voids, which indicates that successive precursor wetting is not a useful technique for inverse opal syntheses. The final product consists of 33-nm crystallites that are Shcherbinaite V_2O_5 (PDF no. 41-1426), but the lithium-ion insertion rate capability of this inverse opal composition was not determined.

2.4.2.2 Lithium Metal Oxide Cathodes

Lithium metal oxides are common lithium-ion cathode compositions because the transition metal atoms in their crystalline lattices undergo reversible redox reactions that require cations be inserted or de-inserted in order to maintain electroneutrality. In the case of $LiCoO_2$, Li ions intercalate between layers of CoO_2 octahedra in a rock-salt crystalline lattice, amounting to a theoretical 136 mAh g^{-1} $LiCoO_2$ [133]. Sols of mixed Li and Co salts precipitate as oxalates within CCT interstices and form spinel-like and then rock-salt crystal structures with increasing calcination temperatures [36]. At a precursor ratio of 1 Li:1 Co, a portion of the Li precursor volatilizes during calcination such that Co_3O_4 phase separates from the desired $LiCoO_2$ polymorph. Adding the Li precursor in stoichiometric excess achieves phase purity, but the inverse opal structure distorts due to crystallite growth at the 700°C temperature that is required to form this phase.

Soft chemical syntheses, like the above example with $LiCoO_2$ inverse opals, can be tuned to strike a delicate balance between achieving the desired crystalline phase and maintaining an ordered macroporous nanostructure. One way to find this balance between crystallinity and structure is to rely on chemical additives that complex the metal salts from solution into a more homogeneous mixture. Sol–gel precursors may seem homogeneous, but they can be inherently otherwise because the components of many precursors naturally oligomerize [134, 135]. Chelating agents, like poly(ethylene glycol) (PEG), therefore play an important role in the sol–gel syntheses of mixed metal oxides. Molecular complexing agents like PEG can minimize phase separation and grain sizes in crystalline materials by electrostatically binding both cation types in close molecular proximity to one another.

For instance, colloidal-crystal-templated $LiCoO_2$ experiences excessive grain growth and sintering at 700°C, even in the presence of PEG as a chelator [36]. However, incorporating hydrogen hexachloroplatinate monohydrate ($H_2PtCl_6{\cdot}H_2O$) into the precursor limits the growth of $LiCoO_2$ to occur only along the c-axis of the $LiCoO_2$ rhombohedral unit cell [136]. None of the resulting macroporous $LiCoO_2$ powders that are formed in this way fit the conventional definition of an inverse opal, but the structure maintains an interconnected macroporous network and nanoscopic walls that support high-rate cycling. For instance, PEG-doped and Pt-doped $LiCoO_2$ deliver almost 40 mAh g^{-1} at 700 mA g^{-1}, whereas bulk $LiCoO_2$ polarizes to negligible specific capacities at a comparable current density.

Lithium nickel oxide is another binary lithium metal oxide that has been fabricated as an inverse opal cathode material [75]. The optimized synthesis of $LiNiO_2$ requires balance between forming the right crystalline phase and maintaining the inverse opal geometry. With a precursor ratio of 1 Li:1 Ni, the inverse opal framework phase separates into NiO and Li_2CO_3 at <700°C in air; above this temperature, the inverse opal structure is altered by grain growth. A stoichiometric excess of Li in the precursor overcomes lithium depletion that occurs by volatilization and achieves phase purity. Likewise, the excess Li prevents Ni^{3+} cations from occupying Li^{+} sites in the layered lattice and thereby keeps Ni^{3+} from becoming a diffusion barrier to Li ions. The degree of cation mixing in the layers of $LiNiO_2$ is monitored by comparing the X-ray diffraction reflections of the (103) and (104) crystal planes such that a larger ratio corresponds to less mixing and therefore better ion transport in the solid state. Based on this information, the ideal calcination conditions for sol–gel-derived $LiNiO_2$ are 600°C for 5 h in O_2. Longer calcinations permit cation mixing, whereas higher temperatures create wormlike macropores.

A 2008 publication about fabricating $LiMn_2O_4$ spinel inverse opal powders emphasizes the importance of choosing metal nitrate precursors that solidify, rather than acetate salts that melt, before polymer CCTs liquefy during thermal processing [137]. Through the lens of the previous work with $LiCoO_2$ [36] and $LiNiO_2$ [75], it is clear that Li volatilizes from 1 Li : 2 Mn precursor solution to result in a mixture of α-Mn_2O_3 and $LiMn_2O_4$ nanocrystallites (5–20 nm) at 700°C. This article also introduces the use of solid-state ^7Li magic angle spinning NMR to distinguish the chemical environment around Li ions in the deposited framework material at increasingly greater calcination temperatures. The chemical bonding information from NMR shows that Li-rich amorphous materials (100 ppm) gradually transform with increasing temperatures into a spinel phase with Li ions in tetrahedral (500 ppm) and octahedral (1,500 ppm) sites. Unlike many metal oxide inverse opal compositions that form granular powders, $LiMn_2O_4$ is highly sintered under these conditions into monoliths as large as $1 \times 10 \times 10$ mm. These monoliths are electrochemically cycled (3.5–4.6 V vs. Li^+/Li) without adding binders or conductive additives. The authors employed aqueous lithium-ion electrolyte because aqueous electrolytes have been reported to have similar effects as organic solvents on $LiMn_2O_4$. It is not clear how their methods avoid electrochemically oxidizing water at such positive chemical potentials. Approximately 70% of the material's discharge capacity is retained when current densities increase by a factor of 40, and the $LiMn_2O_4$ inverse opal morphology exhibits no major structural changes over 30 cycles.

2.4.2.3 Lithium Iron Phosphate Cathodes

Lithium iron(II) phosphate ($LiFePO_4$) is a lithium-ion cathode composition that has been actively studied since Goodenough's initial paper on the material in 1997 [138]. Substituting for cobalt and nickel, both of which are costly strategic metals in common Li cathodes, iron is a relatively inexpensive surrogate and has a low toxicity. Lithium iron phosphates have an olivine structure and tend to suffer from poor electronic and ionic conductivities, although these have been repeatedly improved by the work of Chiang and others [139]. One approach to increase the mobility of electrons and ions in this composition is to create $LiFePO_4$ inverse opal materials with nanoscopic diffusion paths, open macropore networks that minimally impede electrolyte flux, and carbon coatings that boost the material's electronic conductivity. By this rationale, Lu et al. prepared an open macroporous $LiFePO_4$ powder in 2005 by combining a stoichiometric (1 Li: 1 Fe : 1 PO_4) precursor sol with a suspension of poly(styrene-methyl methacrylate-acrylic acid) spheres [140]. After the spheres settle and decompose at 600–800°C in nitrogen, a phase-pure $LiFePO_4$ nanomaterial forms with disordered macropores and ~5 wt.% residual amorphous carbon as an electronic conductor. The authors report that these electrodes discharge approximately 100 mAh g^{-1} at a rate of 5.9 C (theoretical specific capacity = 170 mAh g^{-1}).

From another perspective, $FePO_4$ is an interesting choice for an inverse opal cathode because it can reversibly host Li ions depending on the degree of crystallinity in the $FePO_4$ lattice. Highly ordered macroporous $FePO_4$ inverse opals are prepared by infiltrating PS CCTs with a standard sol–gel precursor that begins to crystallize into hexagonal $FePO_4$ at 500°C [141]. As-prepared samples that are calcined to 400°C have a maximum reversible capacity of 100 mAh g^{-1}, even though they are as X-ray amorphous as unheated samples. As is often the case in

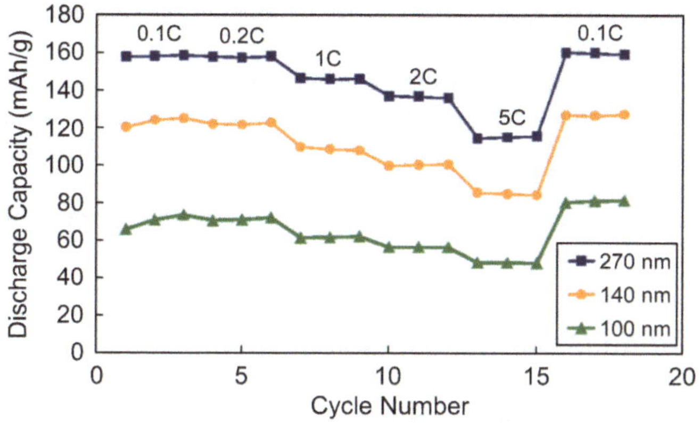

Fig. 2.5 The specific discharge capacity of $LiFePO_4$ inverse opals as a function of the diameter of the PMMA colloidal sphere template used to prepare each (Reprinted with permission from [41]. 2009 American Chemical Society

so-called "amorphous" metal oxides, order exists in small atomic-scale domains that are surrounded by a disordered atomic-scale matrix [142]. Ions conduct through the ion wire of vacancies in the disordered regions; electrons conduct through nucleated nanocrystallites. Optimal mixed ion–electron conduction appears to exist in sol–gel-derived $FePO_4$ that are calcined at 400°C, whereas hexagonal $LiFePO_4$ (600°C) is known to be electrochemically inactive [141]. The specific capacity of the 400°C material decreases by ~65% as the cycling rate increases to 2 C. No electrochemical comparison was made to bulk $LiFePO_4$.

Perhaps the best example of how the inverse opal structure benefits $LiFePO_4$ is recent work by Doherty et al. [41]. The authors synthesize inverse opal $LiFePO_4$ powders by slightly underfilling PMMA CCTs with a sol of iron nitrate, lithium acetate, and phosphoric acid—lithium acetate is a surprising choice because it melts (T_m = 53–56°C) long before PMMA spheres combust. Three different PMMA sphere sizes (100, 140, and 270 nm) are used to elucidate the role of an inverse opal's pore size on its performance at different cycling rates.

Electrochemical processes are dually ionic and electronic, and resistance to either form of charge transport makes it imperative that ions and electrons move through the smallest distances possible. Dense, non-templated surface crusts negate the flux of liquid electrolyte that endows inverse opal materials with high power densities. It is therefore critical for electrochemical applications to underfill CCTs with precursor, as was accomplished in this 2009 work, in order to avoid forming non-templated electrode materials [41].

The phase-pure $LiFePO_4$ inverse opals prepared in this synthesis are increasingly sintered at processing temperatures up to 800°C [41]. Sintering decreases charge-transfer resistance in the electrodes and makes possible discharge rates that are as high as 10 C (93 mAh g^{-1} after 50 cycles). The largest discharge capacities for inverse opal $LiFePO_4$ are achieved by templating the structure with 270-nm PMMA spheres (see Fig. 2.5).

Large spheres ensure that electrode macropore networks are continuous and maintain a flux of liquid electrolyte to all parts of the electrode. Residual amorphous carbon is also key to enhancing the electronic conductivity of the olivine material, although having too much carbon creates a lithium-ion diffusion barrier on the surface of active electrode particles and prevents higher rates of discharge.

2.4.3 Solid-State Lithium-Ion Electrolytes

Electrolytes conduct ionic charge, but not electrons, between electrodes in order to force electricity to reroute through an external circuit (i.e., a device). While liquid electrolytes are very common in conventional lithium-ion batteries, solid-state batteries are conceptually appealing because they never leak, are not flammable, and do not result in SEI diffusion bottlenecks like those that form as organic solvents decompose in the presence of lithium salts [143].

Charge-transfer impedance through the solid phase is one of the major challenges to working with solid electrolytes. Ions diffuse through solids at much slower rates than through liquids, and this effect is exacerbated when working with solid electrolyte materials in the bulk. In that sense, using the inverse opal structure as a solid electrolyte for power-dense lithium-ion batteries is enticing to some because the inverse opal structure's interconnected macropores can be interfilled with three-dimensionally continuous electrode materials such that ions conduct over nanoscopic solid-state diffusion path lengths.

Gamma-lithium aluminate (γ-$LiAlO_2$) is purportedly an ion-conducting matrix that is used for molten carbonate fuel cells [144] and was prepared as a solid-state lithium-ion electrolyte inverse opal in 2003 [145]. Two sol–gel approaches were attempted, and a mixture of γ-$LiAlO_2$ and $LiAl_5O_8$ was present in nanocrystalline inverse opal macropore walls. No electronic or electrochemical measurements were recorded.

A more traditional lithium-ion solid electrolyte composition, $Li_{0.29}La_{0.55}TiO_3$ (LLT), is prepared from alkoxide and acetate precursors as phase-pure inverse opals [146]. Polystyrene CCTs that contain these precursors are first combusted, and the nanocrystallites in the pore walls are subsequently sintered at 1,000°C for several hours in order to reduce charge-transport impedance. According to complex impedance measurements, the inverse opal structure does not greatly diminish the ionic conductivity of inverse opal LLT at room temperature (1.1×10^{-4} S cm^{-1}) as compared to ground and pelletized LLT inverse opals that presumably densify at 25 MPa (2.1×10^{-4} S cm^{-1}). This finding suggests that the inverse opal structure has ample pathways for solid-state ionic transport. As expected, ionic transport is less facile in amorphous LLT inverse opals (~10^{-5} S cm^{-1}) because ions in amorphous solids may only slowly percolate through disordered vacancies rather than through an ordered crystalline lattice [147].

The LLT inverse opal structure can be interfilled with active electrode materials to prepare solid-state electrochemical half-cells, and these mixed battery material composites are closely related to the lithium-ion microbattery design that is reviewed in Sect. 2.5. Solid-state electrolyte LLT inverse opal membranes can be infiltrated with multiple precursor loadings that deposit $LiMn_2O_4$ as a lithium-ion cathode [148]. These composites have been tested as two-electrode cells against Li foil using a dry copolymer separator that contains a lithium imide salt. The specific capacity of the composite (83 mAh g^{-1}) is lower than theorized (148 mAh g^{-1}) at a low discharge rate of 0.016 C, but achieves a volumetric capacity of 220 mAh cm^{-3} that is comparable to measurements of $LiMn_2O_4$ in liquid electrolyte.

If inverse opal solid-state electrolytes have a future in battery applications, one challenge will be to avoid detrimental solid-state chemical reactions that occur at the interface between the inverse opal electrolyte and interfilled electrode material. Based on X-ray diffraction studies of a composite that comprises a NASICON-like $Li_{1.5}Al_{0.5}Ti_{1.5}(PO_4)_3$ inverse opal (LATP) and an interfilled $LiMn_2O_4$ cathode, Li diffuses out of LATP to form a few weight percent of rutile TiO_2 [149]. Forming mixed phases, especially from an electrochemically inert material such as rutile titania, can only reduce the mobility of ions in the electrolyte. Polarization is understandably observed in the constant current cycling profile of the interfilled $LiMn_2O_4$. Rather than seeing two discrete redox events in $LiMn_2O_4$ at 4.0 and 4.1 V, these plateaus are blurred into one. This solid electrolyte composition, however, is redox active at approximately 2.4 V, which could be dangerous if a 4-V battery were discharged to this threshold.

2.5 Lithium-Ion Microbatteries Based on Inverse Opal Nanoarchitectures

Based on the compendium of research that has been reviewed to this point, it is safe to say that nanostructured inverse opal electrodes are capable of power densities that exceed those of bulk materials. The inverse opal structure is also promising as a three-dimensionally interconnected scaffolding upon which battery materials can be deposited to form a three-dimensionally interpenetrated (3D) microbattery. Such a battery would not only have the power density of an individual inverse opal electrode but would also create an energy-dense microbattery architecture that stores energy more compactly than a low-density (<26 vol.%) inverse opal electrode. Distributed sensor applications and microelectromechanical systems (MEMS) need microbatteries that are scalable to fit on miniaturized devices and that provide relatively large energy densities per unit mass, geometrical footprint, and volume [150–153].

Carbon inverse opals have unique properties that make them a good choice as an electrode framework upon which to assemble three-dimensionally interpenetrated (3D) microbatteries. Carbon is readily formed as robust inverse opal monoliths that require no additional conductive additives to transport electronic charge or

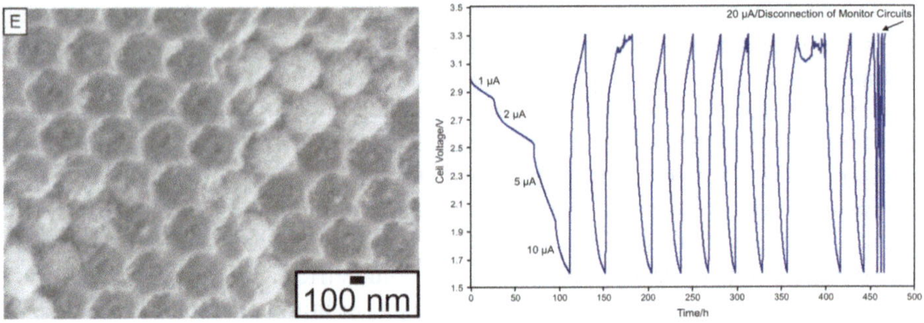

Fig. 2.6 A scanning electron micrograph and galvanostatic cycling profile of a three-dimensionally interpenetrated lithium-ion microbattery that is assembled from a carbon inverse opal anode, an ultrathin poly(phenylene oxide) solid electrolyte coating, and a vanadia aerogel-like cathode ([157]—reproduced by permission of The Electrochemical Society)

binders to maintain the physical form of each electrode [40, 88, 101]. Monolithic inverse opals have been prepared as metals [154] and oxides [11, 137, 155] but are somewhat fragile depending on the degree of sintering between crystallites in macropore walls. Carbon is a standard lithium-ion anode composition that is known to transport and host 100s mAh g^{-1} of electrochemical charge. Importantly, it can also be tailored to have macropore dimensions that are large enough to accommodate other battery nanomaterials within the inverse opal pore network [156, 157]. Based on these characteristics, carbon inverse opals are well suited as lithium-ion anodes in 3D microbatteries.

Once carbon inverse opals are anchored to current collectors, nanoscopic thin-film polymer electrolytes are electropolymerized at all electrifiable interfaces throughout the carbon electrode [156, 157]. Polyphenols conformally deposit as 10–90-nm coatings that self-limit as the growing polymer layers become increasingly resistive [158–162]. These polymers electronically passivate all underlying carbon electrode surfaces, yet modestly transport ions, and withstand dielectric breakdown in 4-V electrochemical environments.

Polyphenol-coated carbon inverse opals are subsequently vacuum infiltrated with neat vanadyl alkoxide liquid to deposit amorphous vanadium oxide as a lithium-ion cathode material within the remaining macroporous void space [156, 157]. Multiple fillings are necessary to maximize the loading of cathode materials. Two forms of vanadia have been tested as cathodes: vanadium oxide xerogels and aerogel-like materials, the latter being formed when pore fluids in the gel are replaced with acetone as a low-surface-tension fluid. In both cases, charge transport through vanadia is likely the rate-determining step during the electrochemical cycling of these 3D microbatteries.

Lithium ions are introduced into the carbon–polymer composite either prior to vanadia infiltration or after the cathode is incorporated into the macropore structure [156, 157]. The dense nature of the fully assembled cell makes ionic transport very slow, and introducing Li ions into the carbon–polymer composite alone is not only much faster but also incorporates Li ions into the nanoscopic polymer electrolyte film. Fully assembled inverse opal microbatteries are electrochemically evaluated completely in the solid state (see Fig. 2.6).

Microbatteries that contain vanadia aerogel-like cathodes discharge substantially more (16 mAh g^{-1}) electrochemical energy than xerogel vanadia cathodes (70 μAh g^{-1}) because the aerogel structure has far more surface area than dense xerogels. Although still a prototype, these microbatteries are the first example of incorporating inverse opal lithium-ion battery materials together to construct a functioning electrochemical cell.

2.6 Summary and Outlook

This chapter illustrates the many compositions of lithium-ion anodes, cathodes, and solid electrolytes that have been fabricated with the inverse opal structure. The essence of this body of work reinforces that ion diffusion through nanostructured solids, particularly those with through-connected macropore structures, delivers power densities that exceed those of bulk electrode materials. Whether highly ordered or otherwise, interconnected macropores minimally impede the flow of liquid electrolyte through inverse opal monoliths, which in part grants substantial electrochemical rate capabilities to some otherwise insulating compositions. Low-density nanostructured materials, like inverse opals, are prone to morphological changes during cycling, specifically in anode compositions that reversibly form Li alloys. Electrically conductive, monolithic carbon inverse opals are critical as binderless electrodes that are responsible for some recent developments in 3D microbatteries.

The inverse opal structure and other nanostructured electrodes will continue to play an important role in lithium-ion batteries as a way to boost electrode power densities. With time, additional research may develop new ways to artfully arrange other compositions of inverse opal lithium-ion electrodes into energy-dense solid-state microbattery configurations. The intentionally periodic and interconnected spherical macropores also make electronically conductive inverse opals an attractive high-surface-area current collector for future work with other lithium-ion insertion nanoparticles and electroactive nanoparticles and coatings.

References

1. Schalkwijk WA, Scrosati B (2002) Advances in lithium-ion batteries. Kluwer/Plenum, New York
2. Lytle JC, Stein A (2006) Recent progress in syntheses and applications of inverse opals and related macroporous materials prepared by colloidal crystal templating. In: Cao G, Brinker CJ (eds) Annual review of nano research, 1st edn. World Scientific, Hackensack
3. Patrissi CJ, Martin CR (1993) Sol-gel based template synthesis and Li-insertion rate performance of nanostructured vanadium pentoxide. J Electrochem Soc 146:3176–3180

4. Owens BB, Passerini S, Smyrl WH (1999) Lithium ion insertion in porous metal oxides. Electrochim Acta 45:215–224
5. Poizot P, Laruelle S, Grugeon S, Dupont L, Tarascon JM (2000) Nano-sized transition metal oxides as negative-electrode materials for lithium-ion batteries. Nature 407:496–499
6. Nordlinder S, Edstrom K, Gustafsson T (2001) The performance of vanadium oxide nanorolls as cathode material in a rechargeable lithium battery. Electrochem Solid State Lett 4:A129–A131
7. Huang H, Yin SC, Kerr T, Taylor N, Nazar LF (2002) Nanostructured composites: a high capacity, fast rate $Li_3V_2(PO_4)_3$/carbon cathode for rechargeable batteries. Adv Mater 14:1525–1528
8. Mui SC, Trapa PE, Huang B, Soo PP, Lozow MI, Wang TC, Cohen RE, Mansour AN, Mukerjee S, Mayes AM, Sadoway DR (2002) Block-copolymer-templated nanocomposite electrodes for rechargeable lithium batteries. J Electrochem Soc 149:A1610–A1615
9. Li N, Mitchell DT, Lee KP, Martin CR (2003) A nanostructured honeycomb carbon anode. J Electrochem Soc 150:A979–A984
10. Sauvage F, Baudrin E, Morcrette M, Tarascon JM (2004) Pulsed laser deposition and electrochemical properties of $LiFePO_4$ thin films. Electrochem Solid State Lett 7:A15–A18
11. Sakamoto J, Dunn B (2002) Hierarchical battery electrodes based on inverted opal structures. J Mater Chem 12:2859–2861
12. Yan H, Blanford CF, Holland BT, Parent M, Smyrl WH, Stein A (1999) A chemical synthesis of periodic macroporous NiO and metallic Ni. Adv Mater 11:1003–1006
13. Stein A (2001) Sphere templating methods for periodic porous solids. Micropor Mesopor Mater 44–45:227–239
14. Thomas A, Goettmann F, Antonietti M (2008) Hard templates for soft materials: creating nanostructured organic materials. Chem Mater 20:738–755
15. Zakhidov AA, Baughman RH, Iqbal Z, Cui C, Khayrullin I, Dantas SO, Marti J, Ralchenko VG (1998) Carbon structures with three-dimensional periodicity at optical wavelengths. Science 282:897–901
16. Su F, Zhao XS, Wang Y, Zeng J, Zhou Z, Lee JY (2005) Synthesis of graphitic ordered macroporous carbon with a three-dimensional interconnected pore structure for electrochemical applications. J Phys Chem B 109:20200–20206
17. Velev OD, Jede TA, Lobo RF, Lenhoff AM (1997) Porous silica via colloidal crystallization. Nature 389:447–448
18. Holland BT, Blanford CF, Stein A (1998) Synthesis of macroporous minerals with highly ordered three-dimensional arrays of spheroidal voids. Science 281:538–540
19. Harris WS (1958) Electrochemical studies in cylic esters. Dissertation, University of California
20. Whittingham MS, Huggins RA (1971) Measurement of sodium ion transport in beta alumina using reversible solid electrodes. J Chem Phys 54:414–416
21. Whittingham MS, Huggins RA (1971) Transport properties of silver beta alumina. J Electrochem Soc 118:1–6
22. Whittingham MS, Huggins RA (1972) Electrochemical preparation and characterization of alkali metal tungsten bronzes, M_xWO_3. In: Solid state chemistry—Proceedings of the 5th materials research symposium sponsored by the Institute for Materials Research, National Bureau of Standards, edited by Roth RS, Schneider SJ Jr, U.S. Department of Commerce and National Bureau of Standards, Washington, D.C., pp 51–62
23. Whittingham MS (1974) The hydrated intercalation complexes of the layered disulfides. Mater Res Bull 9:1681–1689
24. Whittingham MS, Gamble FR Jr (1975) The lithium intercalates of the transition metal dichalcogenides. Mater Res Bull 10:363–371
25. Whittingham MS (1976) Electrical energy storage and intercalation chemistry. Science 192:1126–1127

26. Courtney IA, Dahn JR (1997) Key factors controlling the reversibility of the reaction of lithium with SnO_2 and Sn_2BPO_6 glass. J Electrochem Soc 144:2943–2948
27. Courtney IA, Dahn JR (1997) Electrochemical and in-situ x-ray diffraction studies of the reaction of lithium with tin oxide composites. J Electrochem Soc 144:2045–2052
28. Yu P, Popov BN, Ritter JA, White RE (1999) Determination of the lithium ion diffusion coefficient in graphite. J Electrochem Soc 146:8–14
29. Churikov AV, Ivanishchev AV, Ivanishcheva IA, Sycheva VO, Khasanova NR, Antipov EV (2010) Determination of lithium diffusion coefficient in LiFePO4 electrode by galvanostatic and potentiostatic intermittent titration techniques. Electrochim Acta 55:2939–2950
30. Capiglia C, Saito Y, Kageyama H, Mustarelli P, Iwamoto T, Tabuchi T, Tukamoto H (1999) ^{7}Li and ^{19} F diffusion coefficients and thermal properties of non-aqueous electrolyte solutions for rechargeable lithium batteries. J Power Sources 81–82:859–862
31. Xu K (2004) Nonaqueous liquid electrolytes for lithium-based rechargeable batteries. Chem Rev 104:4303–4417
32. Valoen LO, Reimers JN (2005) Transport properties of $LiPF_6$-based Li-ion battery electrolytes. J Electrochem Soc 152:A882–A891
33. Nyman A, Behm M, Lindbergh G (2008) Electrochemical characterisation and modelling of the mass transport phenomena in $LiPF''_6$-EC-EMC electrolyte. Electrochim Acta 53:6356–6365
34. Pekala RW, Farmer JC, Alviso CT, Tran TD, Mayer ST, Miller JM, Dunn B (1998) Carbon aerogels for electrochemical applications. J Non-Cryst Solids 225:74–80
35. Jurewicz K, Vix-Guterl C, Frackowiak E, Saadallah S, Reda M, Parmentier J, Patarin J, Beguin F (2004) Capacitance properties of ordered porous carbon materials prepared by a templating procedure J. Phys Chem Solids 65:287–293
36. Ergang NS, Lytle JC, Yan H, Stein A (2005) The effect of a macropore structure on cycling rates of $LiCoO_2$. J Electrochem Soc 152:A1989–A1995
37. Doescher MS, Pietron JJ, Denning BM, Long JW, Rhodes CP, Edmonson CA, Rolison DR (2005) Using an oxide nanoarchitecture to make or break a proton wire. Anal Chem 77:7924–7932
38. Fu R, Zheng B, Liu J, Weiss S, Ying JY, Dresselhaus MS, Dresselhaus G, Satcher J, Baumann T (2004) Studies of the chemical and pore structures of the carbon aerogels synthesized by gelation and supercritical drying in isopropanol. J Appl Polymer Sci 91:3060–3067
39. Wang J, Yang X, Wu D, Fu R, Dresselhaus MS, Dresselhaus G (2008) The porous structures of activated carbon aerogels and their effects on electrochemical performance. J Power Sources 185:589–594
40. Lee KT, Lytle JC, Ergang NS, Oh SM, Stein A (2005) Synthesis and rate performance of monolithic macroporous carbon electrodes for lithium-ion secondary batteries. Adv Funct Mater 15:547–556
41. Doherty CM, Caruso RA, Smarsly BM, Drummond CJ (2009) Colloidal crystal templating to produce hierarchically porous $LiFePO_4$ electrode materials for high power lithium ion batteries. Chem Mater 21:2895–2903
42. Goodwin JW, Hearn J, Ho CC, Ottewill RH (1973) The preparation and characterization of polymer lattices formed in the absence of surface active agents. Br Polym J 5:347–362
43. Goodwin JW, Ottewill RH, Pelton R, Vianello G, Yates DE (1978) Control of particle size in the formation of polymer lattices. Br Polym J 10:173–180
44. Zou D, Ma S, Guan R, Park M, Sun L, Aklonis JJ, Salovey R (1992) Model filled polymers. V. Synthesis of crosslinked monodisperse polymethacrylate beads. J Polym Sci Part A Polym Chem 30:137–144
45. Tanrisever T, Okay O, Sonmezoglu IC (1996) Kinetics of emulisifier-free emulsion polymerization of methyl methacrylate. J Appl Polym Sci 61:485–493
46. Xia Y, Gates B, Yin Y, Lu Y (2000) Monodispersed colloidal spheres: old materials with new applications. Adv Mater 12:693–713

47. Stöber W, Fink A, Bohn E (1968) Controlled growth of monodisperse silica spheres in the micron size range. J Colloid Interface Sci 26:62–69
48. van Blaaderen A, van Geest J, Vrij A (1992) Monodisperse colloidal silica spheres from tetraalkoxysilanes: particle formation and growth mechanism. J Colloid Interface Sci 154:481–501
49. Odian G (1991) Principles of polymerization. Wiley, New York
50. Xia Y, Gates B, Li ZY (2001) Self-assembly approaches to three-dimensional photonic crystals. Adv Mater 13:409–413
51. Davis KE, Russel WB, Glantschnig WJ (1991) Settling suspensions of colloidal silica: observations and x-ray measurements. J Chem Soc Faraday Trans 87:411–424
52. Mayoral R, Requena J, Moya JS, López C, Cintas A, Míguez H, Meseguer F, Vázquez L, Holgado M, Blanco A (1997) 3D long-range ordering in an SiO_2 submicrometer-sphere sintered superstructure. Adv Mater 9:257–260
53. Míguez H, Mesguer F, López C, Blanco A, Moya JS, Requena J, Mifsud A, Fornés V (1998) Control of the photonic crystal properties of fcc-packed submicrometer SiO_2 spheres by sintering. Adv Mater 10:480–483
54. Pieranski P (1983) Colloidal crystals. Contemp Phys 24:25–73
55. El-Aasser MS, Robertson AA (1971) An ultracentrifugation technique for the study of latex coalescence. J Colloid Interface Sci 93:504–512
56. Efremov IF (1976) Periodic colloid structures. In: Matijevic E (ed). Surface and Colloid Science, Vol 8, Wiley, New York
57. Tsunekawa S, Barnakov YA, Poborchii VV, Samoilovich SM, Kasuya A, Nishina Y (1997) Characterization of precious opals: AFM and SEM observations, photonic band gap, and incorporation of CdS nano-particles. Micropor Mater 8:275–282
58. Wijnhoven JEGJ, Vos WL (1998) Preparation of photonic crystals made of air spheres in titania. Science 281:802–804
59. Jiang P, Bertone JF, Hwang KS, Colvin VL (1999) Single-crystal colloidal multilayers of controlled thickness. Chem Mater 11:2132–2140
60. Vlasov YA, Bo XZ, Sturm JC, Norris DJ (2001) On-chip natural assembly of silicon photonic bandgap crystals. Nature 414:289–293
61. Lytle JC, Yan H, Ergang N, Smyrl WH, Stein A (2004) Structural and electrochemical properties of three-dimensionally ordered macroporous tin(IV) oxide films. J Mater Chem 14:1616–1622
62. Woodcock LV (1997) Entropy difference between crystal phases. Nature 388:235–237
63. Woodcock LV (1997) Entropy difference between the face-centered cubic and hexagonal close-packed crystal structures. Nature 385:141–143
64. Blanford CF, Carter CB, Stein A (2004) A Method for determining void arrangements in inverse opals. J Microscopy 216:263–287
65. Sanders JV (1968) Diffraction of light by opals. Acta Crystallogr A 24:427–434
66. Balakirev VG, Bogomolov VN, Zhuravlev VV, Kumzerov YA, Petranovskii VP, Romanov SG, Samoilovich LA (1993) Three-dimensional superlattices in opals. Crystallogr Rep 38:348–353
67. Okubo T (1990) Phase transition between liquid-like and crystal-like structures of deionized colloidal suspensions. J Chem Soc Faraday Trans 86:2871–2876
68. Larsen AE, Grier DG (1997) Like-charge attractions in metastable colloidal crystallites. Nature 385:230–233
69. Baughman RH, Dantas SO, Stafstrom S, Zakhidov AA, Mitchell TB, Dublin DHE (2000) Materials with negative compressibilities in one or more dimensions. Science 288:2018–2022
70. Wu JZ, Bratko D, Blanch HW, Prausnitz JM (2000) Effects of three body forces on the phase behavior of charged colloids. J Chem Phys 113:3360–3365
71. Garcia-Santamaria F, Miyazaki HT, Urquia A, Ibisate M, Belmonte M, Shinya N, Meseguer F, Lopez C (2002) Nanorobotic manipulation of microspheres for on-chip diamond architectures. Adv Mater 14:1144–1147

72. Yan H, Blanford CF, Holland BT, Smyrl WH, Stein A (2000) General synthesis of periodic macroporous solids by templated salt precipitation and chemical conversion. Chem Mater 12:1134–1141
73. Yan H, Blanford CF, Lytle JC, Carter CB, Smyrl WH, Stein A (2001) Influence of processing conditions on structures of 3D ordered macroporous metals prepared by colloidal crystal templating. Chem Mater 13:4314–4321
74. Sakamoto JS, Dunn B (2002) Hierarchical battery electrodes based on inverted opal structures. J Mater Chem 12:2859–2861
75. Yan H, Sokolov S, Lytle JC, Stein A, Zhang F, Smyrl WH (2003) Colloidal-crystal-templated synthesis of ordered macroporous electrode materials for lithium secondary batteries. J Electrochem Soc 150:A1102–A1107
76. Nishikawa K, Dokko K, Kinoshita K, Woo S-W, Kanamura K (2009) Three-dimensionally ordered macroporous Ni–Sn anode for lithium batteries. J Power Sources 189:726–729
77. Wijnhoven JEGJ, Zenvenhuizen SJM, Hendriks MA, Vanmaekelbergh D, Kelly JJ, Vos WL (2000) Electrochemical assembly of ordered macropores in gold. Adv Mater 12:888–890
78. Cassagneau T, Caruso F (2002) Semiconducting polymer inverse opals prepared by electropolymerization. Adv Mater 14:34–38
79. Tian S, Wang J, Jonas U, Knoll W (2005) Inverse opals of polyaniline and its copolymers prepared by electrochemical techniques. Chem Mater 17:5726–5730
80. Jiang P, Cizeron J, Bertone JF, Colvin VL (1999) Preparation of macroporous metal films from colloidal crystals. J Am Chem Soc 121:7957–7958
81. Cong H, Cao W (2004) Macroporous Au materials prepared from colloidal crystals as templates. J Colloid Interface Sci 278:423–427
82. Huang L, Wang Z, Sun J, Miao L, Li Q, Yan Y, Zhao D (2000) Fabrication of ordered porous structures by self-assembly of zeolite nanocrystals. J Am Chem Soc 122:3530–3531
83. Wang YJ, Tang Y, Ni Z, Hua WM, Yang WL, Wang XD, Tao WC, Gao Z (2000) Synthesis of macroporous materials with zeolitic microporous frameworks by self-assembly of colloidal zeolites. Chem Lett 5:510–511
84. Meng Q, Fu CH, Einaga Y, Gu ZZ, Fujishima A, Sato O (2002) Assembly of highly ordered three-dimensional porous structure with nanocrystalline TiO2 semiconductors. Chem Mater 14:83–88
85. Subramania G, Constant K, Biswas R, Sigalas MM, Ho K-M (2002) Visible frequency thin film photonic crystals from colloidal systems of nanocrystalline titania and polystyrene microspheres. J Am Ceram Soc 85:1383–1386
86. Miclea TP, Romanov SG, Torres CM, Liang Z, Susha A, Caruso F (2004) Towards 3D metal-dielectric photonic crystal. Optical characterization. Mol Cryst Liq Cryst 415:211–219
87. Wang Z, Ergang NS, Al-Daous MA, Stein A (2005) Synthesis and characterization of three-dimensionally ordered macroporous carbon/titania nanoparticle composites. Chem Mater 17:6805–6813
88. Baumann TF, Satcher JH (2004) Template-directed synthesis of periodic macroporous organic and carbon aerogels. J Non-Cryst Solids 350:120–125
89. Li ZJ, Jaroniec M, Lee YJ, Radovic LR (2002) High surface area graphitized carbon with uniform mesopores synthesized by a colloidal imprinting method. Chem Commun 13:1346–1347
90. Perpall MW, Prasanna K, Perera U, DiMaio J, Ballato J, Foulger SH, Smith DW Jr (2003) Novel network polymer for templated carbon photonic crystal structures. Langmuir 19:7153–7156
91. Yoon SB, Chai GS, Kang SK, Yu J-S, Gierszal KP, Jaroniec M (2005) Graphitized pitch-based carbons with ordered nanopores synthesized by using colloidal crystals as templates. J Am Chem Soc 127:4188–4189
92. Blanco A, Chomski E, Grabtchak S, Ibisate M, John S, Leonard SW, Lopez C, Meseguer F, Miguez H, Mondia JP, Ozin GA, Toader O, van Driel HM (2000) Large-scale synthesis of a

silicon photonic crystal with a complete three-dimensional bandgap near 1.5 micrometres. Nature 405:437–440
93. von Freymann G, John S, Schulz-Dobrick M, Vekris E, Tetreault N, Wong S, Kitaev V, Ozin GA (2004) Tungsten inverse opals: the influence of absorption on the photonic band structure in the visible spectral region. Appl Phys Lett 84:224–226
94. Juarez BH, Garcia PD, Golmayo D, Blanco A, Lopez C (2005) ZnO inverse opals by chemical vapor deposition. Adv Mater 17:2761–2765
95. Wu Q, Ross AD, Gleason KK (2005) Nanoporous organosilicate glass films via chemical vapor deposition onto colloidal crystal templates. Plasma Process Polym 2:401–406
96. Vanysek P (1998) Revision of tables of electrochemical series. CRC Press, Boca Raton
97. Dahn JR, Zheng T, Liu Y, Xue JS (1995) Mechanisms for lithium insertion in carbonaceous materials. Science 270:590–593
98. Yu J-S, Yoon SB, Chai GS (2001) Ordered uniform porous carbon by carbonization of sugars. Carbon 39:1421–1446
99. Kang S, Yu J-S, Kruk M, Jaroniec M (2002) Synthesis of an ordered macroporous carbon with 62 nm spherical pores that exhibit unique gas adsorption properties. Chem Commun 2:1670–1671
100. Yu J-S, Kang S, Yoon SB, Chai GS (2002) Fabrication of ordered uniform porous carbon networks and their application to a catalyst supporter. J Am Chem Soc 124:9382–9383
101. Baumann TF, Satcher JH (2003) Homogeneous incorporation of metal nanoparticles into ordered macroporous carbons. Chem Mater 15:3745–3747
102. Take H, Matsumoto T, Hiwatashi S, Nakayama T, Niihara K, Yoshino K (2004) Preparation and electronic properties of nanoporous carbon inverse opal. Jpn J Appl Phys 43:4453–4457
103. Wang Z, Li F, Ergang NS, Stein A (2006) Effects of hierarchical architecture on electronic and mechanical properties of nanocast monolithic porous carbons and carbon-carbon nanocomposites. Chem Mater 18:5543–5553
104. Aliev AE, Lee SB, Baughman RH, Zakhidov AA (2007) Thermal properties of carbon inverse opal photonic crystals. J Lumin 125:11–17
105. Wang Z, Li F, Ergang NS, Stein A (2008) Synthesis of monolithic 3D ordered macroporous carbon/nano-silicon composites by diiodosilane decomposition. Carbon 46:1702–1710
106. Boukamp BA, Lesh GC, Huggins RA (1981) All-solid lithium electrodes with mixed-conductor matrix. J Electrochem Soc 128:725–729
107. Fu LJ, Yang LC, Shi Y, Wang B, Wu YP (2009) Synthesis of carbon coated nanoporous microcomposite and its rate capability for lithium ion battery. Micropor Mesopor Mater 117:515–518
108. Winter M, Besenhard JO (1999) Electrochemical lithiation of tin and tin-based intermetallics and composites. Electrochim Acta 45:31–50
109. Beaulieu LY, Hatchard TD, Bonakdarpour A, Fleischauer MD, Dahn JR (2003) Reaction of Li with alloy thin films studies by in situ AFM. J Electrochem Soc 150:A1457–A1464
110. Scott RWJ, Yang SM, Chabanis G, Coombs N, Williams DE, Ozin GA (2001) Tin dioxide opals and inverted opals: near-ideal microstructures for gas sensors. Adv Mater 13:1468–1472
111. Acciarri M, Barberini R, Canevali C, Mattoni M, Mari CM, Morazzoni F, Nodari L, Polizzi S, Ruffo R, Russo U, Sala M, Scotti R (2005) Ruthenium(platinum)-doped tin dioxide inverted opals for gas sensors: synthesis, electron paramagnetic resonance, mossbauer, and electrical investigation. Chem Mater 17:6167–6171
112. Wang Z, Fierke MA, Stein A (2008) Porous carbon/tin(iv) oxide monoliths as anodes for lithium-ion batteries. J Electrochem Soc 155:A658–A663
113. Kulinowski KM, Jiang P, Vaswani H, Colvin VL (2000) Porous metals from colloidal templates. Adv Mater 12:833–838
114. Velev OD, Kaler EW (2000) Structured porous materials via colloidal crystal templating: from inorganic oxides to metals. Adv Mater 12:531–534

115. Eradat N, Huang JD, Vardeny ZV, Zakhidov AA, Khayrullin I, Udod I, Baughman RH (2001) Optical studies of metal-infiltrated opal photonic crystals. Synth Metal 116:501–504
116. Reynolds AL, Cassagne D, Jouanin C, Arnold JM (2001) Optical properties of bare, sintered and coated opal-based photonic crystals. Synth Metal 116:453–456
117. Esmanski A, Ozin GA (2009) Silicon inverse-opal-based macroporous materials as negative electrodes for lithium ion batteries. Adv Funct Mater 19:1999–2010
118. Sumida T, Wada Y, Kitamura T, Yanagida S (2002) Electrochemical change of the photonic stop band of the ordered macroporous WO3 films. Chem Lett 31:180–181
119. Colbow KM, Dahn JR, Haering RR (1989) Structure and electrochemistry of the spinel oxides $LiTi_2O_4$ and $Li_{4/3}Ti_{5/3}O_4$. J Power Sources 26:397–402
120. Ohzuku T, Ueda A, Yamamoto N (1995) Zero-strain insertion material of $Li[Li_{1/3}Ti_{5/3}]O_4$ for rechargeable lithium cells. J Electrochem Soc 142:1431–1435
121. Peramunage D, Abraham KM (1998) Preparation of micron-sized $Li_4Ti_5O_{12}$ and its electrochemistry in polyacrylonitrile electrolyte-based lithium cells. J Electrochem Soc 145:2609–2615
122. Sarciaux S, Le Gal La Salle A, Guyomard D, Piffard Y (1998) Electrochemical study of the lithium insertion mechanism Into $Li_4Ti_5O_{12}$. Mol Cryst Liq Cryst 311:63–68
123. Robertson AD, Trevino L, Tukamoto H, Irvine JTS (1999) New inorganic spinel oxides for use as negative electrode materials in future lithium-ion batteries. J Power Sources 81–82:352–357
124. Wang GX, Bradhurst DH, Dou SX, Liu HK (1999) Spinel $Li[Li_{1/3}Ti_{5/3}]O_4$ as an anode material for lithium ion batteries. J Power Sources 83:156–161
125. Chen CH, Vaughey JT, Jansen AN, Dees DW, Kahaian AJ, Goacher T, Thackeray MM (2001) Studies of Mg-substituted $Li_{4-x}Mg_xTi_5O_{12}$ spinel electrodes ($0 < x < 1$) for lithium batteries. J Electrochem Soc 148:A102–A104
126. Ronci F, Reale P, Scrosati B, Panero S, Albertini VR, Perfetti P, di Michiel M, Merino JM (2002) High-resolution in-situ measurements of the $Li_{4/3}Ti_{5/3}O_4$ "Zero-Strain" insertion material. J Phys Chem B 106:3082–3086
127. Sorensen EM, Barry SJ, Jung HK, Rondinelli JR, Vaughey JT, Poeppelmeier KR (2006) Three-dimensionally ordered macroporous $Li_4Ti_5O_{12}$: effect of wall structure on electrochemical properties. Chem Mater 18:482–489
128. Woo S-W, Dokko K, Kanamura K (2007) Preparation and characterization of three dimensionally ordered macroporous $Li_4Ti_5O_{12}$ anode for lithium batteries. Electrochim Acta 53:79–82
129. Bing Z, Yuan Y, Wang Y, Fu Z-W (2006) Electrochemical characterization of a three dimensionally ordered macroporous anatase TiO_2 electrode. Electrochem Solid State Lett 9:A101–A104
130. Li ZH, Zhao TP, Zhan XY, Gao DS, Xiao QZ, Lei GT (2010) High capacity three-dimensional ordered macroporous CoFe2O4 as anode material for lithium ion batteries. Electrochim Acta 55:4594–4598
131. Wang Y, Cao G (2008) Developments in nanostructured cathode materials for high-performance lithium-ion batteries. Adv Mater 20:2251–2269
132. Holland BT, Blanford CF, Do T, Stein A (1999) Synthesis of highly ordered, three-dimensional, macroporous structures of amorphous or crystalline inorganic oxides, phosphates, and hybrid composites. Chem Mater 11:795–805
133. Shao-Horn Y, Croguennec L, Delmas C, Nelson EC, O'Keefe MA (2003) Atomic resolution of lithium ions in $LiCoO_2$. Nature Mater 2:464–467
134. Kallala M, Sanchez C, Cabane B (1993) Structures of inorganic polymers in sol-gel processes based on titanium oxide. Phys Rev E 48:3692–3704
135. Kavan L, Grätzel M (2002) Facile synthesis of nanocrystalline Li4Ti5O12 (spinel) exhibiting fast Li insertion. Electrochem Sol St Lett 5:A39–A42
136. Du K, Zhang H, Qi L (2004) Improved rate performance of $LiCoO_2$-based multi-phase composite due to Pt additive. Electrochim Acta 50:211–216

137. Tonti D, Torralvo MJ, Enciso E, Sobrados I, Sanz J (2008) Three-dimensionally ordered macroporous lithium manganese oxide for rechargeable lithium batteries. Chem Mater 20:4783–4790
138. Padhi AK, Nanjundaswamy KS, Goodenough JB (1997) Phospho-olivines as positive-electrode materials for rechargeable lithium batteries. J Electrochem Soc 144:1188–1194
139. Meethong N, Kao Y-H, Carter WC, Chiang Y-M (2010) Comparative study of lithium transport kinetics in olivine cathodes for Li-ion batteries. Chem Mater 22:1088–1097
140. Lu J, Tang Z, Zhang Z, Shen W (2005) Preparation of LiFePO4 with inverse opal structure and its satisfactory electrochemical properties. Mater Res Bull 40:2039–2046
141. Cui W-J, Liu H-J, Wang C-X, Xia Y-Y (2008) Highly ordered three-dimensional macroporous $FePO_4$ as cathode materials for lithium–ion batteries. Electrochem Comm 10:1587–1589
142. Lytle JC, Rhodes CP, Long JW, Pettigrew KA, Stroud RM, Rolison DR (2007) The importance of combining disorder with order for Li-ion insertion into cryogenically prepared nanoscopic ruthenia. J Mater Chem 17:1292–1299
143. Peled E (1979) The electrochemical behavior of alkali and alkaline earth metals in nonaqueous battery systems—the solid electrolyte interphase model. J Electrochem Soc 126:2047–2051
144. Paetsch LM, Doyon JD, Faroque M (1993) 3rd symposium on carbonate fuel cell technology. In: Shores D, Maru H, Uchida I, Selman JR (eds) 1st edn. Electrochemical Society, Pennington
145. Sokolov S, Stein A (2003) Preparation and characterization of macroporous γ-$LiAlO_2$. Mater Lett 57:3593–3597
146. Dokko K, Akutagawa N, Isshiki Y, Hoshina K, Kanamura K (2005) Preparation of three dimensionally ordered macroporous $Li_{0.35}La_{0.55}TiO_3$ by colloidal crystal templating process. Solid State Ion 176:2345–2348
147. Kanamura K, Akutagawa N, Dokko K (2005) Three dimensionally ordered composite solid materials for all solid-state rechargeable lithium batteries. J Power Sources 146:86–89
148. Hara M, Nakano H, Dokko K, Okuda S, Kaeriyama A, Kanamura K (2008) Fabrication of all solid-state lithium-ion batteries with three-dimensionally ordered composite electrode consisting of $Li_{0.35}La_{0.55}TiO_3$ and $LiMn_2O_4$. J. Power Source 189:485–489
149. Nakano H, Dokko K, Hara M, Isshiki Y, Kanamura K (2008) Three-dimensionally ordered composite electrode between $LiMn_2O_4$ and $Li_{1.5}Al_{0.5}Ti_{1.5}(PO_4)_3$. Ionics 14:173–177
150. Long JW, Dunn B, Rolison DR, White HS (2004) Three-dimensional battery architectures. Chem Rev 104:4463–4492
151. Nathan M, Golodnitsky D, Yufit V, Strauss E, Ripenbein T, Shechtman I, Menkin S, Peled E (2005) Recent advances in three dimensional thin film microbatteries. Mater Res Soc Symp Proc 835:K10.10.11–K10.10.16
152. Baggetto L, Niessen AH, Roozeboom F, Notten PHL (2008) High energy density all-solid-state batteries: a challenging concept towards 3D integration. Adv Funct Mater 18:1–10
153. Lai W, Erdonmez CK, Marinis TF, Bjune CK, Dudney NJ, Xu F, Wartena R, Chiang Y-M (2010) ultrahigh-energy-density microbatteries enabled by new electrode architecture and micropackaging design. Adv Mater 22:E139–E144
154. Denny NR, Han S, Turgeon RT, Lytle JC, Norris DJ, Stein A (2005) Synthetic approaches toward tungsten photonic crystals for thermal emission. In: Proceedings of SPIE—The International Society for Optical Engineering Boston, MA, USA, 6005: 600505/600501–600505/600513
155. Li F, Wang Z, Ergang NS, Fyfe CA, Stein A (2007) Controlling the shape and alignment of mesopores by confinement in colloidal crystals: designer pathways to silica monoliths with hierarchical porosity. Langmuir 23:3996–4004
156. Ergang NS, Lytle JC, Lee KT, Oh SM, Smyrl WH, Stein A (2006) Photonic crystal structures as a basis for a three-dimensionally interpenetrating electrochemical cell system. Adv Mater 18:1750–1753

157. Ergang NS, Fierke MA, Wang Z, Smyrl WH, Stein A (2007) Fabrication of a fully infiltrated three-dimensional solid-state interpenetrating electrochemical cell. J Electrochem Soc 154:A1135–A1139
158. McCarley RL, Thomas RE, Irene EA, Murray RW (1990) Optical, electrical, and electrochemical characteristics of ultrathin poly(phenylene oxide) films: organic dielectrics less than 10 nm thick. J Electroanal Chem 290:79–92
159. McCarley RL, Irene EA, Murray RW (1991) Permeant molecular sieving with electrochemically prepared 6-nm films of poly(phenylene oxide). J Phys Chem 95:2492–2498
160. Long JW, Rhodes CP, Young AL, Rolison DR (2003) Ultrathin, protective coatings of poly (o-phenylenediamine) as electrochemical proton gates: making mesoporous MnO_2 nanoarchitectures stable in acid electrolytes. Nano Lett 3:1155–1161
161. Rhodes CP, Long JW, Doescher MS, Fontanella JJ, Rolison DR (2004) Nanoscale polymer electrolytes: ultrathin electrodeposited poly(phenylene oxide) with solid-state ionic conductivity. J Phys Chem B 108:13079–13087
162. Rhodes CP, Long JW, Rolison DR (2005) Direct electrodeposition of nanoscale solid polymer electrolytes via electropolymerization of sulfonated phenols. Electrochem Solid State Lett 8:A579

Chapter 3
Nano-engineered Silicon Anodes for Lithium-Ion Rechargeable Batteries

Rahul Krishnan, Rahul Mukherjee, Toh-Ming Lu, and Nikhil Koratkar

Abstract Lithium-ion batteries continue finding use in a variety of applications and are the ideal choice for next generation wireless communication devices and hybrid electric vehicles. However, there is a significant need to improve the power density, energy density, and the cycle life of these batteries for which nano-engineered silicon anodes have been realized as an effective approach. These silicon nanostructures provide very high charge storage capacity when compared with traditional carbon-based anodes. Moreover, shorter lithium diffusion distances associated with nanostructured anodes enable the battery to be operated at ultrafast charge/discharge rates which are critical for high-power applications such as in automotives. The purpose of this chapter is to introduce the reader to the concept of nanostructured silicon as an anode material for lithium-ion battery technology and to review the latest developments in the field.

R. Krishnan
Department of Materials Science and Engineering, Rensselaer Polytechnic Institute, Troy, NY 12180, USA

R. Mukherjee
Department of Mechanical, Aerospace and Nuclear Engineering, Rensselaer Polytechnic Institute, Troy, NY 12180, USA

T.-M. Lu
Department of Physics, Applied Physics, and Astronomy, Rensselaer Polytechnic Institute, Troy, NY 12180, USA

N. Koratkar (✉)
Department of Materials Science and Engineering, Rensselaer Polytechnic Institute, Troy, NY 12180, USA

Department of Mechanical, Aerospace and Nuclear Engineering, Rensselaer Polytechnic Institute, Troy, NY 12180, USA
e-mail: koratn@rpi.edu

Y. Abu-Lebdeh and I. Davidson (eds.), *Nanotechnology for Lithium-Ion Batteries*, Nanostructure Science and Technology, DOI 10.1007/978-1-4614-4605-7_3,

3.1 Introduction

Rechargeable lithium-ion batteries are ubiquitous in today's information-rich, mobile society. They are one of the most popular types of battery for portable electronics, with one of the best energy-to-weight ratios, suffering virtually no memory effect, and a slow discharge rate when not in use [1]. Among the various new material concepts for lithium-ion batteries that are being explored by academia, research laboratories, and industry, nanostructured silicon shows exceptional promise as an anode material. Nanostructured silicon in the form of nanowires, nanorods, nanoscoops, and nano-compliant layers have shown a significant improvement in addressing issues related to the charge capacities, charge/discharge rates, and cycle life, thereby improving the performance characteristics of a lithium-ion battery.

Lithium (Li)-based rechargeable batteries were first proposed in the early 1960s and since then the battery has undergone several transformations. Around the 1970s, sulfide-based intercalation compounds [2] and metallic Li were used as the cathode and anode materials, respectively. However, it posed serious safety hazards because of dendritic Li growth from replating during charge-discharge cycling [3]. By the end of 1980s and early 1990s, it was shown that use of an insertion material such as pyrolytic carbon as the anode solved this problem [4, 5]. This led to the rocking-chair or lithium-ion technology. Since Li was present only in its ionic state, the dendrite problem was effectively resolved with this technology. However, the disadvantage of using an insertion material as the anode was that these anodes had higher redox potentials than Li metal. This in turn reduced the cell potential. To overcome this problem, transition metal oxides were employed as the cathode material due to their higher insertion potential with respect to transition metal disulfides [6]. Commercially, the most popular material for the anode is graphite or carbon [7] and the first commercial C/$LiCoO_2$-based lithium-ion battery was introduced by Sony in 1991 [8]. Extensive research on lithium-ion battery concepts is ongoing, with an intention to make these batteries applicable in a wide variety of applications. For instance, electric vehicles will require high-power batteries, mobile phones and other modern communication devices will require a high-energy density from the battery, and Uninterrupted Power Storage (UPS) and Stationary Storage Batteries (SSB) will require long cycle life lithium-ion batteries [9].

3.2 Silicon as Lithium-Ion Battery Anode

Silicon (Si) has the highest known theoretical charge capacity of ~4,200 mAh/g, which is more than ten times that of existing graphite anodes (which have a capacity of ~372 mAh/g corresponding to the formation of LiC_6) [10]. Moreover, silicon is also the second most abundant element on earth. Silicon has thus received a huge attention as a prospective replacement material for use as anodes in a lithium-ion battery.

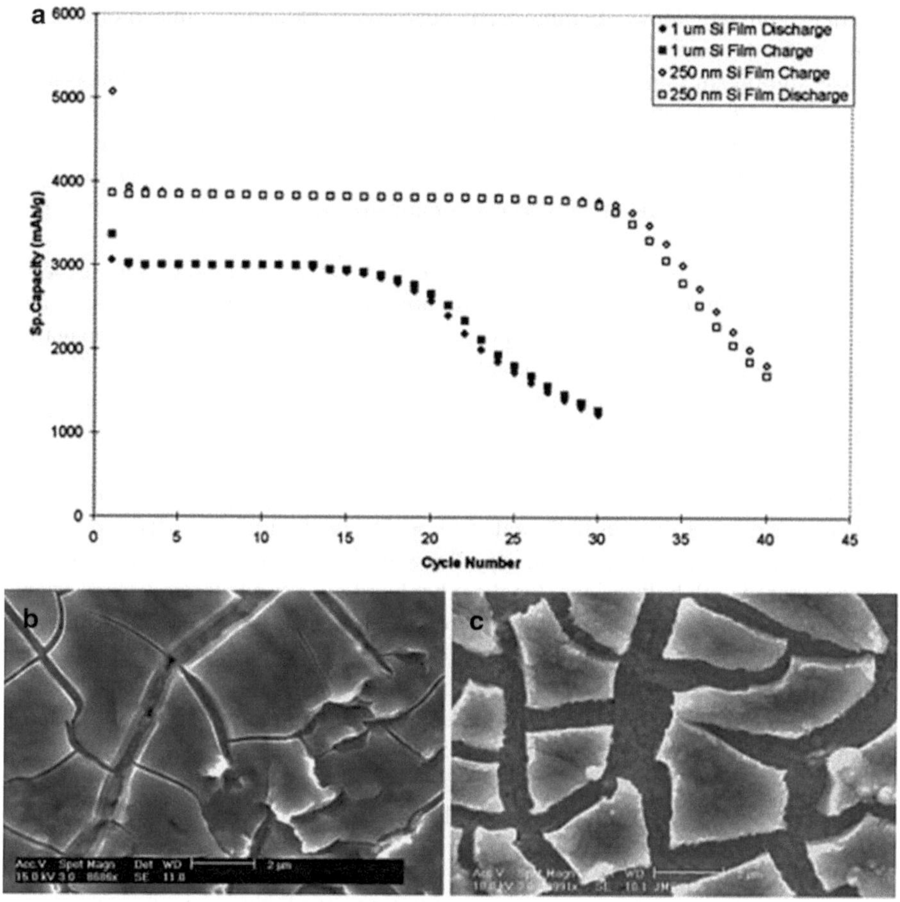

Fig. 3.1 Characterization of amorphous Si films as the battery anode. (**a**) Specific capacity plotted as a function of cycle number. (**b**) Stress-induced cracking of the film after a few cycles. (**c**) Delamination and peeling of the film from the collector electrode after extended cycling (Reprinted with permission from Ref. [12])

However, the insertion and extraction of lithium ions in the Si anode result in a tremendous volume change of Si. Lithium intercalates with silicon to form $Li_{22}Si_5$ alloy [11] and correspondingly shows a volume change in the order of 400%. Stresses associated with such volume changes cause cracking and pulverization of the Si anode resulting in a loss of electrical contact and eventual fading of capacity. Figure 3.1a shows the specific capacity of amorphous Si film anodes of thicknesses 1 μm and 250 nm [12]. The films exhibit near theoretical capacity for a few cycles after which the capacity begins to fade drastically, indicating the onset of the above-mentioned damages. Figure 3.1b–c shows the stress-induced breakup of the film into small islands, causing pileup of these islands to form pillars, which eventually get peeled off the underlying electrode, thereby resulting in loss of electrical contact. Some attempts have been made to address this problem, for example, Ohara et al. tested a Si film deposited on a Ni substrate [13]. The cycle life was significantly improved as Ni develops a passivating layer which acts as an excellent binding agent between the substrate and the silicon film, essentially due to the strong affinity of silicon to oxygen. These cells showed capacities as high as 1,700–2,200 mAh/g for an impressive 750 cycles at a 2C charge/discharge rate.

There are several advantages of using nanostructured silicon anodes over bulk silicon. Nanostructured silicon can better accommodate the strain from Li-Si alloying resulting in better stress tolerance and longer cycle life. This superior resistance to fracture is because cracks do not reach their critical size for propagation as they do in bulk materials [14]. The critical sizes for crack propagation obtained from fracture mechanics calculations are mostly much larger than the dimensions of nanomaterials. They provide enhanced electrode surface area and shorter Li diffusion distances which enable faster charge/discharge rates [10, 14]. It is also known that nanomaterials are more reactive and have reduced energy barriers for alloy formation [15]. This is mainly because most of the atoms are in a higher energy state than in bulk materials owing to the higher surface to volume ratio. Thus, nanostructured silicon can alloy with Li more readily at room temperature compared with bulk silicon. On the other hand, increased electrode surface area also increases the amount of solid electrolyte interphase (SEI) resulting in larger irreversible capacity losses. However, in light of the major advantages of nanomaterials, significant research effort has gone into developing different forms of nano-silicon anodes. In another approach to minimize the volumetric strain in silicon, very thin films of n-type Si (~50 nm thick) were vacuum deposited on a nickel foil and tested under a wide variety of C-rates [16, 17]. These thin films achieved capacities of around 3,500 mAh/g at 2C for 200 cycles. They were also tested at 12C for 1,000 cycles and achieved capacities as high as 3,000 mAh/g. At such high rates, there is a phenomenon of capacity suppression which lasts over the initial 50 cycles or so. The mechanism behind this phenomenon had not been well studied. However it is possible that since the rate of cycling is so high, there is increasing diffusion of Li into the bulk of Si with each cycle until all the Si has been lithiated. The use of doped silicon in these films resulted in higher conductivity and better performance at higher rates. However, the main disadvantage of this approach was the low mass of the active anode material. Increasing the mass of the film was limited by the poor performance of thicker silicon films. To accommodate these Si anodes in practical cells, it is important that their capacities match that of the commercial standard cathodes such as $LiCoO_2$. Yin et al. showed that Si films of thickness greater than 4 μm provide geometric capacities of around 2.6 mAh/cm^2 which match that of the cathode (~2 mAh/cm^2) [18]. These films were grown by electron-beam deposition on a Cu substrate with a concave-convex surface. As a result, the thick Si layer was reported to have good adhesion to Cu, providing capacity matching over a long cycle life. In cases where the cathode capacity is much smaller compared to that of the anode, the cathode is made much thicker in order to match the capacities. The ultimate goal would be to develop high-capacity anodes and cathodes so that they can both be thin.

3.3 Silicon Nanostructures

A study by Cui and coworkers showed that it was possible to overcome the pulverization of silicon-based anodes by incorporating silicon nanowires (Fig. 3.2a, b) [19]. Not only could these anodes accommodate large strain, they also provided a good electronic contact, while the nanoscale wires facilitated short lithium diffusion distances. They achieved the theoretical charge capacity for silicon anodes and observed little fading of capacity, maintaining a discharge capacity close to 75% of the maximum value when tested at a rate of 0.05C for up to 10 cycles (Fig. 3.2c). A vapor-liquid-solid method was incorporated to grow the one-dimensional silicon nanowires directly on a stainless steel substrate that also acted as the current collector. The nanowires were ~89 nm in diameter and on

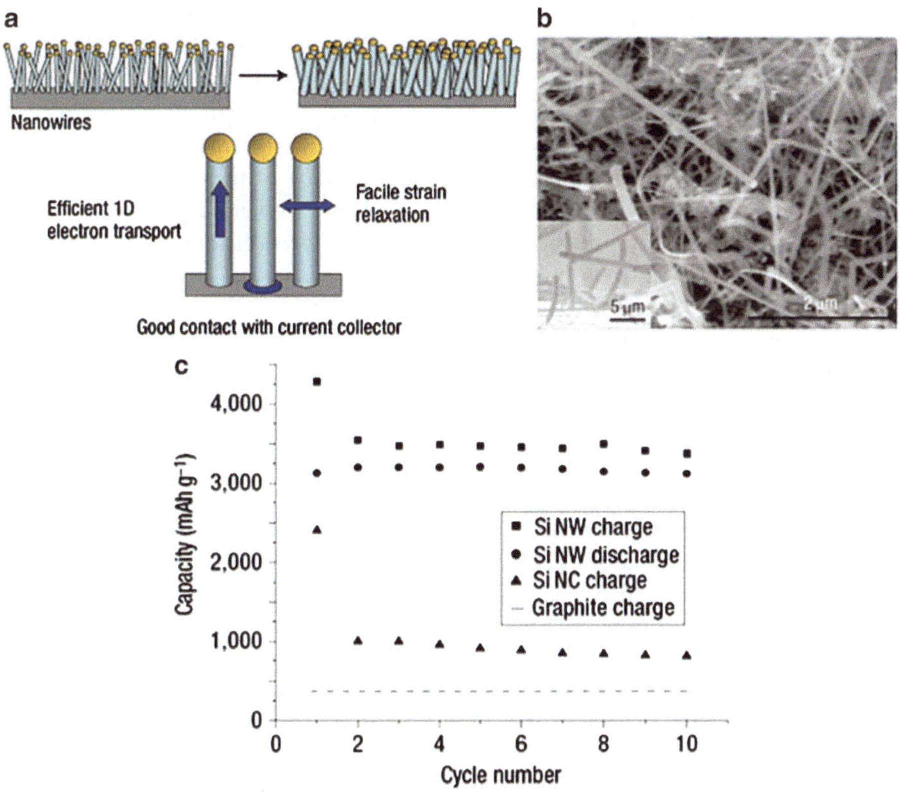

Fig. 3.2 Testing of Si nanowires as the battery anode. (**a**) Concept schematic of Si nanowire electrode assembled on the current collector. (**b**) Scanning electron micrograph of Si nanowires that comprise the device anode. (**c**) Capacity versus cycle number for various electrode configurations. The Si nanowires show stable capacity (~3,500 mAhg^{-1}) without any degradation with increase in the number of cycles (Reprinted with permission from Ref. [19])

subsequent cycling were found to have increased to ~141 nm in diameter. This increase in volume is most certainly associated with alloying and de-alloying of Lithium with silicon. Moreover, the initially crystalline silicon nanowires also underwent a drastic change in their atomic structure and were found to have transformed into amorphous Li_xSi [20]. Each of the nanowire has sufficient space between adjacent nanowires, and hence they can expand in volume freely. This is considered to be the primary reason behind stress alleviation of the nanowire arrays. Nano-mechanics could potentially assist in stress alleviation as well since nanocrystalline materials are generally governed by nanoscale grain-boundary dynamics that provide increased toughness and stress to failure compared to equivalent bulk materials. Finally, other phenomena such as the nature of the phase transition and adhesion effects at the nanoscale could also contribute and these need to be investigated in-depth.

Cui and coworkers [21] also managed to achieve high capacities over extensive cycling by incorporating interconnected silicon hollow nanosphere electrodes. The diffusion-induced stress is greatly reduced by the hollow sphere morphology. This facilitated repeated cycling of the electrodes without delamination or pulverization. The hollow nanosphere was observed under TEM after lithiation and a 240% volume expansion was observed. Interestingly, no fracture was observed in the hollow spheres, indicating good structural stability. Another possibility that could be playing a role in improving the performance of these electrodes is the reduced SEI formation. Aurbach et al. reported that irreversible capacities were primarily contributed by side reactions occurring between the composite electrode and the electrolyte [22]. Formation of SEI thus results in irreversible loss of lithium and contributes to the capacity fading of the cell. However, the hollow sphere geometry has a free surface on the inside of the structure not exposed to the electrolyte. This would reduce the side reactions and hence result in reduced irreversible capacity.

In a later report published by Chan et al., detailed X-ray photoelectron spectroscopy (XPS) analysis on the cycled Si nanowire anode was performed to evaluate the composition of the solid electrolyte interphase (SEI) [23]. The major component of the SEI was reported to be Li_2CO_3. This was in contrast to the SEI study on amorphous thin film anodes [24] where the major component was found to be LiF. The percentage of different molecular species found on the cycled Si nanowire anodes was estimated from the high-resolution XPS survey scans and are shown in Fig. 3.3a. In the same report, Chan et al. also showed that presence of SiO_2 was not the primary reason for the initial irreversible capacity loss. In fact, Si nanowires that were etched with HF to remove the surface oxide showed a larger drop in capacity over 35 cycles (at 0.2C) in comparison with unetched pristine Si nanowires (Fig. 3.3b). Thus, it was established that the surface oxide helped to form a stable SEI film which is essential for the long life of the anode material.

Cui and coworkers have also studied germanium (Ge) nanowires [25] instead of silicon. The room temperature diffusivity of lithium in germanium is 400 times that of silicon which makes it an attractive anode for high-power applications. The same group has also developed crystalline-amorphous core-shell silicon

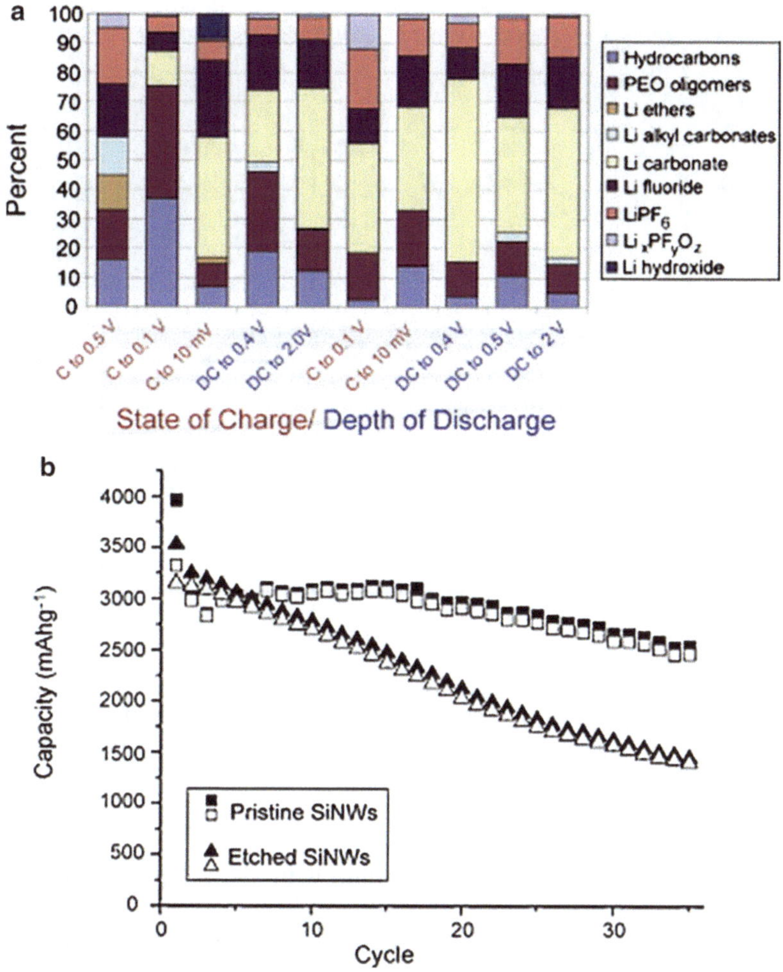

Fig. 3.3 (**a**) Quantitative analysis of molecular species found in the solid electrolyte interphase (SEI) on the surface of silicon nanowires at different potentials obtained from high-resolution X-ray photoelectron spectroscopy (XPS) scans. (**b**) Capacity as a function of cycle number for etched and unetched silicon nanowires showing that surface oxide does not account for the irreversible capacity loss (Reprinted with permission from Ref. [23])

nanowire–based anodes [26]. It was determined by Kumta and coworkers [27] that amorphous silicon is better able to withstand pulverization during cycling as compared to crystalline silicon. Thus, the crystalline-amorphous core shells serve a twofold purpose – the amorphous shells are selectively electrochemically active while the crystalline silicon core provides a stable mechanical support and an efficient electrical conducting pathway. These anodes exhibited excellent electrochemical performance at rates of charging as high as 6.8 A/g and demonstrated

Fig. 3.4 Testing of porous Si particles with nanometer-scale wall thickness. (**a**) Scanning electron micrographs of the porous Si particles indicating a pore wall size of ~40 nm. (**b**) These nanoporous Si structures maintained a charge capacity of greater than 2,800 mAh/g for over a hundred cycles (Reprinted with permission from Ref. [29])

high charge storage capacity (~1,000 mAh/g) with ~90% capacity retention over 100 cycles.

Shimizu et al. have grown high-density epitaxial nanowire arrays by using VLS growth in a porous anodic aluminum oxide (AAO) template [28]. AAO allows for great flexibility in the design of the nanowire diameter and density. In another interesting work, Cho and coworkers fabricated 3D porous bulk silicon particles (Fig. 3.4a) with a pore wall size of ~40 nm from butyl-capped Si gels and SiO_2 nanoparticles [29]. It is to be noted that after annealing and etching, the final product consists of carbon-coated silicon particles (c-Si). CHN (carbon, hydrogen, nitrogen) analysis, a technique to determine the elemental composition of a sample was carried out to determine the percent composition of the carbon coating. The carbon coating accounted for 12 wt% of carbon in the entire electrode sample. These nano-silicon sponges were able to accommodate large strains without pulverization even after 100 cycles and maintained a charge capacity greater than 2,800 mAh/g throughout at a cycling rate of 1C (Fig. 3.4b). The unique shape and structure of these particles not only allows for faster transport of the Li ions through the electrolyte and the electrode but also enables faster intercalation reactions of the Li ions, thus resulting in a large specific capacity.

3.4 Silicon Nanorods by Oblique Angle Deposition

Oblique angle deposition (OAD) with or without substrate rotation is a highly preferred approach for directly patterning the nanostructured surfaces on the current collecting substrates [30–32]. OAD is a physical vapor deposition process in which the flux arrives at a highly oblique angle of incidence (typically >80°) from the substrate normal as shown schematically in Fig. 3.5a. If the substrate is not rotated, oblique nanorods are obtained. On rotating the substrate, a variety of morphologies can be obtained such as vertical nanorods and nanosprings. As the flux arrives, isolated nanorods are formed by the self-shadowing effect. Shadowing effect is a typical characteristic of OAD and proceeds when certain regions of higher initial

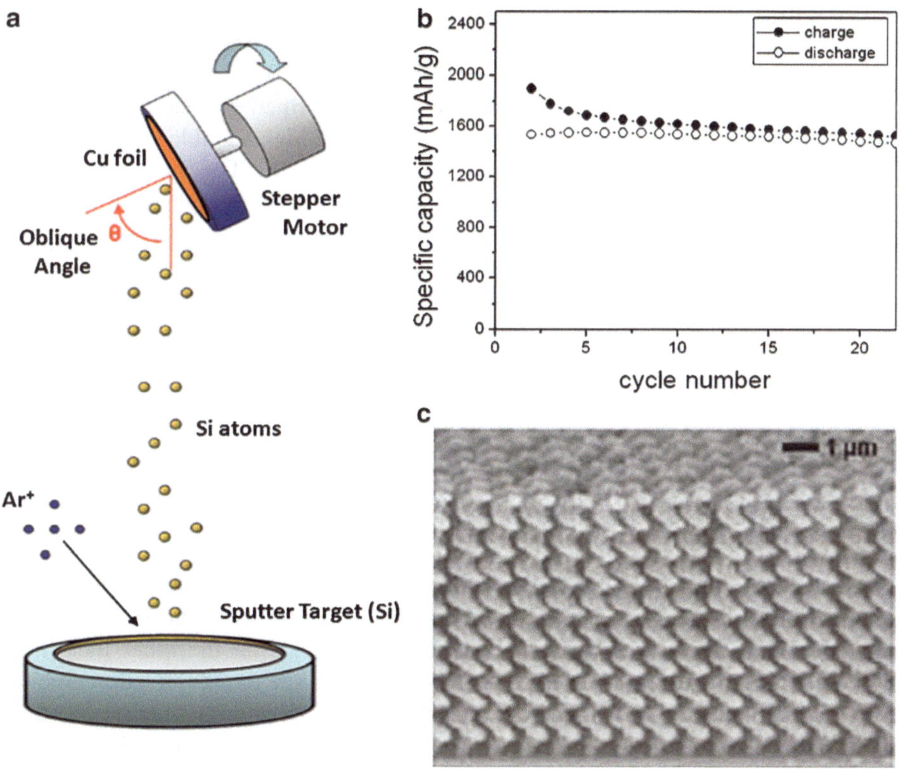

Fig. 3.5 Si nanorod anodes developed by oblique angle deposition. (**a**) Schematic diagram showing the oblique angle deposition (OAD) process with substrate rotation. (**b**) Testing of Si nanorods grown by OAD; the nanorods display a stable capacity of ~1,600 mAh/g which is over four times greater than graphite electrodes. There is an initial loss in the charge and discharge capacities for the first few cycles indicative of probable wettability issues between the electrode and the electrolyte and SEI formation. (**c**) Scanning electron micrograph of an amorphous Si nanospring array grown by OAD. The nanosprings are highly compliant along the axial direction (Reprinted with permission from Ref. [33])

growth shadow the regions behind them by obstructing the flow of incident flux to those areas. By adjusting the deposition rate, incidence angle, and the substrate rotation speed, the shadowing effect can be largely controlled and 3D nanostructures with very large aspect ratio and controllable porosity, shape, and symmetry can be deposited using OAD without the need for expensive lithography and multistep processes.

The primary disadvantage of solution-based growth techniques can be effectively overcome by physical vapor deposition methods such as sputtering and e-beam evaporation. These physical vapor deposition processes are clean and repeatable, have high deposition rates, and can be effectively grown over large areas with very good substrate adhesion. Moreover, these deposition techniques allow for greater control over the deposition characteristics such as dimensions and porosity. Brett and coworkers used OAD (also called glancing angle deposition or GLAD) to grow vertical silicon nano-columns to test them as anodes for lithium-ion batteries [34]. They obtained capacities of ~3,000 mAh/g and their anodes exhibited impressive capacity retention (~83%) after 70 charge/discharge cycles at C-rates of 0.1, 0.2, and 0.5C. Experiments performed by our group with Si nanorods deposited by OAD on Cu foils also show a highly stable capacity that is over fourfold better than graphite anodes [33] (Fig. 3.5b). The initial irreversible capacity loss is due to the formation of a solid electrolyte interphase (SEI) which has a high electronic resistivity. With continued cycling, these nanorods display stable capacities of ~1,600 mAh/g. Our group has demonstrated [35] that silicon nanosprings are highly compliant and are therefore expected to offer enhanced stress resilience. Such silicon nanospring architectures (Fig. 3.5c), which are easily grown by OAD techniques cannot be fabricated using the traditional VLS growth methodologies.

3.5 Nano-compliant Supports

Nanoscale thin film amorphous silicon electrodes display good stability over a large number of cycles. However, these electrodes have insufficient material for a viable battery. Bulk silicon and micron-sized silicon particles on the other hand have shown significant capacity fading and poor cycle life due to pulverization and loss of electrical contact. Thus, there is an urgent need to develop thicker silicon films with improved stress resistance such that the anodes may exhibit stable capacity without deterioration.

One approach to forming stress resistant Si thin films is by using a nanostructured compliant layer (NCL) [36, 37]. The NCL consists of slanted nanorods grown by oblique angle deposition, sandwiched between the film and the substrate (Fig. 3.6a). The technique is all in situ, does not require any lithography steps, and the nanostructured layer can be made from the same material as the deposited film (e.g., Si as in Fig. 3.6b). Our group used this approach [36] to study the stresses in sputter-deposited tungsten films (Fig. 3.6c). The tests showed that the

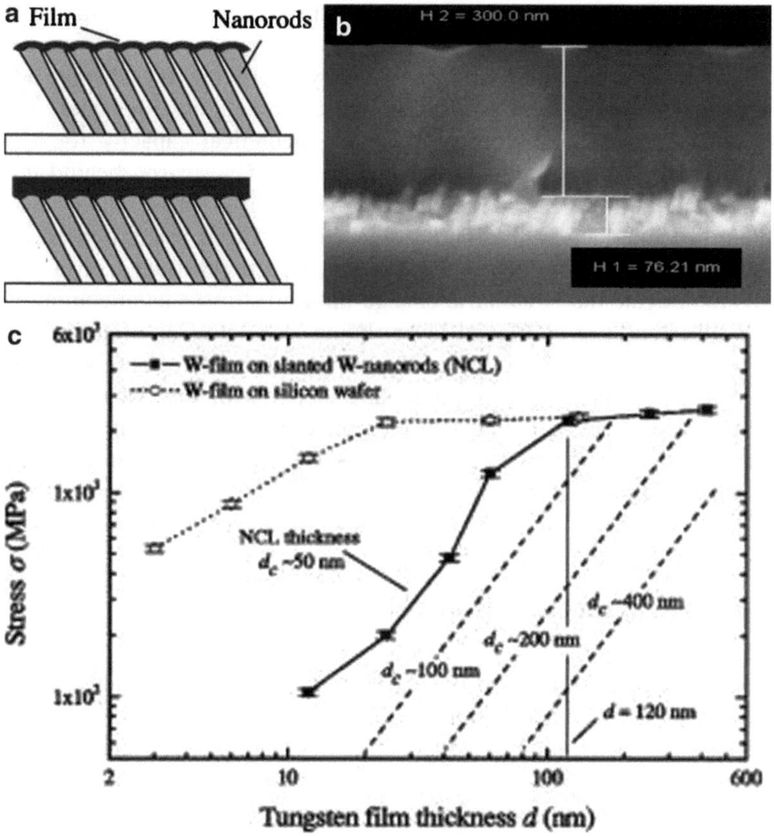

Fig. 3.6 Nano-compliant support structures for Si film anodes. (**a**) Schematic of the nano-compliant-layer (NCL) concept showing a film supported by an array of inclined nanorods. (**b**) Fabrication of a Si film supported on an NCL of inclined Si nanorods by the oblique angle deposition (OAD) technique with substrate rotation. (**c**) Reduction in film stress caused by the NCL. Data shows up to an order of magnitude reduction in stress depending on the film thickness. The data is for a tungsten film supported on a tungsten NCL (Fig. 3.6c is reprinted with permission from Ref. [36])

stress value can be reduced by almost one order of magnitude by incorporating NCL. Stress alleviation in NCL is primarily obtained through [37] initiating a delay in the inception of stress buildup and by toughening the interface to facilitate higher levels of strain energy to be stored in the film before delamination buckling can set in. The NCL thus relieves the stress in continuous films and effectively improves the adhesion, which results in larger critical thickness for delamination buckling and higher quality films. Because of these attributes, the NCL could thus permit the deposition of thicker silicon films in order to achieve mass scalability of the anodes. Such thicker silicon films will exhibit an increased structural stability and will provide for a greater amount of material for lithium insertion/extraction.

The process can be further repeated to form multilayer (or stacked) films with very large thickness [36, 37].

NCL also assists the anode characteristics by improving the substrate adhesion of silicon films. Studies done by Kumta and coworkers have revealed that even though amorphous silicon films exhibit near theoretical capacity for a limited number of cycles, they eventually delaminate [11]. One reason behind this phenomenon could be the formation of amorphous Cu-Si phases that resulted in weakening the interfacial adhesion between Si and Cu. It was proposed by them that an interfacial layer between the Si film and Cu substrate with a very low modulus to mitigate the volumetric change-induced stresses in silicon layer could improve adhesion. The interfacial layer should also act as an effective diffusion barrier to eliminate the possibility of Cu-Si phase formation. The NCL proposed here not only provides a highly compliant support structure which relieves the stress, but by constructing the NCL from a material such as chromium, one could significantly improve adhesion between Si and Cu, while chromium would also act as an efficient diffusion barrier for Si.

3.6 Nanostructured Silicon-Carbon Composites

Some other approaches used to overcome the issues with Si anodes include Si dispersed in an active/inactive matrix, Si mixed with different binders and Si-C composites [38]. Generation of nanocomposites comprising a nanocrystalline Si powder coated with a thin layer of amorphous carbon serving as an interfacial adhesion layer, together embedded in a mechanically softer but more ductile carbon matrix that is relatively inactive in the electrochemical potential window of 0.02–1.2 V has led to moderate success of sustaining reversible capacities on the order of 700 mAh/g [39, 40]. These capacities were obtained at a C-rate of ~0.25C. Carbon-coated silicon particles have also been shown to achieve capacities as high as 1,000 mAh/g when charging and discharging at a constant current of ~0.3 mA/mg [41]. The carbon coating enhances structural stability and also prevents local capacity fading by maintaining electrical contact between the silicon particles. Carbon is known to form a very stable solid electrolyte interphase (SEI) instantly within the first few cycles. This is in contrast to silicon where the SEI breaks up from volumetric expansion exposing new silicon surface for SEI formation. As a result, in silicon, the SEI formation may occur over several cycles resulting in a much larger capacity loss [42]. Thus, by using carbon-coated silicon, it is possible to isolate silicon from the electrolyte resulting in a stable electrode with minimal capacity loss. Similarly, Ng et al. have obtained a reversible capacity of 1,450 mAh/g with spheroidal carbon-coated silicon nanocomposite anode structures at a cycling rate of 100 mA/g [43]. Incorporation of carbon nanotubes has led to reversible capacities as high as 1,000 mAh/g under a constant current density of 250 $\mu A/cm^2$ [44]. However, some studies report [45] that nano-sized Si particles in composites tend to agglomerate after the insertion/

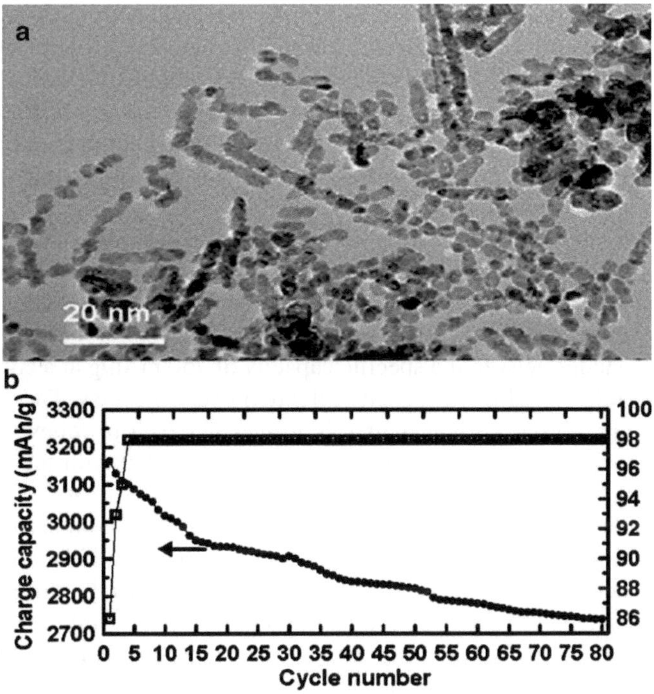

Fig. 3.7 (**a**) Transmission electron microscope (TEM) image of the silicon-carbon core-shell nanowires. (**b**) Capacity performance of the composite nanowires at a rate of 0.2 C (600 mA/g) (Reprinted with permission from Ref. [47])

extraction of Li ions, and this size increase results in poor Li insertion/extraction kinetics. Kwon et al. reported the synthesis of Si quantum dots coated with amorphous carbon [46]. These structures achieved a first charge capacity of 1,257 mAh/g with a coulombic efficiency of 71%. The uniform distribution of Si quantum dots along with the carbon coating prevented aggregation during the cycling. In another article, Kim and Cho reported the synthesis of Si-carbon core-shell nanowires using an SBA-15 template (Fig. 3.7a) [47]. These nanowires were ~6.5 nm in diameter and achieved a capacity of 3,163 mAh/g in the first charge with 87% capacity retention after 80 cycles at a rate of 0.2C (Fig. 3.7b). This performance appears to be better than carbon-silicon core-shell nanowires [48]. This reinforces the need for a carbon-electrolyte interface in Si-C composites. Other composite anode research efforts include silicon nanoparticles trapped in a polymer binder matrix [49], silicon/graphite composite nanowires [50], silicon-graphene composites [51, 52], silicon-carbon nanotube heteronanostructures [53] and silicon nanowire – MWNT mixtures [54]. These composite structures have all shown stable capacities at low rates over tens of cycles. The capacity of silicon-based composites is reduced with respect to pure silicon due to the weight fraction of low-capacity carbon. This reduction in capacity is the trade-off for the minimization in capacity losses. Recently, a hierarchical bottom-up self-assembly technique consisting of dendritic carbon structures coated with silicon nanoparticles has attracted attention [55]. These structures yielded a capacity of ~1,600 mAh/g over 100 cycles at a rate of 1C. The dendrite structure of carbon helps provide efficient electron conduction acting like a conducting backbone. It also provides the appropriate porosity required for the silicon nanoparticles to undergo volumetric expansion.

Studies have further revealed that on anchoring silicon nanoparticles to carbon nanofibers [56], a significant improvement could be achieved. The carbonization process that is carried out to anchor the silicon nanoparticles induces a strong interaction between silicon and carbon through thick amorphous silicon oxide layer. These anodes delivered a specific capacity of 750 mAh/g at a current density of 50 mA/g and showed good capacity retention over 50 cycles. The polymer-like elasticity of the fibrous carbon matrix is further believed to relieve the induced stress by accommodating the volume expansion associated with the alloying and de-alloying of lithium.

In spite of all the aforementioned advances in the field of composite anodes, the key issue of a weak structural interface between carbon and silicon can never be fully eliminated. Silicon and carbon both exhibit drastically different volume changes associated with intercalation and de-intercalation of lithium ions. This in turn makes the composite highly vulnerable to rapid delamination particularly at higher charge/discharge rates due to significant mismatch of the induced strains in silicon and carbon.

In order to address the rapid pulverization in composite electrodes at higher current densities, our group demonstrated that strain-graded nanoscoop anodes [57] provided improved cycle stability at exceptionally high charge/discharge rates. The anode proposed in the strain-graded architecture comprised of amorphous carbon nanorods deposited directly on top of the stainless steel substrate with aluminum and silicon nanoscoops deposited on top of the carbon nanorods in this order. Figure 3.8a shows a schematic of the proposed architecture before and after experiencing volume expansion along with an actual SEM image of the anode. A typical cyclic voltammogram (CV) carried out on the lithium-ion coin cell with the proposed strain-graded carbon-aluminum-silicon anode against lithium metal is shown in Fig. 3.8b. Cyclic voltammetry is an effective tool to study the electrode properties. In the CV, a broad peak was observed at around 0.65 V in the delithiation cycle, while in the lithiation cycle, a peak was observed below 0.2 V. The CV was carried out at a scan rate of 10 mV/s. Such CV testing is also used to ensure that the cell is assembled properly and does not exhibit a resistive behavior. The concept and results of the strain-graded carbon-aluminum-silicon have been discussed in details in the next section of this chapter.

One of the major limitations to using silicon as anode material for lithium-ion batteries arises from the semi conductive nature of silicon. This makes the use of carbon-based conductive additives an attractive field of study as carbon can improve the performance of the battery by enhancing the overall charge transportation mechanism. Wang et al. [58] demonstrated that graphene

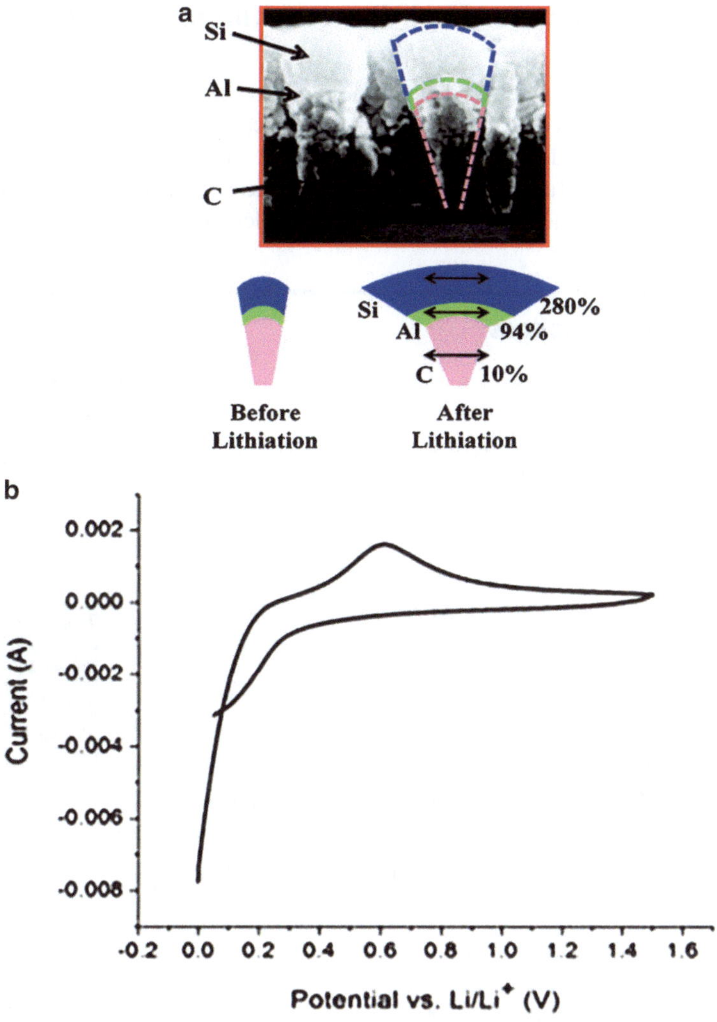

Fig. 3.8 (**a**) Schematic of the C-Al-Si strain-graded nanoscoop architecture before and after volume expansion along with an actual SEM image of the nanoarchitecture. (**b**) Cyclic voltammetry of lithium-ion batteries with C-Al-Si nanoscoop architecture anodes against lithium metal. The CV was carried out at a scan rate of 10 mV/s between 0.05 and 1.5 V (Reprinted with permission from Ref. [57]. Copyright 2010 American Chemical Society)

nanosheets significantly improved the lithium storage capacity of porous single crystalline silicon nanowires. Graphene acts as a conductive additive and the nanosheets cover larger areas of the nanowires, thereby providing greater areas for charge transfer. Silicon nanowires with graphene conductive additives showed an initial charge capacity of 2,347 mAh/g with a capacity retention of 87% after 20 cycles.

3.7 Silicon Anodes for High-Power Lithium-Ion Batteries

The recent past has shown rapid development in the field of hybrid automotives and electric vehicles motivated by the need to reduce our fossil fuel consumption. Supercapacitors provide high power density but lack sufficient energy density [59] as shown in the Ragone plot of energy storage domains (Fig. 3.9a). Lithium-Ion batteries on the other hand have high energy densities and provide the longest

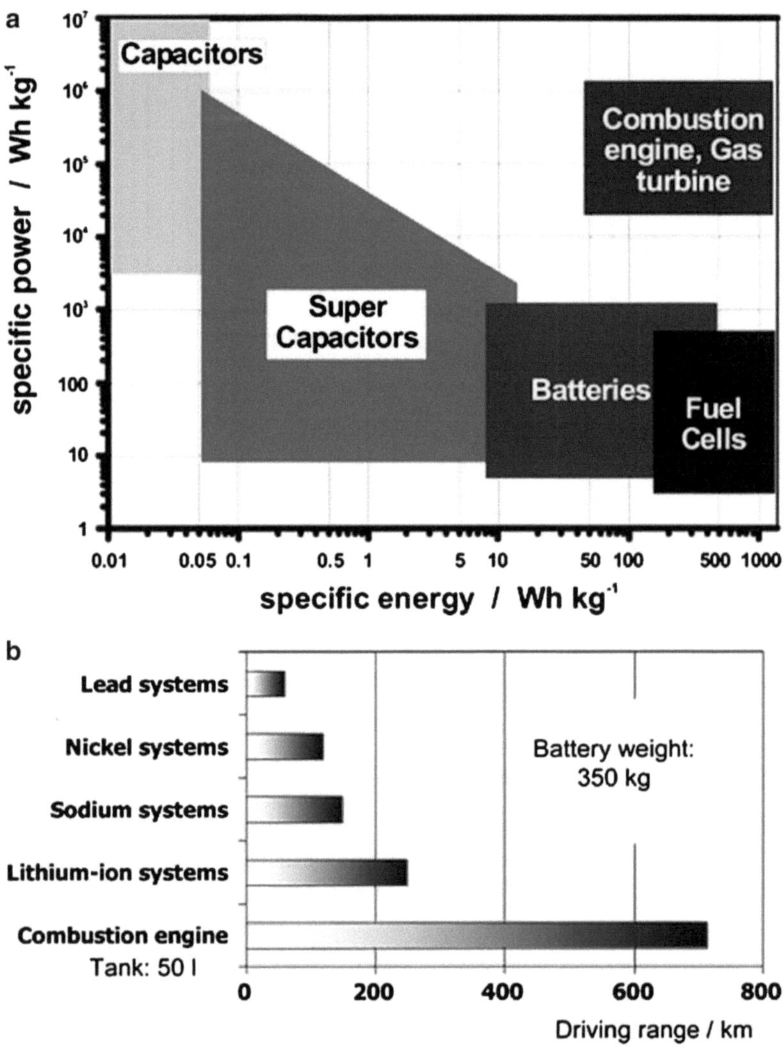

Fig. 3.9 (**a**) Relative energy storage domains for various electrochemical energy systems with respect to power density and energy density. (**b**) Driving range for a vehicle powered by various battery systems in relation to a gasoline-powered vehicle (Reprinted with permission from Ref. [55])

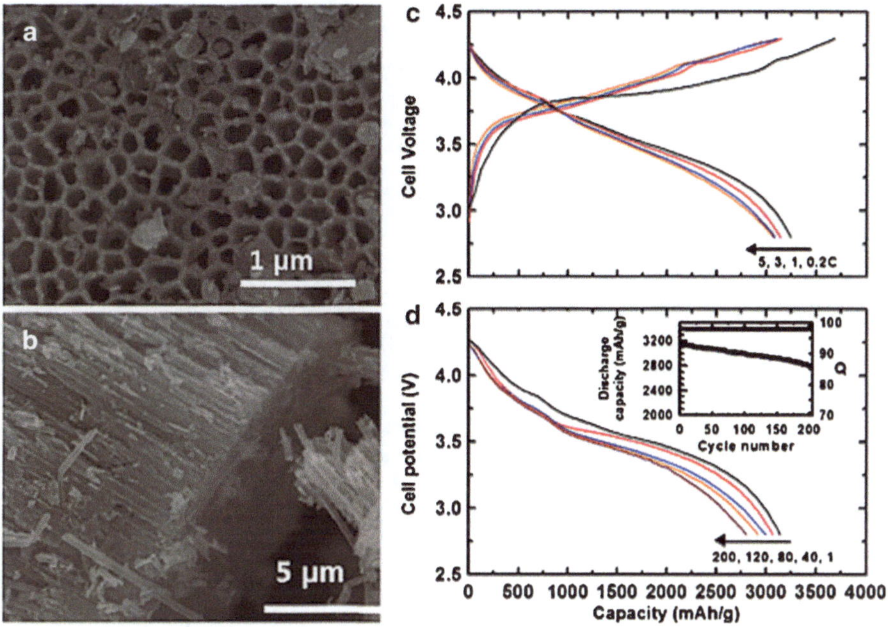

Fig. 3.10 FE-SEM images of the silicon nanotubes. (**a**) Top view and (**b**) side view. (**c**) Cell voltage as a function of capacity at different C-rates up to 5 C. (**d**) Cell potential as a function of capacity for different cycle numbers and inset is the capacity performance of silicon nanotubes at 1 C. The silicon nanotube electrode data is shown from a full cell assembled against $LiCoO_2$ (Reprinted with permission from Ref. [62])

driving range compared to any other battery system powering a vehicle (Fig. 3.9b). High-power batteries designed for electric vehicles should typically withstand massive current demands while providing high energy density. This can be achieved by using highly stress-tolerant nano-silicon-based anode materials that would provide not only high capacities but also high current bursts with a long cycle life.

High-power cathode materials [60, 61] have propelled the development of high-rate capable silicon anodes in the attempt to make equally competent anode materials. Cho, Cui, and coworkers reported carbon-coated silicon nanotubes that were prepared by a chemical deposition using porous alumina membrane templates (Fig. 3.10a, b) [62]. These nanotubes were tested at rates as high as 5C (15 A/g) where they yielded an initial capacity of over 3,000 mAh/g (Fig. 3.10c). They also showed capacity retention of around 89% after 200 cycles when tested at a rate of 1C (Fig. 3.10d). The carbon coating was provided to stabilize the SEI formation over long cycle life.

In another approach, Zhou et al. synthesized $TiSi_2$ nanonets that acted as a conducting network [63]. These nanonets were coated with Si nanoparticles (Fig. 3.11a, b). The conducting backbone network provided efficient electron conduction. These structures were cycled at a current density of 8.4 A/g and yielded

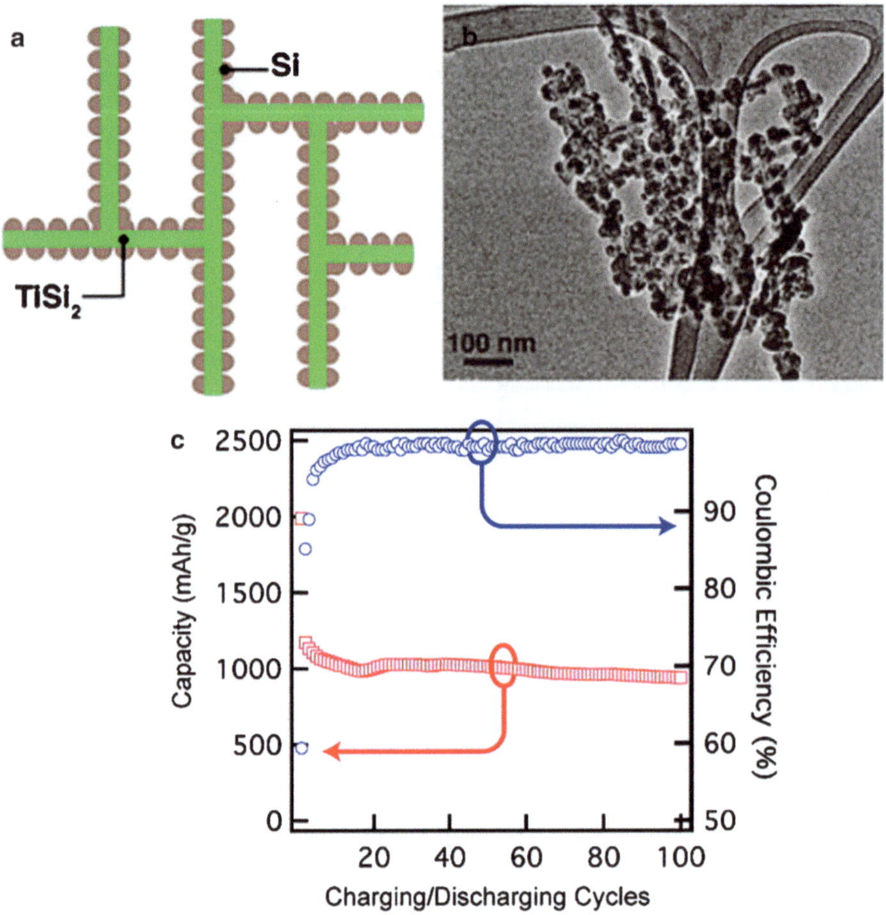

Fig. 3.11 (**a**) Schematic showing the silicon nanoparticles coated around the $TiSi_2$ nanonets. (**b**) TEM image showing the Si/$TiSi_2$ heteronanostructure. (**c**) Capacity performance at a current density of 8,400 mA/g (Reprinted with permission from Ref. [63])

capacities of ~1,000 mAh/g in the range of 100 cycles (Fig. 3.11c). However, the $TiSi_2$ backbone was shown to alloy with Li at a potential of 60 mV. This could result in destruction of the backbone and increase in the capacity losses. To avoid this, it was necessary to modify the charge/discharge voltage window to exclude the $TiSi_2$ reaction potential. Mesoporous TiO_2 has also been tested as a Li battery anode at rates of 10C and 20C. However, the capacities achieved were relatively low (~140 mAh/g and ~100 mAh/g at the respective C-rates [64]).

The carbon-aluminum-silicon functionally strain-graded architecture [57] tested by our group has shown a considerable improvement in capacity at ultrahigh C-rates. The composite nanostructured anode toughened the interface by providing a strain gradation and could facilitate rapid volume changes at rates as high as 100C

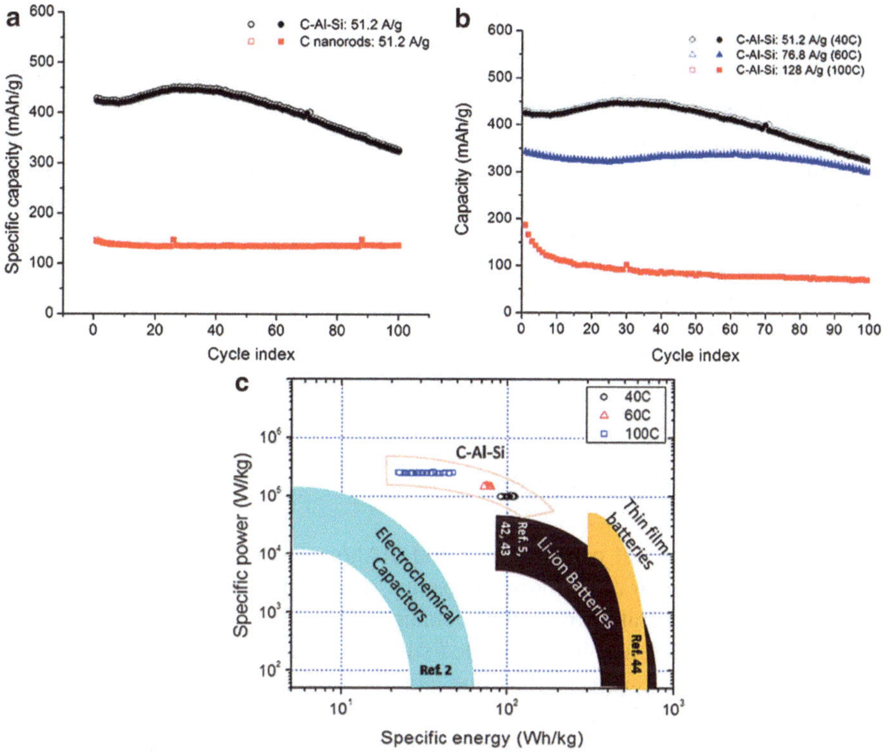

Fig. 3.12 (**a**) Capacity versus cycle index for carbon-aluminum-silicon functionally strain-graded multilayer architecture at 40C, 60C, and 100C over 100 cycles. (**b**) Capacity as a function of cycle index for the C-Al-Si multilayer architecture when compared to a carbon nanorod having the same length and diameter. Both these anodes were tested at a current density of 51.2 A/g. (**c**) C-Al-Si electrode performance shown on the Ragone plot and compared to the performance of electrochemical capacitors [65, 66], lithium-ion batteries [67, 68], and thin film batteries [69]. The gravimetric density calculations were based on the mass of active electrodes only (Reprinted with permission from Ref. [57]. Copyright 2010 American Chemical Society)

without delamination or pulverization. Aluminum, with a volume expansion of ~94%, was chosen since it provided a strain that would be intermediate between carbon (~10%) and silicon (~280%). The composite anode was deposited using DC Magnetron sputtering using oblique angle deposition technique. Ultrahigh rates of 40C, 60C, and 100C were tested to assess the structural stability of the multilayer C-Al-Si nanoarchitecture (see Fig. 3.12a). At a C-rate of 40C, an average capacity of 412 mAh/g was achieved over 100 cycles with a capacity fading of only ~0.2% with every cycle. At 60C, the average capacity over 100 cycles was noted to be ~330 mAh/g, with a ~90% capacity retention after 100 cycles. To further study the role of silicon in the strain-graded architecture, carbon nanorods of the same dimensions (length and diameter) were tested at 40C (see Fig. 3.12b). The average capacity of the carbon nanorods was only 140 mAh/g, almost three times lesser than that of the C-Al-Si multilayer architecture tested at the same rate (the capacity obtained from the carbon nanorods was lesser than the theoretical value of 372 mAh/g due to partial lithiation at such high rates). This confirms the partial lithiation of silicon even at charge/discharge rates (C-rates) as high as 40C (i.e., current density of ~51 A/g) and its contribution to the increased specific capacity. The C-Al-Si nanoscoops were also tested at a C-rate as high as 100C (corresponding to a current density of ~128 A/g). At this accelerated current density, an exceptional power density of ~250 kW/kg was obtained with the C-Al-Si architecture (see Fig. 3.12c) however the energy density declined sharply suggesting that at such ultrahigh C-rates Li^+ is unable to diffuse even partially into the silicon nanoscoops.

3.8 Summary and Outlook

Silicon has a theoretical capacity of 4,200 mAh/g which is more than ten times higher than the theoretical charge capacity of graphite anodes. However, the extensive stress generated by volume changes associated with intercalation and de-intercalation of lithium has prevented silicon anodes from being realized commercially. Research suggests that one possible solution to this problem is to enhance the stress resilience of the electrode by using nanostructured Si electrode architectures. These nanostructures can significantly improve the structural stability, provide a good substrate adhesion, and allow for shorter lithium diffusion distances. Thicker silicon films on nano-compliant structures could also be an alternative in order to develop a toughened interface and prevent rapid delamination of the anode. Carbon-silicon composites have also shown immense promise. Carbon can provide a stable solid electrolyte interface and can also accommodate the stresses resulting from the silicon volume expansions, while silicon enhances the capacity due to larger lithium uptake. Thus, silicon-based composites show great promise for the next generation of lithium-ion battery anode materials. With the demand for hybrid electric vehicles expected to increase by at least 15% by the year 2015 [70], there is increasing research efforts to develop high-power and high-capacity lithium-ion batteries. This has led to several silicon-based nanoarchitectures capable of high-rate charge/discharge. An in-depth study of critical scientific issues is necessary in order to develop lithium ion batteries for practical applications. A detailed understanding of the mechanism of stress buildup in the nanostructure during lithiation and its relaxation during reversed process is essential to predict how fracture can be avoided in nanostructures, especially at ultrahigh discharge rates. This necessitates the development of high-fidelity multiscale models to predict the stress buildup and damage accumulation in nano-engineered electrode structures. Moreover, attempts need to be made to improve the capacity at higher charge/discharge rates in order to make these batteries feasible for high-power applications. Finally, researchers will need to look at ways to extend the cycle life of the electrodes since longevity of the battery will be a prime requirement if lithium-ion batteries are to make an entry into the commercial market.

Acknowledgments We acknowledge funding support from the nano and bio mechanics program of the United States National Science Foundation (award number CMMI 0969895) to N.K. and T.M.L.

References

1. Tarascon J-M, Armand M (2001) Issues and challenges facing rechargeable lithium batteries. Nature 414:359–367
2. Whittingham MS (1976) Electrical energy storage and intercalation chemistry. Science 192:1126–1127
3. Broussely M, Biensan P, Simon B (1999) Lithium insertion into host materials: the key to success for Li ion batteries. Electrochim Acta 45:3–22
4. Murphy DW, DiSalvo FJ, Carides JN, Waszczak JV (1978) Topochemical reactions of rutile related structures with lithium. Mater Res Bull 13:1395–1402
5. Lazzari M, Scrosati B (1980) A cyclable lithium organic electrolyte cell based on two intercalation electrodes. J Electrochem Soc 127:773–774
6. Mizushima K, Jones PC, Wiseman PJ, Goodenough JB (1980) Li_xCoO_2 ($0 < x < -1$): a new cathode material for batteries of high energy density. Mater Res Bull 15:783–789
7. Endo M, Kim C, Nishimura K, Fujino T, Miyashita K (2000) Recent development of carbon materials for Li ion batteries. Carbon 38:183–197
8. Nagaura T, Towaza K (1990) Prog Batt Solar Cells 9:209
9. Guo Y-G, Hu J-S, Wan L-J (2008) Nanostructured materials for electrochemical energy conversion and storage devices. Adv Mater 20:2878–2887
10. Patil A, Patil V, Shin D, Choi J, Paik D, Yoon S-J (2008) Issue and challenges facing rechargeable thin film lithium batteries. Mater Res Bull 43:1913–1942
11. Wang W, Datta MK, Kumta PN (2007) Silicon-based composite anodes for Li-ion rechargeable batteries. J Mater Chem 17:3229–3237
12. Maranchi JP, Hepps AF, Evans AG, Nuhfer NT, Kumta PN (2006) Interfacial properties of the a-Si/Cu:active–inactive thin-film anode system for lithium-ion batteries. J Electrochem Soc 153:A1246–A1253
13. Ohara S, Suzuki J, Sekine K, Takamura T (2003) Li insertion/extraction reaction at a Si film evaporated on a Ni foil. J Power Sources 119:591–596
14. Graetz J, Ahn CC, Yazami R, Fultz B (2003) Highly reversible lithium storage in nanostructured silicon. Electrochem Solid State Lett 6:A194–A197
15. Gao B, Sinha S, Fleming L, Zhou O (2001) Alloy formation in nanostructured silicon. Adv Mater 13:816–819
16. Ohara S, Suzuki J, Sekine K, Takamura T (2004) A thin film silicon anode for Li-ion batteries having a very large specific capacity and long cycle life. J Power Sources 136:303–306
17. Takamura T, Ohara S, Uehara M, Suzuki J, Sekine K (2004) A vacuum deposited Si film having a Li extraction capacity over 2000 mAh/g with a long cycle life. J Power Sources 129:96–100
18. Yin J, Wada M, Yamamoto K, Kitano Y, Tanase S, Sakai T (2006) Micrometer-scale amorphous Si Thin-film electrodes fabricated by electron-beam deposition for Li-ion batteries. J Electrochem Soc 153:A472–A477
19. Chan CK, Peng H, Lin G, McIlwrath K, Zhang XF, Huggins RA, Cui Y (2008) High-performance lithium battery anodes using silicon nanowires. Nat Nanotechnol 3:31–35

20. Chan CK, Ruffo R, Hong SS, Huggins RA, Cui Y (2009) Structural and electrochemical study of the reaction of lithium with silicon nanowires. J Power Sources 189:34–39
21. Yao Y, McDowell MT, Ryu I, Wu H, Liu N, Hu L, Nix WD, Cui Y (2011) Interconnected silicon hollow nanospheres for lithium-ion battery anodes with long cycle life. Nano Lett 11(7):2949–2954
22. Aurbach D, Nimberger A, Markovsky B, Levi E, Sominski E, Gedanken A (2002) Nanoparticles of SnO produced by sonochemistry as anode materials for rechargeable lithium batteries. Chem Mater 14(10):4155–4163
23. Chan CK, Ruffo R, Hong SS, Cui Y (2009) Surface chemistry and morphology of the solid electrolyte interphase on silicon nanowire lithium-ion battery anodes. J Power Sources 189:1132–1140
24. Lee YM, Lee JY, Shim H-T, Lee JK, Park J-K (2007) SEI layer formation on amorphous Si thin electrode during precycling. J Electrochem Soc 154:A515–A519
25. Chan CK, Zhang XF, Cui Y (2008) High capacity Li ion battery anodes using Ge nanowires. Nano Lett 8:307–309
26. Cui L-F, Ruffo R, Chan CK, Peng H, Cui Y (2009) Crystalline-amorphous core – shell silicon nanowires for high capacity and high current battery electrodes. Nano Lett 9:491–495
27. Nanda J, Datta M, Remillard JT, O'Neill A, Kumta PN (2009) In situ Raman microscopy during discharge of a high capacity silicon–carbon composite Li-ion battery negative electrode. Electrochem Commun 11:235–237
28. Shimizu T, Xie T, Nishikawa J, Shingubara S, Senz S, Gosele U (2007) Synthesis of vertical high-density epitaxial Si(100) nanowire arrays on a Si(100) substrate using an anodic aluminum oxide template. Adv Mater 19:917–920
29. Kim H, Han B, Choo J, Cho J (2008) Three-dimensional porous silicon particles for use in high-performance lithium secondary batteries. Angew Chem 47:10151–10154
30. Zhao Y-P, Ye D-X, Wang G-C, Lu T-M (2002) Novel nano-column and nano-flower arrays by glancing angle deposition. Nano Lett 2:351–354
31. Kim S-Y, Karabacak T, Lu T-M, Koratkar N (2006) Water electrolysis activated by Ru nanorod array electrodes. Appl Phys Lett 88:263106-1–263106-3
32. Teki R, Karabacak T, Lu T-M, Koratkar N (2006) Enhanced photoemission from nanostructured surface topologies. Appl Phys Lett 89:193116-1–193116-3
33. Teki R, Datta MK, Krishnan R, Parker TC, Lu T-M, Kumta PN, Koratkar N (2009) Nanostructured silicon anodes for lithium ion rechargeable batteries. Small 5:2236–2242
34. Fleischauer MD, Li J, Brett MJ (2009) Columnar thin films for three-dimensional microbatteries. J Electrochem Soc 156:A33–A36
35. Liu DL, Ye DX, Khan F, Tang F, Lim BK, Picu RC, Wang G-C, Lu T-M (2003) Mechanics of patterned helical Si springs on Si substrate. J Nanosci Nanotechnol 3:492–495
36. Karabacak T, Senkevich JJ, Wang G-C, Lu T-M (2005) Stress reduction in sputter deposited films using nanostructured compliant layers by high working-gas pressures. J Vac Sci Technol A 23:986–990
37. Karabacak T, Picu RC, Senkevich JJ, Wang G-C, Lu T-M (2004) Stress reduction in tungsten films using nanostructured compliant layers. J Appl Phys 96:5740–5746
38. Kasavajjula U, Wang C, Appleby AJ (2007) Nano- and bulk-silicon-based insertion anodes for lithium-ion secondary cells. J Power Sources 163:1003–1039
39. Datta M, Kumta PN (2006) Silicon and carbon based composite anodes for lithium ion batteries. J Power Sources 158:557–563
40. Datta MK, Kumta PN (2007) Silicon, graphite and resin based hard carbon nanocomposite anodes for lithium ion batteries. J Power Sources 165:378–378
41. Liu W, Wang J, Wu H, Shieh D, Yang M, Wu N (2005) Electrochemical characterizations on Si and C-coated Si particle electrodes for lithium-ion batteries. J Electrochem Soc 152:A1719–A1725

42. Winter M, Appel WK, Evers B, Hodal T, Moller K-C, Schneider I, Wachtler M, Wagner MR, Wrodnigg GH, Besenhard JO (2001) Studies on the anode/electrolyte interface in lithium ion batteries. Monatshefte fur Chemie 132:473–486
43. Ng SH, Wang J, Wexler D, Konstantinov K, Guo ZP, Liu HK (2006) Highly reversible lithium storage in spheroidal carbon-coated silicon nanocomposites as anodes for lithium-ion batteries. Angew Chem Int Ed 45:6896–6899
44. Wang W, Kumta PN (2007) Reversible high capacity nanocomposite anodes of Si/C/SWNTs for rechargeable Li-ion batteries. J Power Sources 172:650–658
45. Li H, Huang X, Chen L, Zhou G, Zhang Z, Yu D, Mo YJ, Pei N (2000) The crystal structural evolution of nano-Si anode caused by lithium insertion and extraction at room temperature. Solid State Ion 135:181–191
46. Kwon Y, Park G-S, Cho J (2007) Synthesis and electrochemical properties of lithium-electroactive surface-stabilized silicon quantum dots. Electrochim Acta 52:4663–4668
47. Kim H, Cho J (2008) Superior lithium electroactive mesoporous Si@Carbon core – shell nanowires for lithium battery anode material. Nano Lett 8:3688–3691
48. Cui L-F, Yang Y, Hsu C-M, Cui Y (2009) Carbon – silicon core – shell nanowires as high capacity electrode for lithium ion batteries. Nano Lett 9:3370–3374
49. Guo J, Wang C (2010) A polymer scaffold binder structure for high capacity silicon anode of lithium-ion battery. Chem Commun 46:1428–1430
50. Xu W, Flake JC (2010) Composite silicon nanowire anodes for secondary lithium-ion cells. J Electrochem Soc 157:A41–A45
51. Chou S-L, Wang J-Z, Choucair M, Liu H-K, Stride JA, Dou S-X (2010) Enhanced reversible lithium storage in a nanosize silicon/graphene composite. Electrochem Commun 12:303–306
52. Lee JK, Smith KB, Hayner CM, Kung HH (2010) Silicon nanoparticles–graphene paper composites for Li ion battery anodes. Chem Commun 46:2025–2027
53. Wang W, Kumta PN (2010) Nanostructured hybrid silicon/carbon nanotube heterostructures: reversible high-capacity lithium-ion anodes. ACS Nano 4:2233–2241
54. Chan CK, Patel RN, O'Connell MJ, Korgel BA, Cui Y (2010) Solution-grown silicon nanowires for lithium-ion battery anodes. ACS Nano 4:1443–1450
55. Magasinski A, Dixon P, Hertzberg B, Kvit A, Ayala J, Yushin G (2010) High-performance lithium-ion anodes using a hierarchical bottom-up approach. Nat Mater 9:353–358
56. Gómez-Cámer JL, Morales J, Sánchez L (2011) Anchoring Si nanoparticles to carbon nanofibers: an efficient procedure for improving Si performance in Li batteries. J Mater Chem 21:811–818
57. Krishnan R, Lu T-M, Koratkar N (2011) Functionally strain-graded nanoscoops for high power Li-ion battery anodes. Nano Lett 11(2):377–384
58. Wang X-L, Han W-Q (2010) Graphene enhances Li storage capacity of porous single-crystalline silicon nanowires. ACS Appl Mater Interfaces 2(12):3709–3713
59. Winter M, Brodd RJ (2004) What are batteries, fuel cells, and supercapacitors? Chem Rev 104:4245–4270
60. Kang K, Meng YS, Breger J, Grey CP, Ceder G (2006) Electrodes with high power and high capacity for rechargeable lithium batteries. Science 311:977–980
61. Kang B, Ceder G (2009) Battery materials for ultrafast charging and discharging. Nature 458:190–193
62. Park M-H, Kim MG, Joo J, Kim K, Kim J, Ahn S, Cui Y, Cho J (2009) Silicon nanotube battery anodes. Nano Lett 9:3844–3847
63. Zhou S, Liu X, Wang D (2010) Si/$TiSi_2$ heteronanostructures as high-capacity anode material for Li ion batteries. Nano Lett 10:860–863
64. Jung H-G, Oh S-W, Ce J, Jayaprakash N, Sun YK (2009) Mesoporous TiO_2 nano networks: anode for high power lithium battery applications. Electrochem Commun 11:756–759
65. Simon P, Gogotsi Y (2008) Materials for electrochemical capacitors. Nat Mater 7:845–854

66. Lee SW, Yabuuchi N, Gallant BM, Chen S, Kim B-S, Hammond PT, Shao-Horn Y (2010) High-power lithium batteries from functionalized carbon-nanotube electrodes. Nat Nanotechnol 5:531–537
67. Chen H, Armand M, Demailly G, Dolhem F, Poizot P, Tarascon J-M (2008) From biomass to a renewable Li$X$$C_6O_6$ organic electrode for sustainable Li-ion batteries. ChemSusChem 1:348–355
68. Wu X-L, Jiang L-Y, Cao F-F, Guo Y-G, Wan L-J (2009) $LiFePO_4$ nanoparticles embedded in a nanoporous carbon matrix: superior cathode material for electrochemical energy-storage devices. Adv Mater 21:2710–2714
69. Dudney NJ (2008) Thin film micro-batteries. Electrochem Soc Interface 17:44–48
70. Scrosati B, Garche J (2010) Lithium batteries: status, prospects and future. J Power Sources 195:2419–2430

Chapter 4
Tin-Based Anode Materials for Lithium-Ion Batteries

Fabrice M. Courtel and Yaser Abu-Lebdeh

Abstract Tin and its compounds constitute a new class of high-capacity anode materials that can replace graphitic carbon in current lithium-ion batteries. In the case of the two most studied, tin metal and tin oxide, it was shown that the inevitable volume expansion during electrochemical alloying with lithium can be mitigated using many strategies including formation of nanofilms, nanoparticles, nanocomposites, and nanostructures. It was demonstrated that high reversible capacities can be obtained and this was highlighted by the successful commercialization of a lithium-ion battery with a Sn/Co/C nanocomposite (Nexelion^{TM}).

4.1 Introduction

Most commercial lithium-ion batteries (LIBs) use graphitic carbon as the anode material due to its low cost, long cycle life, and very stable capacity [1]. However, the reversible electrochemical intercalation of lithium ions in its structure leads to a graphite intercalated compound with a composition of one lithium for six carbons (LiC_6, see Fig. 4.1a) that results in capacities limited to only 372 mAh g^{-1} and 830 mAh mL^{-1}. Alternative anode materials are hence currently being investigated with the focus on those based on mechanisms other than intercalation. One of the most common mechanisms is metals or metalloids that can electrochemically and reversibly alloy with lithium such as tin, silicon, antimony, aluminum, etc. [1–6]. These metals react with a large number of lithium atoms which leads to much higher capacities than graphite: tin (993 mAh g^{-1}, 7,313 mAh mL^{-1}), silicon

F.M. Courtel
Atomic Energy of Canada Limited, Chalk River, ON K0J 1J0, Canada

Y. Abu-Lebdeh (✉)
National Research Council of Canada, Ottawa, ON K1A 0R6, Canada
e-mail: Yaser.Abulebdeh@nrc.gc.ca

Y. Abu-Lebdeh and I. Davidson (eds.), *Nanotechnology for Lithium-Ion Batteries*, Nanostructure Science and Technology, DOI 10.1007/978-1-4614-4605-7_4,

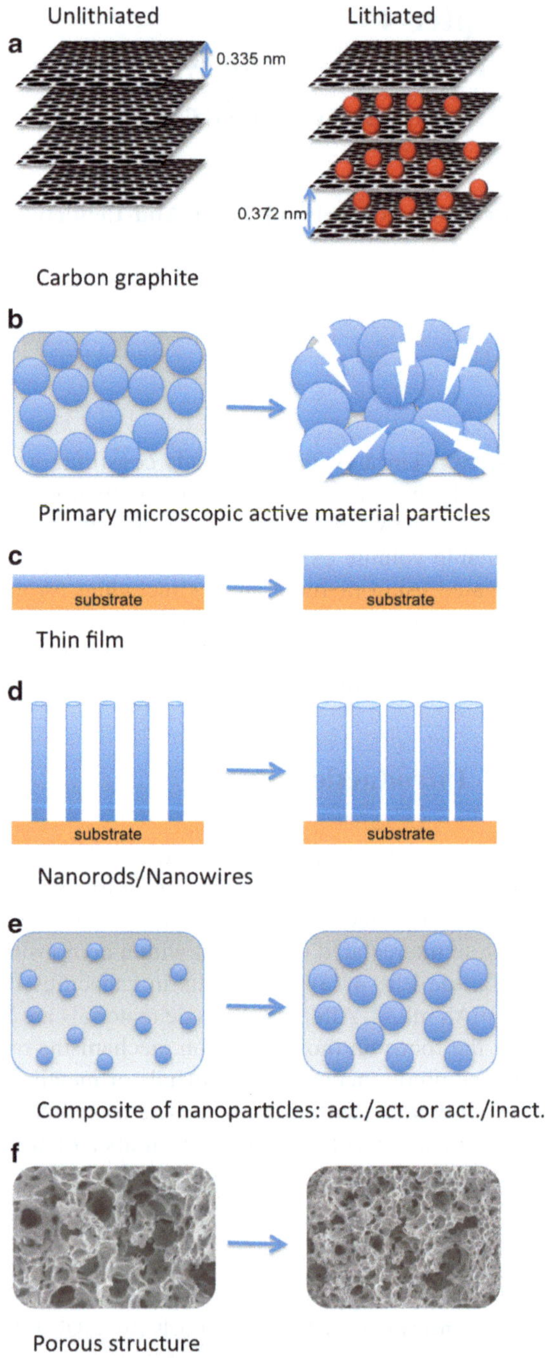

Fig. 4.1 Lithiation process of different forms of anode materials: (**a**) carbon graphite, (**b**) microscopic active material particles, (**c**) thin film, (**d**) nanorods or nanowires, (**e**) composite of nanoparticles (act./act. or act./inact.), and (**f**) porous structure

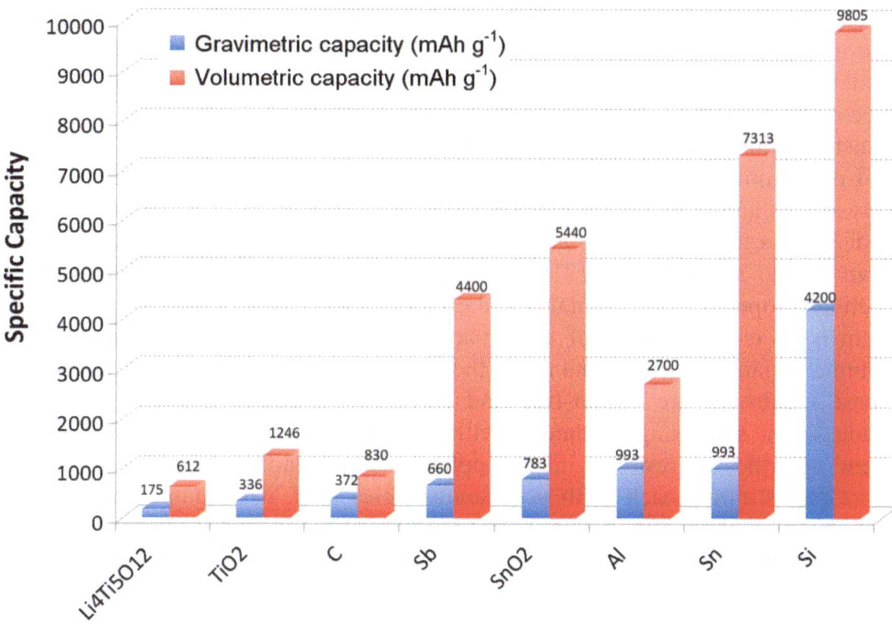

Fig. 4.2 Graph of theoretical gravimetric and volumetric capacities of various anode materials

(4,200 mAh g^{-1}, 9,805 mAh mL^{-1}), antimony (660 mAh g^{-1}, 4,220 mAh mL^{-1}), and aluminum (994 mAh g^{-1}, 2,684 mAh mL^{-1}) [1–6]. Work on lithium alloys started with Dey in 1971 [7] who studied the electrochemical formation of lithium alloys with tin, zinc, lead, and aluminum. Metallic lithium alloys have been widely used for high-temperature thermal batteries for military applications [8]. Extensive work has been done with aluminum which shows the advantage of having a flat lithiation and delithiation potential profile at 0.26 and 0.45 V versus Li/Li^+, respectively [9]. Unlike graphite where lithium is intercalated with only 10% volume change (see Fig. 4.1a), in the case of aluminum alloying reaction, the volume change associated with the formation of LiAl is about 100% [1]. This was found to be detrimental for a bulk electrode which simply pulverizes and leads to the end of its cycle life [10], as illustrated in Fig. 4.1b. However, a 100-nm thin film has shown a reversible capacity of 800 mAh g^{-1} and a lithium diffusion coefficient of 6×10^{-12} cm^2 s^{-1} (see Fig. 4.1c) [9].

Figure 4.2 compares the specific capacities (gravimetric and volumetric) of different anode materials. Anodes showing high values are silicon (Si), silicon monoxide (SiO), tin (Sn), and tin dioxide (SnO_2). They have the advantage of not being too expensive to produce (at least for battery grade) and are widely abundant on earth crust. Tin exists in two different crystalline phases, tin β (a.k.a. white tin) which is metallic and is the most stable form between room temperature up until the melting point (230°C). Below 13.2°C (or lower in presence of impurities), the most stable phase is tin α (a.k.a. gray tin) which is an insulator and could cause problems for low temperature performance especially due to changes in density between the two structures: 7.30 g cm^{-3} for the body-centered tetragonal crystal structure of the β phase versus 5.77 g cm^{-3} for the face-centered cubic diamond structure of the α phase [11]. However, this phase transformation is a very slow process.

Like aluminum and silicon, tin also suffers from a large volume change between its unlithiated and lithiated states, about 260% [1]. This volume change is due to the fact that a large amount of lithium alloys with tin while both atoms have similar atomic radii [12]. As shown in Fig. 4.1c–f, a few strategies have been developed to counterbalance the effect of this large volume change. One of them is to use thin films of a few hundred of nanometers. In that case, the volume expansion occurs mainly in the direction normal to the film plane [13]. These electrodes are binder-free and carbon-free; however, they are usually made via a sputtering technique with low active material mass that makes them of little interest for many applications primarily electric vehicles. A more realistic approach is to use nanoparticles [14], nanocomposites [3, 15], and binders that can better accommodate the volume change [15, 16]. Nanoparticles are known to usually better accommodate the volume change due to faster stress relaxations which improves the cycling performance of the materials. In addition, nanoparticles/nanocomposites provide a much shorter diffusion path for the lithium ions, which also improves the rate capability of these electrodes [3]. However, it is also known that nanoparticles are usually more prone to surface reactions such as those associated with the formation of the solid electrolyte interface (SEI), which usually increases the irreversible capacity and lower the battery performance. In addition, due to the large surface to volume ratio of nanoparticles, a low tap density is obtained, which reduces the volumetric capacity of the electrode [3] and also gives rise to nanoeffects that little is known about their influence (thermodynamic, transport) on the interface (space charge) that might have a positive or negative effect on the battery performance.

4.2 Tin Oxide

Tin oxide has been widely studied in the last decade as it has been found to be attractive because of its high reversible capacity, 783 mAh g^{-1}. It reacts with lithium according to the two following reactions:

$$SnO_2 + 4Li^+ + 4e^- \rightarrow Sn + 2Li_2O \qquad 711\,\text{mAh g}^{-1} \tag{4.1}$$

$$\begin{gathered} Sn + xLi^+ + xe^- \leftrightarrow Li_xSn(0 \le x \le 4.4) \\ 993\,\text{mAh g}^{-1}\text{of Sn or } 783\,\text{mAh g}^{-1}\text{of SnO}_2 \end{gathered} \tag{4.2}$$

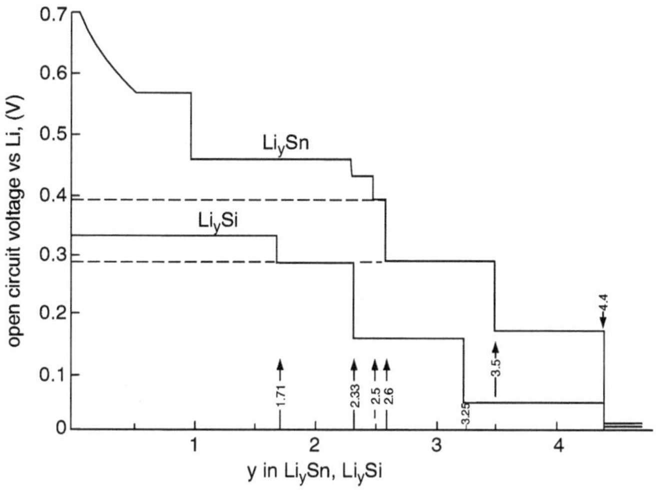

Fig. 4.3 Composition dependence of the potential in the Li–Sn and Li–Si systems [1]

Tin oxide has a theoretical reversible capacity of 783 mAh g^{-1}. An irreversible reaction occurs prior to the $SnLi_{4.4}$ formation: the reduction of SnO_2 to Sn and the formation of a matrix of Li_2O. However, Li_2O is not decomposable which means that a large irreversible capacity of 711 mAh g^{-1} is associated with this reaction. It is important to remember that for a full lithium-ion cell, the cathode is the lithium reservoir in the battery, which means that this irreversible capacity is non-recoverable and lower the total capacity of the battery. Nevertheless, Li_2O has the advantage of providing a matrix that helps in accommodating the volume changes associated with the alloying/de-alloying reaction of tin with lithium. Sn then further reacts with lithium; this reaction is associated with a volume change of about 260% [1]. Tin alloys with lithium in a multistep reaction (multiphase transitions) to form the Li_xSn ($0 \leq x \leq 4.4$), as shown in Fig. 4.3. With up to 4.4 Li per Sn atoms, Sn has a theoretical gravimetric capacity of 993 mAh g^{-1} which is 2.5 times higher than graphite. Its volumetric capacity is even more impressive, 7,300 mAh mL^{-1}, which is about nine times higher than graphite.

As reported by Brousse et al. [4] and Courtney et al. [17], the size of the particles plays a crucial role in the performance of tin oxide, as smaller particles are able to better accommodate the absolute volume change. As an illustration of the size effect, Fig. 4.4 shows the cycling performance of SnO_2 nanoparticles with sizes ranging from 3 to 8 nm. Pristine SnO_2 nanoparticles or composites usually show a medium capacity around 350 to 450 mAh g^{-1} for a rate around C/5 [18–22]. Using a layer-by-layer technique, SnO_2 nanotubes (50 wt%) supported on carbon nanotubes provided a specific capacity of 450 mAh g^{-1} [23]. As shown by Fig. 4.5, Yang et al. prepared 3D flower-shaped SnO_2 nanostructures; those structures exhibit a high capacity of 670 mAh g^{-1} at a rate of about C/8 [24].

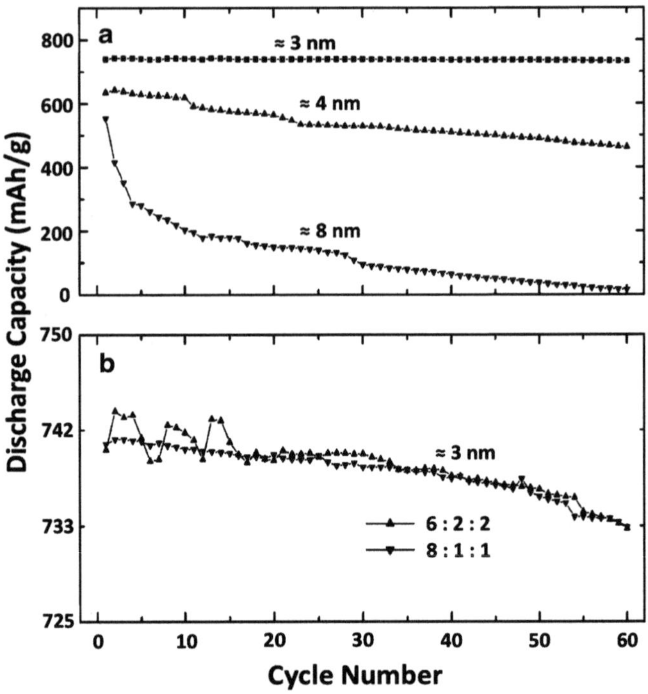

Fig. 4.4 Cycle-life performance of the SnO_2 nanoparticles as a function of (**a**) size (≈3, ≈4, and ≈8 nm) and (**b**) electrode composition (SnO_2:binder:carbon black in a weight ratio of 6:2:2 and 8:1:1). The charge cutoff voltage was 1.2 V (Reprinted with permission from [14]. Copyright (2011) American Chemical Society)

SnO_2 templated structures have greatly improved the reversible capacity by providing free space for the expanding structure limiting the bulk volume expansion. Kim et al. [25] and Yim et al. [26] prepared 3D porous SnO_2 (see Figs. 4.6 and 4.7, respectively). They reported a capacity of 600 mAh g^{-1} after 55 cycles [26]. A capacity value of 775 mAh g^{-1} was obtained at C by Kim et al. which is almost the theoretical reversible capacity of SnO_2 [25]. At higher rates, a capacity up to 730 mAh g^{-1} was obtained at 10 C, which is 93% of the theoretical reversible capacity. Some SnO_2/carbonaceous-material composites [16, 27, 28], such as SnO_2/graphene composites, showed very interesting capacity values, capacity retention, and rate capabilities. This is discussed in great details in chapter six in this book [29–33].

Schoonman et al. have shown that a nanosized silicon-doped tin oxide have a reduced irreversible capacity and an increased reversible capacity [34, 35]. The 26% silicon-doped sample even showed a capacity over the theoretical value which was attributed to excess of defects in the space-charge regions and to the inhomogeneity at the grain boundaries. It involved a higher Li^+ diffusion coefficient at the interfacial region between particles of the two phases.

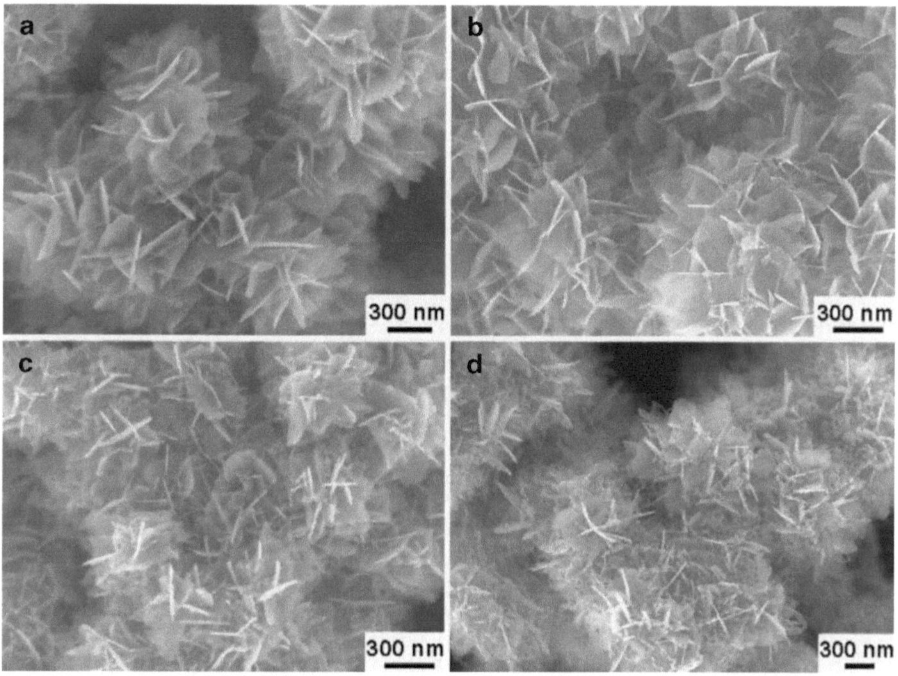

Fig. 4.5 SEM micrographs of the SnO_2 nanostructures prepared after calcinations in air at different temperatures: (**a**) as-synthesized nanoplates obtained by hydrothermal treatments; (**b**) after heating in air at 300°C, (**c**) 400°C, and (**d**) 700°C, respectively (Reprinted from Acta Materialia [24], Copyright (2011), with permission from Elsevier)

4.3 Tin-Based Composites

The main issue with tin oxide is the large irreversible capacity of the first cycle (711 mAh g^{-1}), and this is not avoidable. In addition, once tin oxide has been reduced, it has been reported by many groups that due to their high mobility, $SnLi_{x\ (0 \leq x \leq 4.4)}$ nanoparticles tend to aggregate to form bigger particles which are detrimental for the electrode performance [36, 37]. This aggregation is due to the fact that tin is a very ductile metal (melting point 230°C), which suggests a good mobility of the atoms at room temperature. So each time tin metal is formed, the nanoparticles will have a tendency to aggregate. This aggregation is evidenced by the growth of the peak either in the differential capacity versus voltage curves or in the cyclic voltammograms, as a function of cycle number [17]. The aggregation involves the destruction of the $SnLi_{x\ (0 \leq x \leq 4.4)}/Li_2O$ homogenous composite. Figure 4.8 shows the tin aggregation over cycling in the case of a $Sn_3O_2(OH)_2$ electrode. The key point here is to suppress the aggregation by embedding the nanoparticles in a stronger matrix than Li_2O.

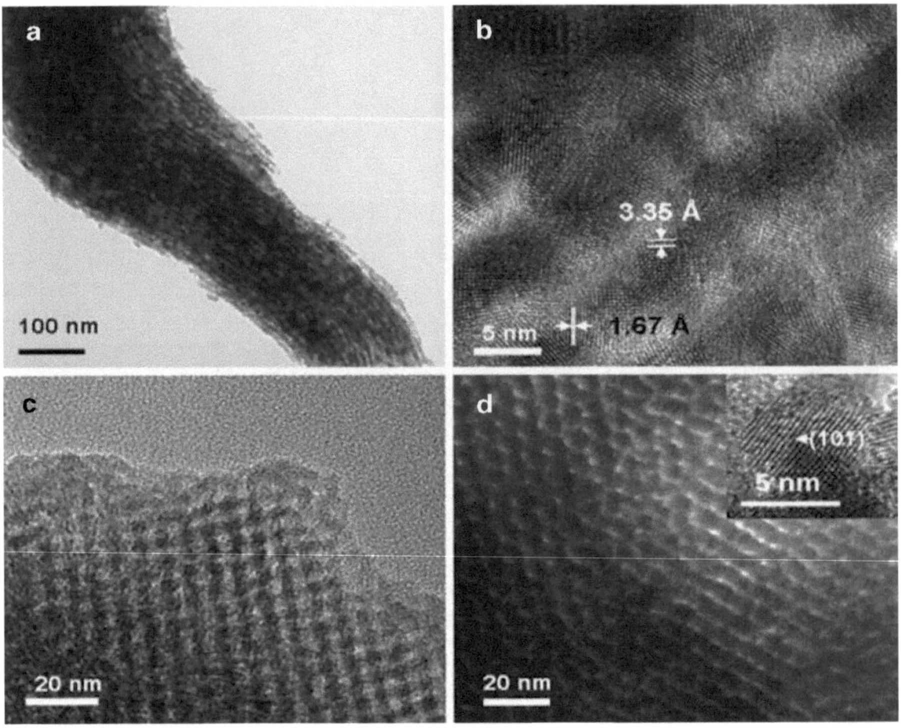

Fig. 4.6 TEM micrographs of SnO_2 nanowires (**a**) along the [110] direction of the p6mm structure; (**b**) HREM image of (a). TEM images of mesoporous SnO_2 along the (**c**) [110] and (**d**) [100] directions ([25] Reproduced by permission of The Royal Society of Chemistry)

Tin metal is attractive since, in theory, no irreversible capacity should be observed. In addition, tin has a high lithium diffusion coefficient, for example, LiSn shows a value of 8×10^{-8} cm^2 s^{-1} [38] and $Li_{4.4}Sn$ has a value of 5.9×10^{-7} cm^2 s^{-1} [39]. These values are quite comparable with those of $Li_4Ti_5O_{12}$ [40] and graphite [41, 42]. However, tin metal is rarely used by itself because the volume change associated with the alloying reaction is too large, even as nanoparticles. Tin is usually used in composites made of either carbon and/or another metal, such as Sn-M or Sn-M-C, with M being an inactive material toward lithium. For instance, as shown by Fig. 4.9, a Sn/C composite made of tin nanoparticles of 200 nm embedded into a calcined polystyrene (PS) resin showed a stable capacity of 500 mAh g^{-1} when cycled at C/13 [43]. The nanocomposite was prepared by infiltrating (tetraethyl)tin into the PS resin. A similar composite made of tin nanoparticles of about 10–100 nm embedded into a carbon matrix provided a capacity of about 500 mAh g^{-1} after 200 cycles at 0.8 C, which represents 96% of the theoretical capacity of the composite [3]. Figure 4.10 shows the TEM micrograph and battery performance of the nanocomposite. The composite was prepared by infiltration of (tributylphenyl)tin into an organic resorcinol–formaldehyde gel.

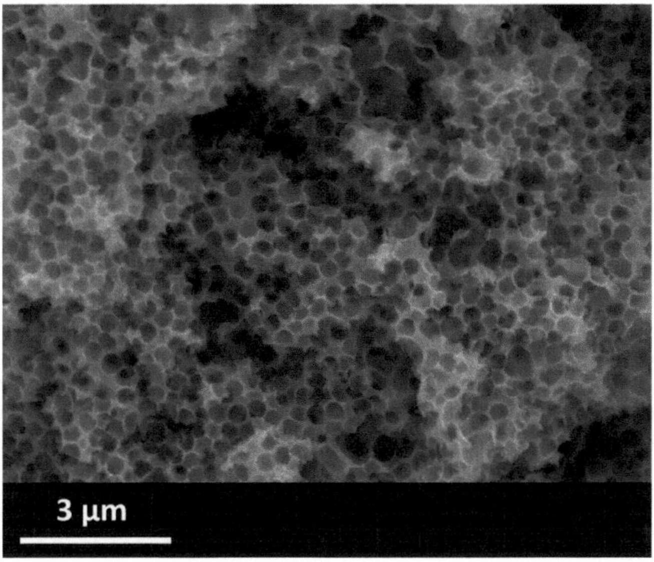

Fig. 4.7 SEM micrograph of the macroporous SnO_2/C composite after the calcination of the template under argon at 260°C for 3 h, and 600°C for 3 h (Reprinted from Journal of Power Sources [26], Copyright (2011), with permission from Elsevier)

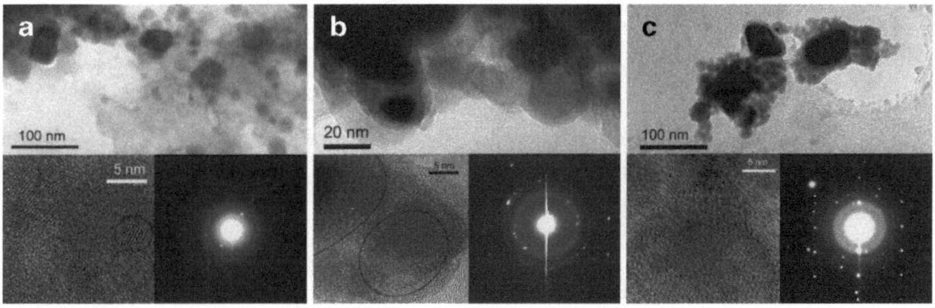

Fig. 4.8 Low magnification TEM micrographs, HRTEM micrographs, and SADP of $Sn_3O_2(OH)_2$ electrodes after (**a**) 1, (**b**) 20, and (**c**) 40 cycles (Reproduced by permission of the Electrochemical Society [37])

This composite also shows good rate capability: from 410 mAh g^{-1} at 1.5 C to 350 mAh g^{-1} at 2 C and 200 mAh g^{-1} at 5 C [3].

As shown by Fig. 4.11a, the Sn/C composite prepared by Hassoun and Scrosati has been tested in full lithium–sulfur battery with a Li_2S/C composite cathode and a gel electrolyte [44]. As shown by Fig. 4.11b, at C/20, a capacity of about 500 mAh g^{-1} was obtained after 35 cycles with the same mass of active materials on both electrodes. At higher rate, C/5, a capacity of about 170 mAh g^{-1} was obtained, as shown by Fig. 4.11c.

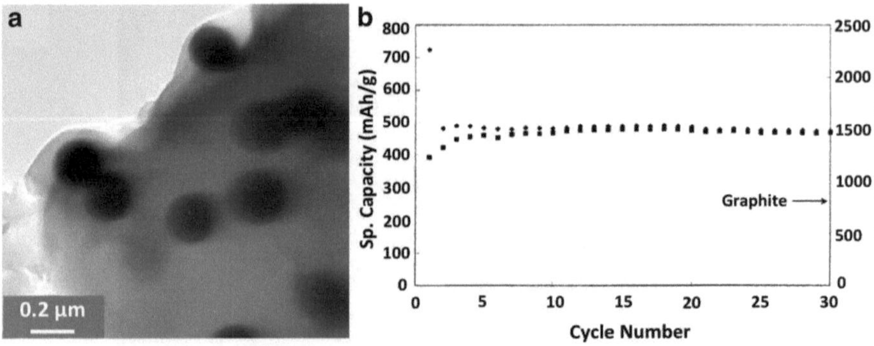

Fig. 4.9 (**a**) TEM micrographs of the Sn/C composite obtained from polystyrene resin infiltrated with tetraethyltin corresponding to a Sn:C composition of 1:1 after heat treatment for 5 h in UHP Ar at 600°C; (**b**) plot of capacity as a function of cycle number of the same composite (current rate: 100 mA/cm^2; potential: 0.02–1.2 V) (Reproduced by permission of the Electrochemical Society [43])

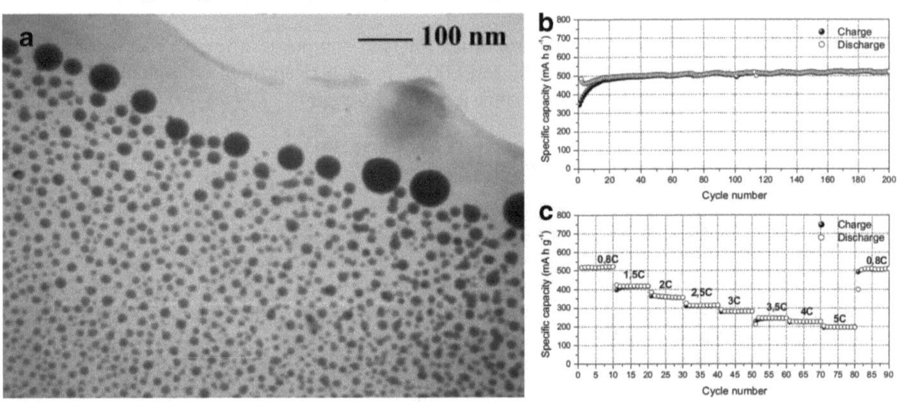

Fig. 4.10 (**a**) TEM micrograph of a Sn/C composite material, (**b**) cycling graph of a half-cell having Sn/C composite electrode and lithium metal as a counter electrode (cells were cycled at a 0.8 C rate), (**c**) cycling graphs a half-cell at various rates (Reprinted with permission of John Wiley & Sons, Inc. [3])

Another strategy to accommodate the volume expansion associated with the alloying reaction is to use a foam-like structure, as explored by Morishita et al. [45]. The free volume left by the dissolved MgO template gives room for volume expansion. A stable capacity of 500 mAh g^{-1} is obtained at C/20. However, a large irreversible capacity is observed for the first cycle, due to oxidized tin at the surface and also due to the non-graphitic carbon generated by the polymer. Unfortunately, no rate capability behavior has been reported. Using an electrodeposited mesoporous tin anode, a good rate capability was obtained by Nara et al. [46]: from 650 mAh g^{-1} at C to 550 mAh g^{-1} at 2 C and 350 mAh g^{-1} at 10 C (Fig. 4.12).

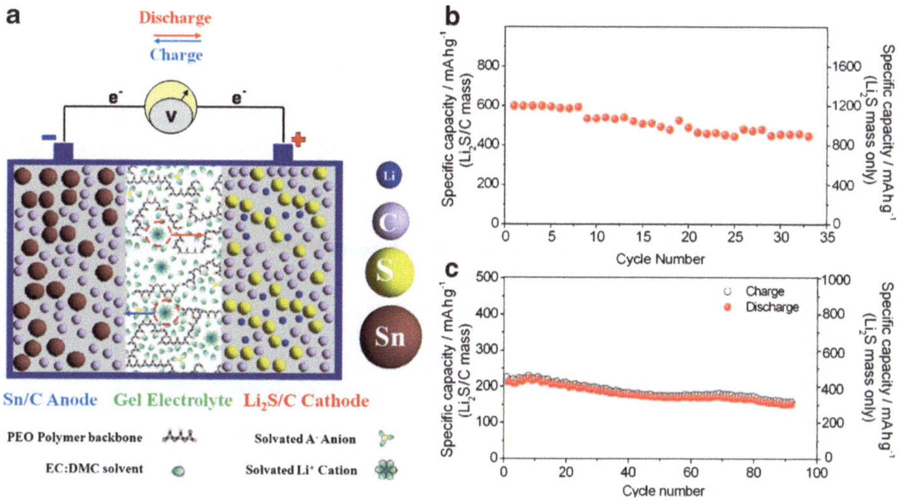

Fig. 4.11 (**a**) Sketch of the Sn/C–Li_2S/C polymer battery developed by Hassoun and Scrosati. The battery is formed by a Sn/C composite anode, a poly(ethylene oxide)-based gel polymer electrolyte, and a Li_2S/C cathode [44]. (**b**) Cycling graph of the full cell cycled at C/20 (0.2–4 V) and (**c**) at C/5 (0.2–4.6 V). The operating temperature was 25°C. The capacity is shown both in terms of Li_2S/C mass and in terms of Li_2S active material mass only (Reprinted with permission of John Wiley & Sons, Inc. [44])

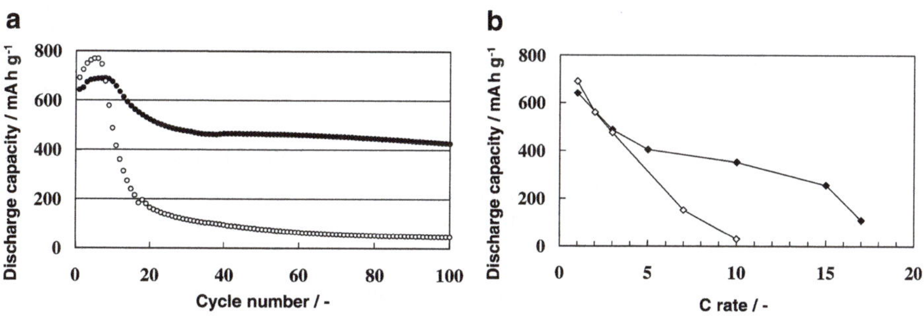

Fig. 4.12 The discharge capacities of the mesoporous Sn (●) and the dense Sn anodes (○). The charge–discharge current density was 994 mA g^{-1} (1 C rate) in the potential range of 0.01–1 V versus Li/Li^+. The discharge capacity at the first cycle of the mesoporous Sn anode (■) and the dense Sn anode electrodeposited with a conventional bath (□). The charge–discharge current density was varied from 0.994 to 16.9 A g^{-1} (1–17 C rate) in the potential range of 0.01–1 V versus Li/Li^+ (Reprinted from Chemistry Letters [46], Copyright (2011), with permission from the Chemical Society of Japan)

Using the same concept of free space in a composite made of 37.6 wt% of tin encapsulated in carbon nanotubes (CNTs), a very stable capacity of about 500 mAh g^{-1} was obtained [47]. This capacity value represents 100% of the theoretical capacity of the composite. Using a similar approach, Yu et al. also prepared a composite made of 66 wt% of tin encapsulated in CNTs [48]. A capacity

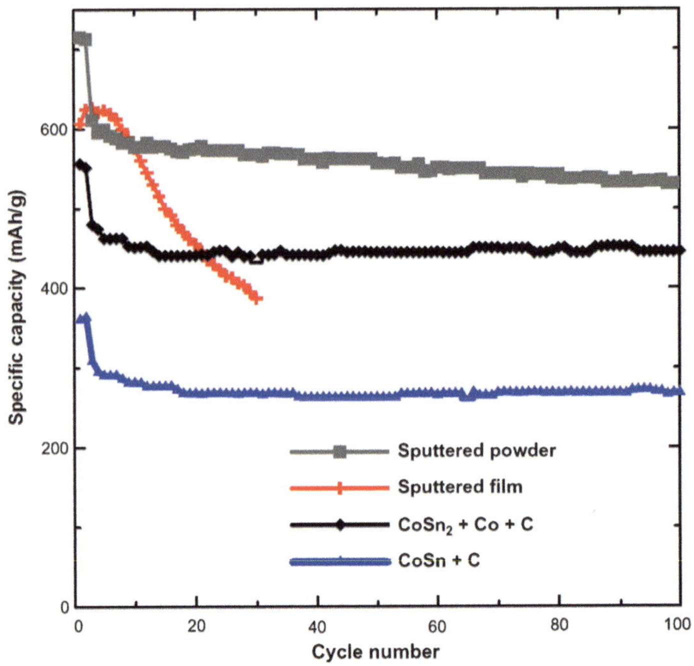

Fig. 4.13 Specific capacity versus cycle number for sputtered and attrited materials. Also included is data for the sputtered film deposited directly onto copper foil (cycled as a thin film). Cells were charged and discharged between 0.005 and 2.5 V for the first two cycles at C/10. The following cycles were made between 0.005 and 1.2 V at C/5. Testing was done at 30°C (Reprinted from Electrochemistry Communications [60], Copyright (2011), with permission from Elsevier)

of 650 mAh g^{-1} was obtained at C/2 after 140 cycles, which represents about 84% of the composite capacity. Their material also shows a good rate capability with a capacity of 570 mAh g^{-1} at 2 C and 295 mAh g^{-1} at 10 C, which represent 74% and 38% of the theoretical capacity. Sn nanoparticles can also be encapsulated in a graphene structure (see Chap. 6 for more information).

Another attractive way to manage the volume change associated with the alloying/de-alloying reaction of tin and lithium is to prepare amorphous nanocomposites made of an active and an inactive elements along with some graphitic carbon, such as Sn-Mn-C [49], Sn-Fe-C [50–52], Sn-Co-C [53–55], Sn-Ni-C [56], or Sn-Cu-C [57, 58]. The intimate mixture of the inactive element and the graphite acts as a matrix to accommodate the volume change during the reaction of Sn with lithium. The most attractive composite in terms of performance and rate capability is the Sn-Co-C composite. Cobalt is inactive toward carbon that prevents the formation of carbides. A stable reversible capacity of 600 mAh g^{-1} at C/12 was obtained with a composition of $Sn_{0.42}Co_{0.34}C_{0.24}$ prepared by magnetron sputtering [59]. With a different composition, $Sn_{0.30}Co_{0.30}C_{0.40}$, prepared by high-energy ball milling, Ferguson et al. obtained a capacity of 450 mAh g^{-1} at C/5 (see Fig. 4.13) [60]. $Sn_{0.31}Co_{0.28}C_{0.41}$, also prepared using high-energy ball milling,

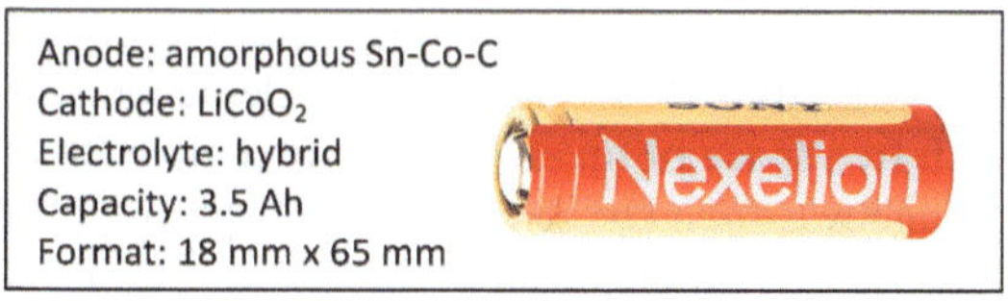

Fig. 4.14 An image of the 18650 Nexelion™ Sony battery that utilizes a tin compound as an anode material ([66])

showed a stable capacity of 500 mAh g^{-1} for rates going from C/14 to C/4 and up to 250–200 mAh g^{-1} at C [55, 61]. For a lower carbon content, with a composition of $Sn_{0.40}Co_{0.40}C_{0.20}$, a reversible capacity of 400 mAh g^{-1} was obtained at C/4 [54]. Other groups obtained similar data [62–64]. Even though no oxide is being used in this case, an irreversible capacity of about 150 mAh g^{-1} is usually observed for the first cycle; this was attributed to a structural reorganization of the Sn-Co alloy following the first de-alloying reaction [61]. Taking into account the mass of inactive cobalt, at C/4 the real measured gravimetric capacity of a $Sn_{0.31}Co_{0.28}C_{0.41}$ anode is about 350 mAh g^{-1}, which is slightly higher than that of a graphite anode. However, when considering the volumetric capacity, an increase of about 650–600% is obtained, considering a Sn/Co density of 14.3 g cm^{-3} [65].

Recently Sony showed an interest in this tin-based amorphous Sn-Co-C composite anode [66–69] and has already commercialized this new generation of lithium-ion battery called Nexelion. The 18650 format is made of a $LiCoO_2$ cathode and a hybrid electrolyte. It provides a capacity of 3.5 Ah. They report a capacity increase of about 30% compared to their conventional battery. This battery shows good rate capability and temperature performance [66, 67]. Other companies such as Panasonic are following the same path [70] (Fig. 4.14).

4.4 Summary and Outlook

In this chapter, we have reviewed the work on tin, tin alloys, and tin dioxide for use as anode materials for next generation lithium-ion batteries. The interest in tin is obvious due to its abundance in the earth crust and its ability to deliver very high capacities when it electrochemically alloys with lithium. It was however shown that this reaction is accompanied by large changes in volume of the electrode, very detrimental to the battery performance. In the case of tin oxide, the problem is exacerbated by the loss of a large capacity to the reduction of the tin oxide to the elemental tin before it reacts with lithium. In this chapter, we have summarized a wide variety of methods that were suggested to overcome or mitigate these issues in order to reach high reversible capacities. It was demonstrated that the use of nanoparticles or nanostructured tin oxide or nanocomposite of tin oxide and a carboneous material could lead to reversible capacities around 700 mAh g^{-1}. In the case of elemental tin, the formation of nanocomposites with free space or tin encapsulation in nanostructures to accommodate the volume change was found to be a successful approach. The most successful strategy was adopted by Sony where an amorphous nanocomposite of tin with cobalt and carbon gave capacities 30% higher than graphite and used in a commercial lithium-ion battery. The use of nanotechnology principles to make better materials has proven to be successful and can be generalized to other lithium alloyable metals with much higher capacities.

References

1. Nazri G-A, Pistoia G (2004) Lithium batteries: science and technology. Kluwer, Boston, 708 p
2. Hatchard TD, Dahn JR (2004) In situ XRD and electrochemical study of the reaction of lithium with amorphous silicon. J Electrochem Soc 151(6):A838–A842
3. Derrien G, Hassoun J, Panero S, Scrosati B (2007) Nanostructured Sn-C composite as an advanced anode material in high-performance lithium-ion batteries. Adv Mater 19(17):2336–2340
4. Brousse T, Crosnier O, Santos-Peña J, Sandu I, Fragnaud P, Schleich DM (2002) Recent progress in the development of tin-based negative electrodes for Li-ion batteries. In: Kumagai N, Komaba S (eds) Materials chemistry in lithium batteries. Research Signpost, Kerala
5. Hassoun J, Derrien G, Panero S, Scrosati B (2008) A nanostructured Sn-C composite lithium battery electrode with unique stability and high electrochemical performance. Adv Mater 20 (16):3169–3175
6. Chan CK, Peng H, Liu G, McIlwrath K, Zhang XF, Huggins RA, Cui Y (2008) High-performance lithium battery anodes using silicon nanowires. Nat Nanotechnol 3(1):31–35
7. Dey AN (1971) Electrochemical alloying of lithium in organic electrolytes. J Electrochem Soc 118(10):1547–1549
8. Huggins RA (2009) Advanced batteries: materials science aspects. Springer, New York
9. Hamon Y, Brousse T, Jousse F, Topart P, Buvat P, Schleich DM (2001) Aluminum negative electrode in lithium ion batteries. J Power Sources 97–98:185–187
10. Rao BML, Francis RW, Christopher HA (1977) Lithium-aluminum electrode. J Electrochem Soc 124(10):1490–1492
11. Callister WD (2007) Materials science and engineering: an introduction, 7th edn. Wiley
12. CRC Handbook of Chemistry and Physics (2010–2011). CRC Press. http://www.hbcpnetbase.com. Accessed in January 2012
13. Lewis RB, Timmons A, Mar RE, Dahn JR (2007) In situ AFM measurements of the expansion and contraction of amorphous Sn-Co-C films reacting with lithium. J Electrochem Soc 154(3): A213–A216
14. Kim C, Noh M, Choi M, Cho J, Park B (2005) Critical size of a nano SnO_2 electrode for Li-secondary battery. Chem Mater 17(12):3297–3301
15. Courtel FM, Bertin E, Saari D, Abu-Lebdeh Y, Davidson IJ (2010) Use of water soluble binders for nano-SnO_2 and nano-SnO_2/carbon composite anodes. In: The 15th international meeting on lithium batteries – IMLB 2010. ECS, Montreal, 27 June –2 July 2010
16. Chou S-L, Wang J-Z, Zhong C, Rahman MM, Liu H-K, Dou S-X (2009) A facile route to carbon-coated SnO2 nanoparticles combined with a new binder for enhanced cyclability of Li-ion rechargeable batteries. Electrochim Acta 54(28):7519–7524

17. Courtney IA, Dahn JR (1997) Key factors controlling the reversibility of the reaction of lithium with SnO_2 and Sn_2BPO_6 glass. J Electrochem Soc 144(9):2943–2948
18. Ng SH, dos Santos DI, Chew SY, Wexler D, Wang J, Dou SX, Liu HK (2007) Polyol-mediated synthesis of ultrafine tin oxide nanoparticles for reversible Li-ion storage. Electrochem Commun 9(5):915–919
19. Chen Y-C, Chen J-M, Huang Y-H, Lee Y-R, Shih HC (2007) Size effect of tin oxide nanoparticles on high capacity lithium battery anode materials. Surf CoatTechnol 202(4–7):1313–1318
20. Wang Y, Su F, Lee JY, Zhao XS (2009) Crystalline carbon hollow spheres, crystalline carbon/SnO_2 hollow spheres, and crystalline SnO_2 hollow spheres: synthesis and performance in reversible Li-ion storage. Chem Mater 18(5):1347–1353
21. Subramanian V, Burke WW, Zhu H, Wei B (2008) Novel microwave synthesis of nanocrystalline SnO_2 and its electrochemical properties. J Phys Chem C 112(12):4550–4556
22. Courtel FM, Baranova EA, Abu-Lebdeh Y, Davidson IJ (2010) In situ polyol-assisted synthesis of nano-SnO_2/carbon composite materials as anodes for lithium-ion batteries. J Power Sources 195(8):2355–2361
23. Du N, Zhang H, Chen B, Ma X, Huang X, Tu J, Yang D (2009) Synthesis of polycrystalline SnO_2 nanotubes on carbon nanotube template for anode material of lithium-ion battery. Mater Res Bull 44(1):211–215
24. Yang R, Gu Y, Li Y, Zheng J, Li X (2010) Self-assembled 3-D flower-shaped SnO_2 nanostructures with improved electrochemical performance for lithium storage. Acta Mater 58(3):866–874
25. Kim H, Cho J (2008) Hard templating synthesis of mesoporous and nanowire SnO_2 lithium battery anode materials. J Mater Chem 18:771–775
26. Yim C-H, Baranova EA, Courtel FM, Abu-Lebdeh Y, Davidson IJ (2011) Synthesis and characterization of macroporous tin oxide composite as an anode material for Li-ion batteries. J Power Sources 196(22):9731–9736
27. Li J, Zhao Y, Wang N, Guan L (2011) A high performance carrier for SnO_2 nanoparticles used in lithium ion battery. Chem Commun 47(18):5238–5240
28. Chen JS, Cheah YL, Chen YT, Jayaprakash N, Madhavi S, Yang YH, Lou XW (2009) SnO_2 nanoparticles with controlled carbon nanocoating as high-capacity anode materials for lithium-ion batteries. J Phys Chem C 113(47):20504–20508
29. Du Z, Yin X, Zhang M, Hao Q, Wang Y, Wang T (2010) In situ synthesis of SnO_2/graphene nanocomposite and their application as anode material for lithium ion battery. Mater Lett 64(19):2076–2079
30. Yao J, Shen X, Wang B, Liu H, Wang G (2009) In situ chemical synthesis of SnO2-graphene nanocomposite as anode materials for lithium-ion batteries. Electrochem Commun 11(10):1849–1852
31. Wang Z, Zhang H, Li N, Shi Z, Gu Z, Cao G (2010) Laterally confined graphene nanosheets and graphene/SnO_2 composites as high-rate anode materials for lithium-ion batteries. Nano Res 3(10):748–756
32. Paek S-M, Yoo E, Honma I (2009) Enhanced cyclic performance and lithium storage capacity of SnO_2/graphene nanoporous electrodes with three-dimensionally delaminated flexible structure. Nano Lett 9(1):72–75
33. Wang X, Zhou X, Yao K, Zhang J, Liu Z (2011) A SnO2/graphene composite as a high stability electrode for lithium ion batteries. Carbon 49(1):133–139
34. Huang H, Kelder EM, Chen L, Schoonman J (1999) Preparation and structure of silicon doped tin oxide composites using an advanced ultrasonic spray method. Solid State Ion 120(1–4):205–210
35. Schoonman J (2003) Nanoionics. Solid State Ion 157(1–4):319–326
36. Courtney IA, McKinnon WR, Dahn JR (1999) On the aggregation of tin in SnO composite glasses caused by the reversible reaction with lithium. J Electrochem Soc 146(1):59–68

37. Kim J-H, Jeong G-J, Kim Y-W, Sohn H-J, Park CW, Lee CK (2003) Tin-based oxides as anode materials for lithium secondary batteries. J Electrochem Soc 150(11):A1544–A1547
38. Wang J, Raistrick ID, Huggins RA (1986) Behavior of some binary lithium alloys as negative electrodes in organic solvent-based electrolytes. J Electrochem Soc 133(3):457–460
39. Anani A, Crouch-Baker S, Huggins RA (1987) Kinetic and thermodynamic parameters of several binary lithium alloy negative electrode materials at ambient temperature. J Electrochem Soc 134(12):3098–3102
40. Kavan L, Prochazka J, Spitler TM, Kalbac M, Zukalova M, Drezen T, Gratzel M (2003) Li insertion into $Li_4Ti_5O_{12}$ (spinel). J Electrochem Soc 150(7):A1000–A1007
41. Takami N, Satoh A, Hara M, Ohsaki T (1995) Structural and kinetic characterization of lithium intercalation into carbon anodes for secondary lithium batteries. J Electrochem Soc 142(2):371–379
42. Guyomard D, Tarascon JM (1992) Li metal-free rechargeable $LiMn_2O_4$/carbon cells: their understanding and optimization. J Electrochem Soc 139(4):937–948
43. Kim I-s, Blomgren GE, Kumta PN (2004) Sn/C composite anodes for Li-ion batteries. Electrochem Solid State Lett 7(3):A44–A48
44. Hassoun J, Scrosati B (2010) A high-performance polymer tin sulfur lithium ion battery. Angew Chem Int Ed 49(13):2371–2374
45. Morishita T, Hirabayashi T, Okuni T, Ota N, Inagaki M (2006) Preparation of carbon-coated Sn powders and their loading onto graphite flakes for lithium ion secondary battery. J Power Sources 160(1):638–644
46. Nara H, Fukuhara Y, Takai A, Komatsu M, Mukaibo H, Yamauchi Y, Momma T, Kuroda K, Osaka T (2008) Cycle and rate properties of mesoporous tin anode for lithium ion secondary batteries. Chem Lett 37(2):142–143
47. Wang Y, Wu M, Jiao Z, Lee JY (2009) Sn@CNT and Sn@C@CNT nanostructures for superior reversible lithium ion storage. Chem Mater 21(14):3210–3215
48. Yu Y, Gu L, Zhu C, van Aken PA, Maier J (2009) Tin nanoparticles encapsulated in porous multichannel carbon microtubes: preparation by single-nozzle electrospinning and application as anode material for high-performance Li-based batteries. J Am Chem Soc 131(44):15984–15985
49. Beaulieu LY, Dahn JR (2000) The reaction of lithium with Sn-Mn-C intermetallics prepared by mechanical alloying. J Electrochem Soc 147(9):3237–3241
50. Mao O, Dunlap RA, Dahn JR (1999) Mechanically alloyed Sn-Fe(-C) powders as anode materials for Li-ion batteries: I. The Sn_2Fe-C system. J Electrochem Soc 146(2):405–413
51. Mao O, Dahn JR (1999) Mechanically alloyed Sn-Fe(-C) powders as anode materials for Li-ion batteries: II. The Sn-Fe system. J Electrochem Soc 146(2):414–422
52. Mao O, Dahn JR (1999) Mechanically alloyed Sn-Fe(-C) powders as anode materials for Li-ion batteries: III. Sn_2Fe:$SnFe_3C$ active/inactive composites. J Electrochem Soc 146(2):423–427
53. Dahn JR, Mar RE, Abouzeid A (2006) Combinatorial Study of $Sn_{1-x}Co_x$ ($0 < x < 0.6$) and $[Sn_{0.55}Co_{0.45}]_{1-y}C_y$ ($0 < y < 0.5$) alloy negative electrode materials for Li-ion batteries. J Electrochem Soc 153(2):A361–A365
54. Hassoun J, Ochal P, Panero S, Mulas G, Bonatto Minella C, Scrosati B (2008) The effect of CoSn/$CoSn_2$ phase ratio on the electrochemical behaviour of $Sn_{40}Co_{40}C_{20}$ ternary alloy electrodes in lithium cells. J Power Sources 180(1):568–575
55. Hassoun J, Mulas G, Panero S, Scrosati B (2007) Ternary Sn-Co-C Li-ion battery electrode material prepared by high energy ball milling. Electrochem Commun 9(8):2075–2081
56. Mukaibo H, Sumi T, Yokoshima T, Momma T, Osaka T (2003) Electrodeposited Sn-Ni alloy film as a high capacity anode material for lithium-ion secondary batteries. Electrochem Solid State Lett 6(10):A218–A220
57. Wolfenstine J, Campos S, Foster D, Read J, Behl WK (2002) Nano-scale Cu_6Sn_5 anodes. J Power Sources 109(1):230–233

58. Thorne JS, Sanderson RJ, Dahn JR, Dunlap RA (2010) Combinatorial study of the Sn-Cu-C system for Li-ion battery negative electrode materials. J Electrochem Soc 157(10): A1085–A1091
59. Todd ADW, Mar RE, Dahn JR (2007) Tin-transition metal-carbon systems for lithium-ion battery negative electrodes. J Electrochem Soc 154(6):A597–A604
60. Ferguson PP, Todd ADW, Dahn JR (2008) Comparison of mechanically alloyed and sputtered tin-cobalt-carbon as an anode material for lithium-ion batteries. Electrochem Commun 10(1):25–31
61. Hassoun J, Panero S, Mulas G, Scrosati B (2007) An electrochemical investigation of a Sn-Co-C ternary alloy as a negative electrode in Li-ion batteries. J Power Sources 171(2):928–931
62. Huang L, Cai J-S, He Y, Ke F-S, Sun S-G (2009) Structure and electrochemical performance of nanostructured Sn-Co alloy/carbon nanotube composites as anodes for lithium ion batteries. Electrochem Commun 11(5):950–953
63. Chen Z, Qian J, Ai X, Cao Y, Yang H (2009) Preparation and electrochemical performance of Sn-Co-C composite as anode material for Li-ion batteries. J Power Sources 189(1):730–732
64. Lee S-I, Yoon S, Park C-M, Lee J-M, Kim H, Im D, Doo S-G, Sohn H-J (2008) Reaction mechanism and electrochemical characterization of a Sn-Co-C composite anode for Li-ion batteries. Electrochim Acta 54(2):364–369
65. Mashimo T, Tashiro S (1994) Synthesis of the WC-type tantalum nitride by mechanical alloying. J Mater Sci Lett 13(3):174–176
66. Sony Corporation (2011) Sony, the market for notebook PC; development of a tin-based amorphous anode, for high-capacity rechargeable lithium-ion battery 3.5 Ah: the "Nexelion" (trans). Available from http://www.sony.co.jp/SonyInfo/News/Press/201107/11-078/
67. Inoue H (2006) High capacity negative-electrode materials next to carbon; Nexelion. In: International meeting on lithium batteries, Biarritz
68. Mizutani S, Inoue H (2005) Negative active material and method for production thereof, non-aqueous electrolyte secondary cell using the same. Patent number: 2005-0208378
69. Kawakami S, Asao M (2005) Electrode material for anode of rechargeable lithium battery, electrode structural body using said electrode material, rechargeable lithium battery using said electrode structural body, process for producing said electrode structural body, and process for producing said rechargeable lithium battery. Patent number: 6,949,312
70. PanasonicNewsBureau (2009) Panasonic develops high-capacity lithium-ion battery cells that can power laptops and electric vehicles. Available from http://panasonic.co.jp/corp/news/official.data/data.dir/en091225-3/en091225-3.html. Accessed January 2012

Chapter 5
Beyond Intercalation: Nanoscale-Enabled Conversion Anode Materials for Lithium-Ion Batteries

Fabrice M. Courtel, Hugues Duncan, and Yaser Abu-Lebdeh

Abstract The use of transition metal oxides as anode materials in lithium-ion batteries offers great advantages over graphitic carbon due to their ability to deliver much higher specific capacities. The mechanism with which they electrochemically react with lithium was found to be peculiar and termed "conversion" to distinguish it from other mechanisms such as intercalation, insertion, and alloying. In this chapter, we have reviewed the behavior of a wide variety of transition metal oxides in lithium-ion batteries and the effect of structure/property relationship on their performance. It was found that a key enabler to the electrochemical reactivity of transition metal oxides is the nanosize effect and essentially the formation of nanoparticles and nanocomposites.

5.1 Introduction

In the recent years, there has been an increasing demand for high-power lithium-ion batteries (LIBs) for applications in portable electronics, plug-in hybrid vehicles (PHEVs), and other electric vehicles (EVs). While many new high-voltage and/or safer positive electrode (referred to as cathode) materials have been successfully developed and commercialized in the last years, such as carbon-coated nano-$LiFePO_4$, [1–3] $LiNi_{1/3}Co_{1/3}Mn_{1/3}O_2$ [4, 5], and $LiNi_{0.8}Co_{0.15}Al_{0.05}O_2$ [5, 6], the replacement of the carbon graphite negative electrode (referred to as anode) has been relatively less successful. Graphite has a high electronic conductivity

F.M. Courtel
Atomic Energy of Canada Limited, Chalk River, ON K0J 1J0, Canada

H. Duncan
Lawrence Berkeley National Laboratory, Berkeley, CA 94720, USA

Y. Abu-Lebdeh (✉)
National Research Council of Canada, Ottawa, ON K1A 0R6, Canada
e-mail: Yaser.Abu-Lebdeh@nrc.gc.ca

Y. Abu-Lebdeh and I. Davidson (eds.), *Nanotechnology for Lithium-Ion Batteries*, Nanostructure Science and Technology, DOI 10.1007/978-1-4614-4605-7_5,

(σ_a: 2.6 × 10^4 S cm^{-1} and σ_c: 2 × 10^2 S cm^{-1} [7]), is not too costly, and has a good capacity retention and very good cycle life. However, it is limited by its theoretical gravimetric and volumetric capacities: 372 mAh g^{-1} and 830 mAh mL^{-1}, respectively. Recently, Sony has prompted an interest in the Sn-Co-C alloy composite anode [8–15] and is now manufacturing this new generation of LIBs called Nexelion^{TM} [16]. This battery leads to much higher gravimetric and volumetric capacities but to a lower long-term reversibility compared to graphite. Sony reported a capacity increase of about 30% compared to their conventional battery, good rate capability, and good temperature performance [8]. Others, such as Toshiba, chose a safer and "zero-strain" [17] insertion anode material, the nano-$Li_4Ti_5O_{12}$ (a.k.a. LTO) [18–21] in the super-charge ion battery (SCiB^{TM}) with an output voltage of 2.4 V. LTO has a very good cycle life and high rate capability that allows quick charges/discharges [22], which is of interest for PHEV applications [18]. In addition, its high delithiation voltage (1.5 V vs. Li/Li^+) makes it safer than other anode materials [18–21]; indeed, there is no risk of lithium plating and presumably no SEI formation. However, its drawback is the low theoretical specific capacity (175 mAh g^{-1}/612 mAh mL^{-1}) and high lithium insertion voltage. Another alternative to the common anode material is to use elements, preferably nanometric in size, that can alloy with lithium, such as aluminum, silicon, or tin; they provide high specific capacities usually ranging from 1,000 to 4,000 mAh g^{-1} (2,700 to 9,800 mAh mL^{-1}). However, there is a large volume change between their unlithiated and their lithiated state: 100% for LiAl [23–25], 300–400% for Li_xSi ($x = 3.5$ or 4.2) [23, 26–29], and 260% for $Li_{4.4}Sn$ [23, 30, 31]. That gives rise to mechanical stresses that lead to cracks, eventual disintegration of the electrode, and failure of the LIB [32]. The use of nanoparticles [29] or nanocomposites [33] that can provide faster stress relaxations along with a binder that can accommodate the volume change, such as sodium carboxymethyl cellulose [34] or styrene-butadiene rubber [35], usually mitigate this issue. In addition, it is widely accepted now that anode materials that are prone to large volume changes during battery cycling warrant the use of ionic polymeric binders instead of conventional PVDF. This leads to better performance by allowing better lithium-ion transport from the electrolyte to the particles of the active material, more uniform coverage of the binder to the surface of the particles, and more uniform SEI formation at the surface of the particles [36]. In the last decade, transition metal oxides (TMOs), sulfides (TMSs), nitrides (TMNs), phosphides (TMPs), or fluorides (TMFs) (also referred as TMXs) have shown their potential as the next generation of anode materials for LIBs. Despite the fact that TMXs, with TM having the lowest oxidation number (e.g., Fe^{2+}, Mn^{2+}), lack interstitial sites for lithium insertion and the fact that none of these 3d transition metals (TMs) alloy with lithium, they happen to be active toward lithium [37, 38]. As shown by Fig. 5.1, TMOs provide capacities ranging from 650 to 1,000 mAh g^{-1} (3,300 to 5,000 mAh mL^{-1}) which is two to three times higher than graphite. However, the delithiation potential of these materials is higher than graphite, typically between 1.2 and 2.5 V versus Li/Li^+.

This chapter exposes the processes by which TMXs can be reversibly lithiated and delithiated and the factors that influence their performance, such as capacity

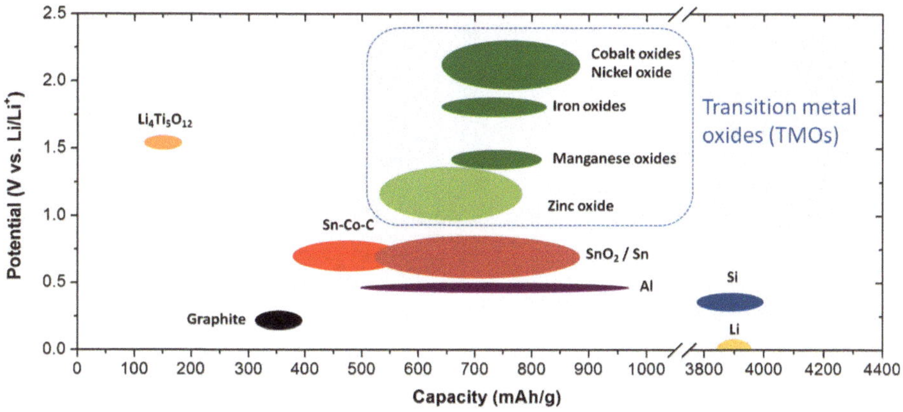

Fig. 5.1 Delithiation potential versus gravimetric specific capacity of anode materials

values, rate capability, and capacity retention. The discussion is mainly focused on TMOs going from chromium oxide to zinc oxide, and from simple oxides to mixed oxides such as spinel AB_2O_4. The last part of this chapter goes over the other TMXs introduced above.

5.2 Lithiation and Delithiation Mechanism

The use of TMOs as anode materials for LIBs is recent, but their use as cathode materials is not new: Li-CuO primary cells have been in use since 1980 [39, 40], and more recently, CuO has been used as a cathode material [41]. At room temperature, the simplified and nonreversible reaction was believed to be Eq. 5.1 with Cu_2O as an intermediate:

$$CuO + 2Li^+ + 2e^- \rightarrow Cu + Li_2O \tag{5.1}$$

The experimental cell voltage was about 1.5 V versus Li/Li$^+$ [39]. Even though in 1981 it has already been shown that the reaction of Fe_2O_3/Fe_3O_4 with lithium was reversible at 420°C [42], it was only in 2000 that Poizot et al. demonstrated that this reaction could be reversible at room temperature [37].

Unlike graphite, $Li_4Ti_5O_{12}$, or cathode materials which undergo a reversible intercalation/insertion reaction with Li$^+$, most TMXs go through a reversible conversion reaction with Li$^+$. While for the intercalation/insertion process less than one electron per metal is transferred (≈0.5 for $LiCoO_2$, 1 for $LiFePO_4$, 0.6 for $Li_4Ti_5O_{12}$, and 0.17 for graphite), for the conversion reaction, often a transfer of more than two electrons per metal occurs. As shown by Eq. 5.2, during the lithiation

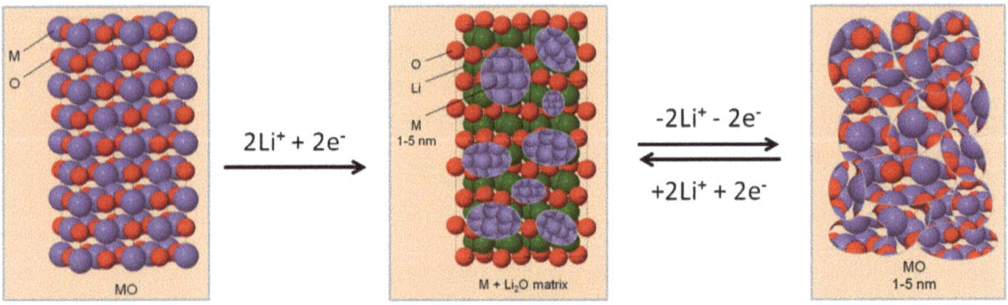

Fig. 5.2 Reversible conversion reaction of MO with lithium

step of the conversion reaction, the TMO is reduced into its metallic state and a decomposable matrix of Li_2O is formed:

$$MO_x + 2xLi^+ + 2xe^- \leftrightarrow M + xLi_2O \tag{5.2}$$

As shown by Fig. 5.2, after the first lithiation, a composite made of metal nanoparticles (typically 1–5 nm) embedded into a Li_2O matrix is formed [37]. During the delithiation process, nanoparticles of the metal oxide (typically 1–5 nm) are formed back, however, not necessarily with the same crystalline structure as the starting oxide. The same conversion reaction occurs with SnO_2 [43] and Sb_2O_3 [44]; however, in these cases, the Li_2O matrix is not decomposable and the subsequent process with lithium will be a reversible alloying reaction between lithium and tin or lithium and antimony. The reversible conversion reaction with lithium is not unique to TMOs; indeed, TMSs, TMNs, TMPs, and TMFs undergo the same reaction. The reaction can be generalized as:

$$MX_y + zLi^+ + ze^- \leftrightarrow M + zLiX_{y/z(y=z \text{ for } X=F; y=z/2 \text{ for } X=O,S;\ y=z/3 \text{ for } X=N,P)} \tag{5.3}$$

The theoretical cell voltage (E) of a half-cell, using lithium metal as counter electrode, is calculated from the Gibbs free energy of formation of the metal oxide nanoparticle and the decomposable Li_2O matrix (ΔG_f) using the Nernst equation shown in Eq. 5.4 [45]. (E is also known as the electromotive force, emf):

$$\Delta G_f = -nFE \tag{5.4}$$

where n is the number of electron exchanged, F the Faraday constant, and E the theoretical cell voltage also known as the electromotive force (emf). ΔG_f is easily calculated using Eq. 5.5:

$$\Delta G_f = n\Delta G_f^\circ(Li_2O) - \Delta G_f^\circ(MO) \tag{5.5}$$

Table 5.1 Theoretical half-cell voltages (E or emf) calculated using the Gibbs free energy of formation ($\Delta_f G$) and theoretical specific capacities of several TMOs [45]

Metal oxide	E or emf (V vs. Li/Li^+)	Theoretical specific capacity (mAh g^{-1})
MgO	−0.00354	1,330
Al_2O_3	0.180	1,577
TiO_2 (rutile)	0.608	1,342
TiO_2 (anatase)	0.625	1,342
SiO_2	0.694	1,784
V_2O_3	0.945	1,073
MnO	1.032	756
Cr_2O_3	1.085	1,058
ZnO	1.252	659
Mn_2O_3	1.431	1,018
FeO	1.61	746
Fe_2O_3	1.631	1,007
MnO_2	1.708	1,233
CoO	1.802	715
NiO	1.954	718
Cu_2O	2.147	375
CuO	2.248	674

Theoretical half-cell voltages for a variety of metal oxides (sulfides, nitrides, and fluorides) and their theoretical specific capacities are shown in Table 5.1 [45].

These values assume that the metal will not alloy with Li, which is the case for all metals except silicon, aluminum, magnesium, and zinc. Furthermore, one should be aware that the Nernst equation is accurate for bulk material only, which means that the surface/interface energy contributions that become significant when nanoparticles are used have been neglected. Taking the latter into account would change the emf values by only 100 mV [45, 46]. Even though Al_2O_3, TiO_2, and SiO_2 exhibit a positive emf and thus their conversion reaction should be spontaneous, there are kinetic limitations that prevent the reaction from taking place. These oxides either are unreactive toward lithium, such as MgO and Al_2O_3, or undergo an insertion reaction, such as TiO_2.

The working potential (U) of the electrode in a half-cell is mostly determined by the emf and is also influenced by the lithiation or the delithiation overpotential (η) according to the following equation:

$$U = \text{emf} \pm \eta(-\text{if lithiation and} + \text{if delithiation}) \tag{5.6}$$

For TMOs that do react with lithium, a large overpotential is observed for both reactions (lithiation and delithiation) due to slow kinetics associated with solid-state reactions. Thus, a large potential window is usually required, typically 10 mV to 3 V versus Li/Li^+. Due to these kinetic limitations, only metal oxides that show an emf above 1 V versus Li/Li^+ have been experimentally proven to be reduced before the electrodeposition of lithium metal on the metal oxide occurs [45].

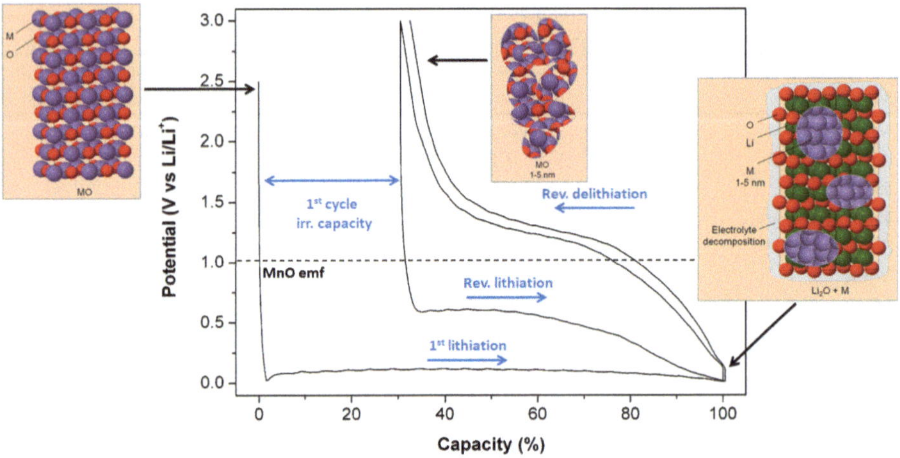

Fig. 5.3 Typical voltage profile of a half-cell made of a lithium metal counter electrode and a TMO working electrode (also applicable for TMSs, TMNs, TMPs, and TMFs). The half-cell was cycled between 10 mV and 3 V at 75 mA g^{-1}. Only the first two cycles are shown

TMXs, for which the metal is at its lowest oxidation state (e.g., MnO, FeO, CoO, NiO, Cu_2O, ZnO), directly undergo the reversible conversion reaction. However, in the case of a higher oxidation state and especially for compounds that exhibit a marked ionic character (i.e., TMOs and TMFs) or a covalent character (i.e., TMPs), due to the presence of vacancies, lithium insertion is going to occur before the reversible conversion reaction [38]. The lithium insertion induces the reduction of the TM to a lower oxidation state, and often, when the lowest (or the most stable) oxidation state is reached, the reversible conversion reaction occurs. Here are some examples of materials that follow this path: MnO_2 [47], Co_3O_4 [48, 49], Mn_3O_4 [50], CuO [51], CuS [52], Cu_3P[53], NiP_3 [54], Fe_2O_3 [55], and FeP [56].

The key point of the reversible conversion reaction of TMOs is the formation of nanoparticles of TM embedded and homogeneously distributed in a Li_2O matrix during the first lithiation. Figures 5.3 and 5.4 show typical voltage profiles of MnO, CoO, NiO, and FeO, respectively, as a function of the state of charge of the anode material in half-cell conditions.

MnO, FeO, CoO, and NiO all undergo a reversible conversion reaction, and their voltage profiles are very similar, except for the position of the lithiation and delithiation plateaus. As shown by Fig. 5.3, MnO exhibits a first lithiation plateau at 0.2 V versus Li/Li^+ where MnO particles are disintegrated into nanoparticles of manganese embedded into a Li_2O matrix. In this particular case, MnO particles were prepared by decomposition of manganese acetate at 800°C under argon. The working electrode composition is 80 wt% MnO, 5 wt% graphite, 5 wt% Super S carbon, and 10 wt% sodium carboxymethyl cellulose. In the case of CoO, starting from well crystallized CoO particle of 100–200 nm aggregated in grains of 1–2 μm (see Fig. 5.4a and b), after the first lithiation (plateau at 0.9 V vs. Li/Li^+), the

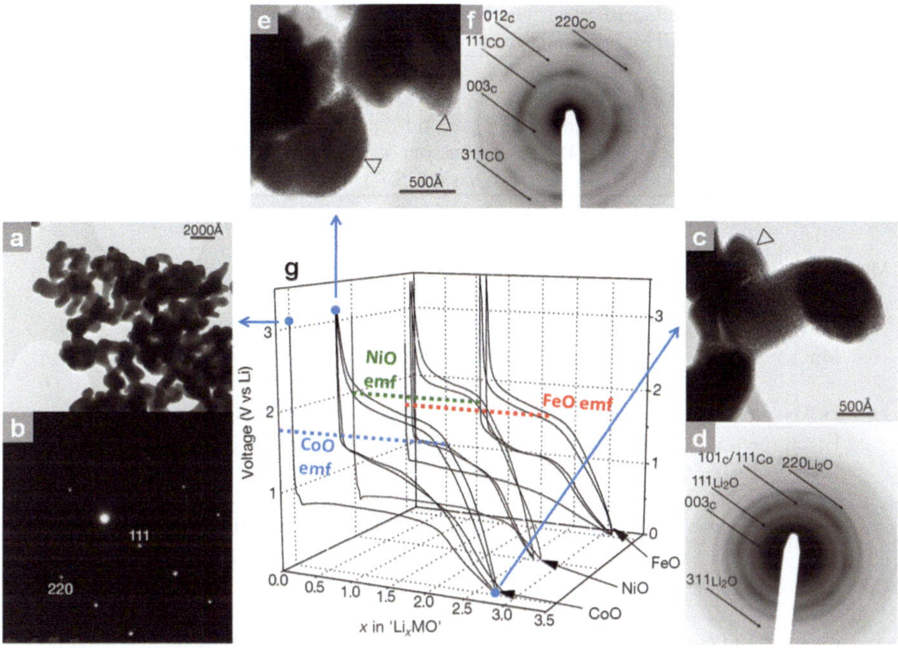

Fig. 5.4 (**a**) TEM micrograph of an uncycled CoO electrode with (**b**) its corresponding SAED patterns (taken along the [22-4] direction in the reciprocal space); (**c**) TEM micrograph of a fully lithiated CoO electrode, with (**d**) its corresponding SAED pattern (negative form); (**e**) TEM micrograph of a delithiated CoO electrode with (**f**) its corresponding SAED pattern (negative form); the subscript c in the hkl notation refers to carbon. (**g**) Voltage profile of a half-cell made of a lithium metal counter electrode and CoO, NiO, and FeO working electrodes, cycled between 0.01 and 3 V at a rate of C/5 (Reprinted by permission from Macmillan Publishers Ltd: Nature, [37], copyright 2011)

particles are totally disintegrated and are then composed of cobalt nanoparticles dispersed in Li_2O (see Fig. 5.4c and d). However, the overall shape of the starting particle/grain is preserved. The preservation of the shape has also been observed for carbon-coated Fe_3O_4 nanospindles [57] and three-dimensionally ordered mesoporous (3-DOM) Cr_2O_3 [58]. As shown in the scheme in Fig. 5.3, the electrode material is covered by a gel-like layer (also called organic layer, polymeric layer, or solid electrolyte interface); the empty triangle on Fig. 5.4c illustrates this layer. As mentioned before, a large overpotential is observed for TMOs; in the case of MnO, it is about 0.8 V, and for CoO, it is about 0.9 V. As explained by Poizot et al., it is believed that the size confinement of the TM nanoparticles enhances their electrochemical activity toward the decomposition of the Li_2O matrix [37]; a phenomenon explained by Maier in terms of a nano-effect resulting from increased surface energies of confined space [59]. The oxidation of the TM particles does not occur at a microscopic scale [37]; the particles must be nanometric and in close contact with Li_2O for this already slow solid-state oxidation reaction to occur [23].

Table 5.2 Summary table of electromotive forces, reduction and oxidation potentials, and theoretical capacities of TMOs ranging from chromium oxide to zinc oxide

TM	Phase	EMF[c] [45] (V)	E first reduction (V vs. Li/Li^+)	E second reduction (V vs. Li/Li^+)	E oxidation (V vs. Li/Li^+)	Number of Li per metal	Theoretical specific capacity [45] (mAh g^{-1})
Cr	Cr_2O_3	1.085	0.15 [60]	0.3–0.15[a] [60]	0.8–1.9[b] [60]	3	1,058
Mn	MnO_2	1.708	0.4[a] [61]	0.4–0.6[b] [61]	0.5–1.4[b][61]	4	1,233
	Mn_2O_3	1.431	1.4/0.35[a] [50]	0.6[b] [50]	1.0–1.4[b] [50]	3	1,018
	Mn_3O_4	–	1.1/0.4 [a] [50]	0.6 [50]	1–1.4[b] [50]	2.7	1,008
	MnO	1.032	0.2 [here]	0.5–0.6[b] [here]	1.0–1.4[b] [here]	2	756
Fe	Fe_2O_3	1.631	1.5/0.75[a] [62]	1.0/1.5[a][62]	1.5–2.0[b] [62]	3	1,007
	Fe_3O_4	–	0.7 [63]	0.8/1.0 [63]	1.5–2.0[b][63]	2.7	926
	FeO	1.61	0.6–0.7 [37]	0.9–1.1 [37]	1.5–1.8[b] [37]	2	746
Co	Co_3O_4	–	1.0 [48]	1.25 [48]	2.1 [48]	2.7	891
	CoO	1.802	1.0 [37]	1.2–1.8[b][37]	2.0–2.3[b][37]	2	715
Ni	NiO	1.954	0.6 [64]	1.0–1.4[b] [64]	1.5–2.2 [b] [64]	2	718
Cu	CuO	2.248	2.2/1.5/0.8 [65]	2.2/1.5/0.8 [66]	1.0–1.5/2.4 [65]	2	674
	Cu_2O	2.147	1.1–1.2 [67]	0–1.5[b] [67]	1.5[b]/2.5 [68]	1	375
Zn	ZnO	1.252	0.3–0.4 [69]	0.9–0.5[b] [70]	from 1.2 [70]	2	659

[a]2 plateaus
[b]Slope
[c]EMF: Electromotive force

As shown in Fig. 5.3, the delithiation plateau of MnO also shows the polarization of the electrode material; it ranges from 1.0 to 1.4 V versus Li/Li^+. In the case of CoO, the position of the plateau is much higher; it ranges between 1.8 and 2.1 V versus Li/Li^+. In general, the delithiation plateau is not affected by the cycle number because the metal-Li_2O composite is already at a nanoscale after the first lithiation. At the end of the delithiation, TMO nanoparticles are obtained, as shown by Fig. 5.4e and f. In most cases, subsequent lithiation plateaus show a much lower overpotential, due to the ease of reducing nanoparticles (1–5 nm) compared to bigger particles ($\approx$100 nm). In the case of MnO, the overpotential of the second lithiation is only 0.4 V. However, as shown in Figs. 5.3 and 5.4, the reduction and oxidation overpotentials are responsible for the large hysteresis that negatively affects the efficiencies of the batteries. Similar effect was also observed in the case of TMXs. Table 5.2 summarizes the lithiation and delithiation potentials for different TMOs, with TMs ranging from chromium to zinc (zinc is included even though it is a post-transition metal).

The composition and the nature of the gel-like layer at the surface of the material vary with the state of lithiation; it is usually thicker when the material is lithiated

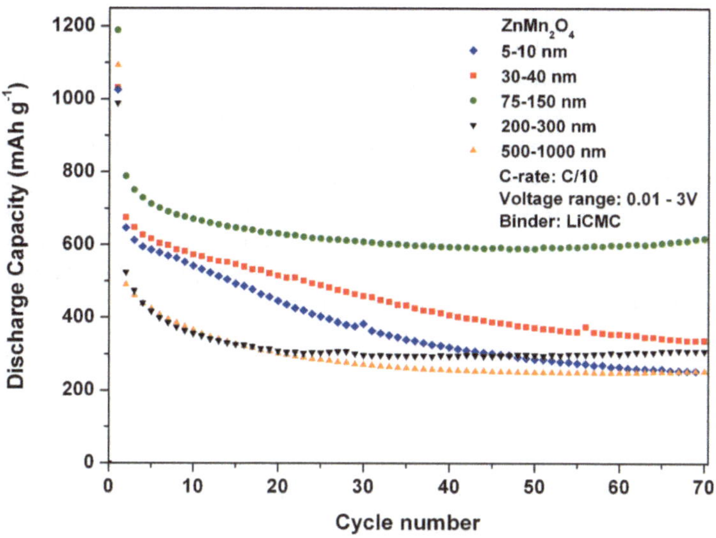

Fig. 5.5 Discharge capacities of $ZnMn_2O_4$ electrodes as a function of cycle number for electrodes made of particles of $ZnMn_2O_4$ having different sizes (size measured via TEM or SEM). Lithium carboxymethyl cellulose was used as binder and the electrodes tested in half-cells and cycled at C/10

which makes this layer unstable [50, 59, 60, 61]. In addition, this layer is known to grow with cycling time and with the increase of temperature and prolonged cycling [23]. Interestingly enough, this unstable layer is able to reversibly store lithium to some extent via an interfacial charging mechanism [45, 46, 72]. This storage capability has been observed for CoO [23] and also for other TMOs such as $ZnMn_2O_4$ [73]. The phenomenon is enhanced by an increase in temperature or prolonged cycling [23, 73].

Even though "nano" is the key point of the reversibility of the conversion reaction of TMOs (and TMXs in general), the capacity retention of these electrode materials is very sensitive to the particle size and morphology of the material. As shown by Poizot et al. in the case of CoO, an optimum particle size of 2 μm is observed, whereas for Cu_2O, it is 1 μm [37]. In the case of spherical particles of Co_3O_4, a crystallite size ranging between 40 and 60 nm seems to be the optimum [74, 75]. Figure 5.5 shows the battery performance of the spinel TMO $ZnMn_2O_4$ prepared via a coprecipitation method [73]. In this particular case, electrodes made of particles of $ZnMn_2O_4$ having a size of 75–150 nm show the highest and the most stable capacity at 690 mAh g^{-1} after 70 cycles [73].

Recently, Maier has suggested a new mechanism for interfacial ion storage at the nanoscale level [76]. This was illustrated in lithium-ion batteries by a study of the storage mechanism in nano-RuO_2 (30–200 nm) where high capacities reaching 1,110 mAh g^{-1} were obtained at unusually high coulombic efficiencies reaching 98% [72]. It was shown that lithium ions can be stored at the interface of the

Ru/Li_2O nanocomposite (between 0 and 1.2 V) with capacities reaching 120 mAh g^{-1} at C/5. The Ru/Li_2O nanocomposite was formed by a conversion reaction prior to an intercalation reaction that formed $LiRuO_2$. The formation of a SEI (5 nm) was observed during charging, similar to other metal oxides (e.g., CoO), but it was suggested to be less plausible in this case because of its formation and dissolution at a different potential above 1.2 V [77]. The storage at the M/Li_2O interface was explained to be capacitive-like in nature where lithium ions occupy interstitial sites in Li_2O close to the boundary compensated by the electron sitting on the metal surface site. A similar mechanism was suggested to take place for a Ti/Li_2O nanocomposite as shown by ab initio calculations [77].

5.3 Metal Oxides

This section presents the properties and performance of TMOs which undergo a reversible conversion reaction with lithium. TMOs go from chromium oxide to zinc oxide, and all potentials are given versus Li/Li^+.

5.3.1 Chromium

According to the emf calculations reported by Maier et al. [45], Cr_2O_3 is the first TMOs from the periodic table which undergoes a conversion reaction with lithium [60] but also means that it is one of the most difficult to reduce. The first lithiation shows a plateau at 0.15 and at 0.3 V for the subsequent lithiations, whereas the delithiation potential ranges from 0.8 to 1.9 V [60]. Even though Cr_2O_3 has a large theoretical specific capacity of 1,058 mAh g^{-1}, micrometric particles always show very poor capacity retention. Particles with size that ranges from 500 nm to 1.5 μm show a first discharge capacity of 1,100 mAh g^{-1}, which is slightly higher than the theoretical capacity. Unfortunately, the second cycle usually shows a specific capacity of 250 mAh g^{-1} or less [60]. Even using nanoparticles (50 nm) prepared via a hydrothermal method is not sufficient to improve significantly the capacity retention and the long-term performance of the electrode [60]. At first, it seemed that this poor capacity was due to the conversion reaction mechanism, which was believed to proceed, according to Dupont et al., as follows: $Cr_2O_3 \rightarrow CrO \leftrightarrow Cr \leftrightarrow CrO$ [78]. Cr_2O_3 was prepared using a pulsed laser deposition technique that produced particles of 50 nm and more in size. In this case, no Cr_2O_3 crystalline phase was observed after the first cycle. However, other groups observed the recovery of Cr_2O_3 after the delithiation [60]. It is worth mentioning that one of the main issues with Cr_2O_3 is its poor electronic conductivity ($\approx 2 \times 10^{-7}$ S cm^{-1}) [71]. This observation applies for all TMOs which exhibit typical insulator or semiconductor behavior with their bandgaps ranging from 3 to 4 eV [79]. In addition, the formation of a very thick and unstable gel-like layer of 30–90 nm on the surface of the nanoparticles is responsible for a capacity loss of about 250 mAh g^{-1} [60]. These issues have been circumvented by making a carbon/ Cr_2O_3 nanocomposite coated with hard carbon; it provided a reversibility of 80%. The reported capacity values are 800 mAh g^{-1} (per gram of composite) for the first cycle and a reversible capacity of 650 mAh g^{-1} after 20 cycles [60, 80]. However, long-term cycling performance has not been reported yet. Doping was another path that has been investigated in order to overcome the low electronic conductivity. Carbon-coated Cr_2O_3 nanoparticles which were previously doped with 2% of Ni have shown a reversibility of almost 90% with an initial capacity of 900 mAh g^{-1} and a reversible capacity of 800 mAh g^{-1} after 20 cycles [71]. The doping increased the electronic conductivity from 2×10^{-7} S cm^{-1} to 3×10^{-4} S cm^{-1}, and the carbon coating improved the stability of the gel-like layer by decreasing its thickness from 30–90 nm to 15–30 nm [60, 71]. More recently, 3-DOM Cr_2O_3 made of 10-nm thick walls and 10-nm pores was successfully prepared, but unfortunately, the lack of electrical conductivity and of carbon coating leads to very poor performance [58]. Other strategies have been investigated, such as 100–200-nm thin films [81] or the growth of a 500-nm porous layer of Cr_2O_3/Mn-Fe-Cr-O directly onto stainless steel [78, 81]. However, these strategies greatly diminished the capacity density of the electrodes.

5.3.2 *Manganese*

Four different manganese oxides are available as anode material for lithium-ion batteries, ranging from Mn^{4+} to Mn^{2+}: MnO_2 [47], Mn_2O_3 [82, 83], Mn_3O_4 [83, 84], and MnO [50, 61, 86–89]. As shown by Table 5.2, they all show high theoretical capacities, starting from 1,233 mAh g^{-1} for MnO_2 to 755 mAh g^{-1} for MnO, which are four to two times greater than graphite. Manganese is an appealing metal; it is not too expensive, abundant, and nontoxic. In addition, the delithiation (oxidation) potential is quite low compared to other TMOs (see Table 5.2). However, the main issue with manganese is the irreversible capacity associated with the first cycle. Like Cr_2O_3, manganese oxides are the most difficult TMOs to reduce; a potential of about 0.2–0.4 V is usually needed [38]. An interesting study was done by Fang et al. that compared the behavior of the four manganese oxides having the same nanorod morphology with a diameter ranging from 100 to 200 nm [50]. Figure 5.6 shows the SEM micrographs and the voltage profiles of these four oxides. The irreversible capacity represents 71.8% for MnO_2, 58.26% for Mn_2O_3, 52.7% for Mn_3O_4, and 45% for MnO.

As previously mentioned, MnO_2 goes through a lithium insertion process before being reduced to manganese metal according to the following path: $MnO_2 \leftrightarrow LiMnO_2 \rightarrow Li_2MnO_2 \rightarrow Mn$ [50, 90]. The large irreversible capacity is explained by the nonreversibility of some of these reactions. Mn_2O_3 is irreversibly reduced into Mn_3O_4 before being further reduced. MnO_2 and Mn_2O_3 do not usually show good capacity retention. The third manganese oxide, Mn_3O_4, goes through an insertion

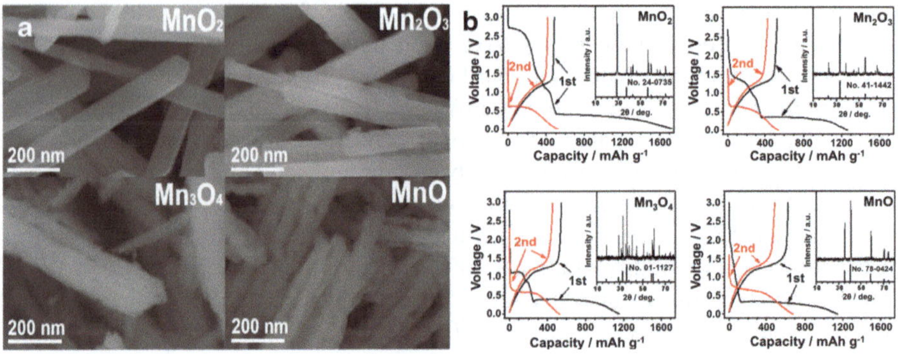

Fig. 5.6 (**a**) SEM micrographs and (**b**) voltage profile of MnO_2, Mn_2O_3, Mn_3O_4, and MnO nanorods prepared via the same method (Reprinted from Electrochemistry Communications [50]. Copyright (2011), with permission from Elsevier)

reaction before being reduced to manganese metal: $Mn_3O_4 \rightarrow LiMn_3O_4 \rightarrow MnO \leftrightarrow Mn$. After the different insertion steps, all manganese oxides show the same first reduction voltage plateau at 0.2–0.4 V, typical of the conversion reaction. An advantage with using manganese oxides is the low voltage of the oxidation plateau of the conversion reaction that ranges from 1.0 to 1.4 V; this allows a higher output cell voltage. Two different reversible processes with lithium have been reported in the literature for Mn_3O_4 [50, 84]. According to Fang et al., first an insertion of lithium occurs to form $LiMn_3O_4$ as shown by Eq. 5.6. A second reaction with lithium then takes place to obtain MnO, as shown by Eq. 5.7. It has been shown that once MnO is obtained, reactions (5.6) and (5.7) are not reversible anymore. From MnO, the reversible conversion reaction occurs, as shown by Eq. 5.8; the latter observation has been demonstrated by high-resolution TEM and selected area electron diffraction (SAED) measurements [50]. A second path has also been proposed by Gao et al. where they state that reaction shown by Eq. (5.9) is totally reversible [84]:

$$Mn_3O_4 + Li^+ + e^- \rightarrow LiMn_3O_4 \quad (5.6)$$

$$LiMn_3O_4 + Li^+ + e^- \rightarrow 3MnO + Li_2O \quad (5.7)$$

$$MnO + 2Li^+ + 2e^- \leftrightarrow Mn + Li_2O \quad (5.8)$$

$$Mn_3O_4 + 8Li^+ + 8e^- \leftrightarrow 3Mn + 4Li_2O \quad (5.9)$$

As expected, MnO shows the lowest irreversible capacity according to reference [50]; however, due to the high reversibility of the conversion reaction, one could expect a very low irreversible capacity. MnO usually exhibits a first-cycle capacity over 1,100 mAh g^{-1} [50, 61, 86–89], which is much higher than its theoretical capacity. Like Cr_2O_3, this is due to the formation of a thick gel-like layer on the manganese nanoparticles. A reversible capacity of about 65–70% of the first-cycle capacity is usually observed, which corresponds to the theoretical capacity of MnO. MnO usually exhibits good capacity retention with or without carbon additives. Zhong et al. reported reversible capacities for a MnO/carbon composite of 650 mAh g^{-1} [91].

5.3.3 *Iron*

Iron oxides are attractive because of the low cost of iron, its environmental friendliness, and its experimental oxidation potential of 1.8 V, which is much lower than cobalt or nickel oxides. Three different iron oxides are available as anode material for lithium-ion batteries, ranging from Fe^{3+} to Fe^{2+}: Fe_2O_3, Fe_3O_4, and FeO. In the 1980s, it was shown by Thackeray and Coetzer that Fe_2O_3 and Fe_3O_4 could react reversibly with a Li/Al alloy in a LiCl-KCl molten salt mixture at 420°C [42]. In this process, α-Fe_2O_3 and Fe_3O_4 were reduced to Fe^0. When discharged at room temperature down to 1 V, the lithium initial insertion in Fe_2O_3 and Fe_3O_4 leads to the creation of lithiated phases $Li_xFe_2O_3$ and $Li_xFe_3O_4$ ($0 \leq x \leq 2$)[92]. It is only recently that the full reduction of iron oxides at room temperature was demonstrated by discharging them down to 5 mV [37]. FeO just goes through the conventional reversible conversion reaction, $FeO + 2Li^+ + 2e^- \leftrightarrow Fe + Li_2O$; this first lithiation occurs at 0.6–0.7 V and yields a first discharge capacity of 1,150 mAh g^{-1} [37]. The oxidation plateau is observed around 1.6 V, and the subsequent reduction plateaus are around 1 V (see Table 5.2). FeO exhibits a reversible capacity of 745 mAh g^{-1}, fading to 300 mAh g^{-1} after 50 cycles. Additional plateaus between 3 and 1 V are observed when using Fe_2O_3 or Fe_3O_4 corresponding to lithium intercalation/deintercalation. In contrast to FeO, which has not been widely investigated, there has been a lot of interest in Fe_2O_3 and Fe_3O_4 as anode materials for lithium-ion batteries. A lot of effort has been put to control the size and morphology of iron oxide particles. It has been reported that micron-sized iron oxides show poor capacity retention [57, 93]. For example, in the case of Fe_2O_3, nanorods [94, 95], nanoflakes [96], and nanotubes [62] were synthesized, while in the case of Fe_3O_4, nanowires [97] and nanospindles [57] were prepared and tested. In both cases, the most stable capacities were obtained when these nanomaterials were coated with carbon, which is explained by the fact that the solid electrolyte interphase (SEI) is created on the carbon rather than on the material itself. After a first discharge capacity of 1,300 mAh g^{-1}, carbon-coated Fe_2O_3 nanotubes yielded a stable capacity of 700 mAh g^{-1} after 150 cycles, compared to 450 mAh g^{-1} for uncoated nanotubes [62]. The same phenomenon was observed for carbon-coated Fe_3O_4 nanowires that showed a stable reversible capacity of 800 mAh g^{-1} over 50 cycles, while their uncoated counterpart performed more poorly, with a reversible capacity of 400 mAh g^{-1} after 50 cycles [97].

Another approach has recently been investigated, that is, growing Fe_3O_4 nanoparticles directly on graphene (exfoliated graphite), and good stability was obtained [98]. The highly conductive graphene sheets act as an electronic conductor [99] and also as support for the Fe_3O_4 nanoparticles. The reported discharge capacities are 1,400 mAh g^{-1} for the first discharge and 900 mAh g^{-1} for the second discharge, slightly increasing with cycling. The rate capability is also very good with a capacity over 600 mAh g^{-1} at rate slightly over C (1,050 mA g^{-1}). Composite electrodes using single-wall carbon nanotubes (SWCNTs) were shown to improve the cycling performance of Fe_3O_4 [91]. The composite was prepared by the reduction of a precursor of FeOOH nanorods in presence of SWCNTs to yield Fe_3O_4 and prepared as a binder-free electrode with only 5 wt% of SWCNTs. Extremely high capacities were reported, reaching over 1,000 mAh g^{-1} at 1 C rate.

5.3.4 *Cobalt*

Cobalt oxides, CoO and Co_3O_4, have both been shown to be active toward lithium [23, 37]. They have theoretical capacities over 700 mAh g^{-1}: 715 mAh g^{-1} for CoO and 890 mAh g^{-1} for Co_3O_4. Like other TMOs, cobalt oxides go through a reversible conversion reaction with lithium. The first reduction reaction occurs at about 1.0 V and the subsequent one at about 1.2–1.8 V [17, 38]. The oxidation occurs at potentials ranging from 2.2 to 2.4 V [37], which is too high and is the main drawback for this particular metal oxide. It will lead to a low output cell voltage when used in a full lithium-ion battery; as shown by Badway et al., a $Co_3O_4/LiCoO_2$ cell has an output voltage of only 2 V [100]. However, since cobalt oxides provide a generally high and stable capacity, it is one of the first transition metal oxides to attract interest. While large particles (over 2 μm) of CoO and Co_3O_4 deliver high and stable capacities [38], the performance of nanosized cobalt oxide will depend greatly on their morphology. Co_3O_4 nanowires [101], nanotubes [102], nanoneedles [103], and hollow microspheres [104] give stable capacities above 700 mAh g^{-1} at low cycling rate. CoO nanowires [105] also yield stable capacities over 700 mAh g^{-1} at 1 C. However, Co_3O_4 nanoplatelets [106] showed capacities with drastic drop or large variations depending on heat treatment. Composites of Co_3O_4 with graphene sheets are useful in preventing metal oxide particles agglomeration and the stacking of the graphene sheets [107]. The metal oxide is grown *in situ* on graphene oxide sheets which are then reduced to produce evenly distributed Co_3O_4 nanoparticles. This leads to a stable capacity of 800 mAh g^{-1} over 30 cycles.

5.3.5 *Nickel*

NiO has been reported as an anode material for lithium-ion batteries. It goes through the conventional reversible conversion reaction with lithium: NiO + $2Li^+$ + $2e^-$ $\leftrightarrow$ Ni + Li_2O [37, 64, 108]. The first lithiation occurs at about 0.6 V, whereas the subsequent lithiation plateaus slope from 1.0 to 1.4 V. Like cobalt oxides, NiO is in general less attractive due to its higher delithiation potential that ranges from 1.5 to 2.2 V, which would provide a lower cell voltage. In addition, NiO usually shows very poor capacity retention [108–112] unless it is nanostructured [64, 113, 114]. Thin films made of nanoparticles of 30–100 nm particles showed a first discharge capacity ranging from 800 to 850 mAh g^{-1} and a reversible capacity of about 450–500 mAh g^{-1} after 100 cycles with cycling rates ranging from 10 to 80 μA cm^{-2} [113]. Some of the best performance were obtained with NiO nanowalls of 100-nm height and less than 40 nm thick, showing a capacity of 640 mAh g^{-1} after 80 cycles at 1.25 C (900 mA g^{-1}) [64]. However, an irreversible loss of about 40% occurs after the first cycle. Similar data were reported for mesoporous carbon-encapsulated NiO nanocomposite [115].

5.3.6 Copper

As previously mentioned in this chapter, copper oxide has been used as early as 1980 as cathode for primary lithium batteries [39, 40], and more recently as cathode for lithium-ion batteries [41]. Its use as an anode was investigated by Poizot et al. in 2000 and the two phases CuO and Cu_2O have been found to be active toward lithium [37]. Like manganese oxides, copper oxides generally show a large irreversible capacity that ranges from 30% to 50% of the first discharge capacity. In addition, due to the higher molecular weight of copper, and its lower oxidation numbers, copper oxides have lower theoretical specific capacities than other TMOs: 674 mAh g^{-1} for CuO and 375 mAh g^{-1} for Cu_2O. It has been shown that a nanosized material does not always provide the best performance, as shown by Grugeon et al.; 1-μm-sized CuO and Cu_2O provide more stable capacities (400 mAh g^{-1}) than 150-nm particles [116]. However, usually, nanostructured CuO or Cu_2O shows better capacity retention than micrometric particles. For example, nanowires having a diameter of 100 nm provided a first discharge capacity of 1,040 mAh g^{-1} and a reversible capacity of 650 mAh g^{-1} after 100 cycles, which represents an irreversibility of 37% [117]. CuO nanoribbon provided similar results with a reversible capacity of 600 mAh g^{-1} after 280 cycles at C/4 and an irreversibility of about 31% [118]. Good rate capability was also reported: 507 mAh g^{-1} at C/3.4, 417 mAh g^{-1} at C/1.7, and 332 mAh g^{-1} at C/0.85. However, one should be careful with these values since the density of the electrode was only 0.4 mg cm^{-2} [118]. CuO was also studied as urchin-like nanostructures of 100 nm [65]. A stable capacity of 600 mAh g^{-1} was obtained after 50 cycles at C/4.5 along with an irreversible capacity of only 25% of the first discharge capacity.

Since an oxidation number of +1 exists for copper, CuO behaves differently from other TMOs; it first goes through an insertion mechanism with mixed oxidation state, that is, $Cu^{2+}{}_{1-x}Cu^{+}{}_{x}O_{(1-x)/2}(0<x<0.4)$ up to Cu_2O, and then the conventional reversible conversion reaction occurs. Figure 5.7 shows the typical voltage profile of a CuO anode.

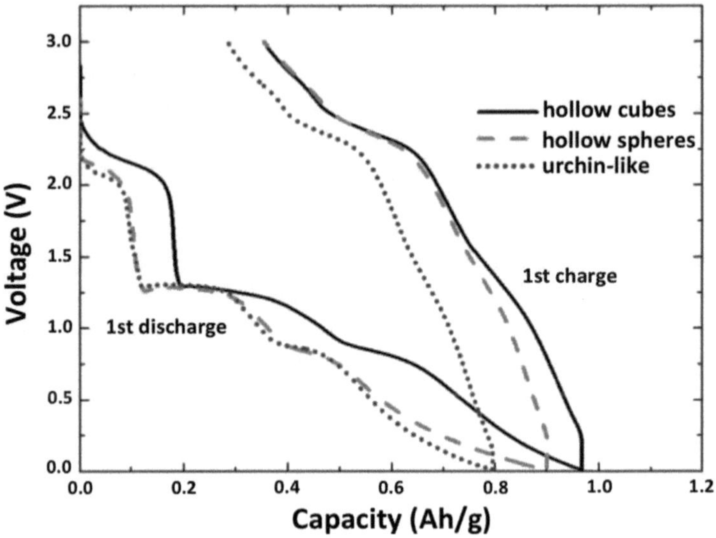

Fig. 5.7 Voltage profile of CuO hollow nanostructures (Reprinted with permission of John Wiley & Sons, Inc. [65])

5.3.7 *Zinc*

Zinc is a post-transition metal, and like the TMOs, ZnO has an interstice-free würtzite structure that does not allow any lithium insertion. As most TMOs, ZnO reacts with lithium through a reversible conversion reaction [119]. However, as a post-TMO, ZnO behaves quite differently since the *in situ* made zinc nanoparticles further react with lithium by forming a ZnLi alloy [120–122]. As most alloying reaction with lithium, Zinc particles undergo a large volume change during this process, about 163% [119]. This expansion is responsible for the large irreversible capacity observed for the first few cycles. Usually, reversible capacities ranging from 500 mAh g^{-1} [123] to 350 mAh g^{-1} [119] or less [73] with very poor capacity retention upon cycling are observed.

5.4 Mixed Oxides: AB_2O_4

Several compounds containing two TMOs and having a spinel-like structure have been investigated, with A having an oxidation state of +2 and B of +3. Three series have been studied, the manganite (AMn_2O_4), ferrite (AFe_2O_4), and cobaltite (ACo_2O_4).

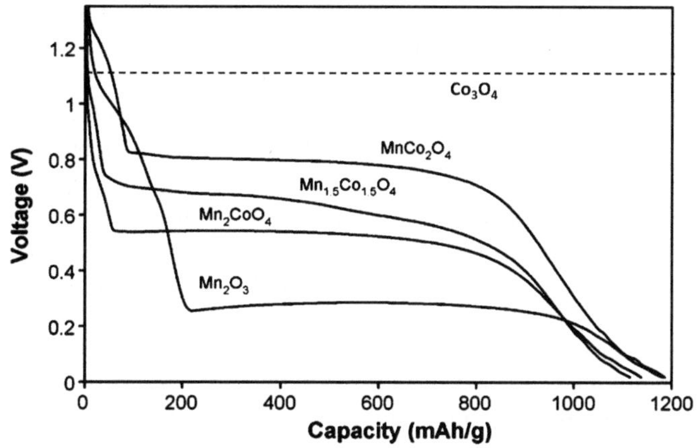

Fig. 5.8 Voltage profile of the first lithiation of $Mn_xCo_{3-x}O_4$ spinel series (with x = 1, 1.5, or 2) and Mn_2O_3 prepared via an ethanol dehydration procedure that precipitates gel-like citrate precursors annealed at 800°C for 24 h. The cells were discharged at 1 C rate (Reprinted from Electrochimica Acta [124]. Copyright (2011), with permission from Elsevier. The voltage plateau value of Co_3O_4 has been taken from reference [125])

In general, their performance is pretty close to the performance of their simple oxides (i.e., Mn_3O_4, Fe_3O_4, and Co_3O_4). Interestingly enough, one would expect two lithiation plateaus corresponding to the two TMs, but only one plateau is experimentally observed for the first lithiation, as shown by Lavela et al. [124] for the $Mn_xCo_{3-x}O_4$ series (x = 0, 1, 1.5, or 2) and Mn_2O_3. Figure 5.8 shows the potential of the first lithiation plateau of this series. It seems that the potential is controlled by the TM content. As expected, Mn_2O_3 exhibits the lowest potential plateau that then shifts to higher values as more cobalt was added into the spinel structure up to Co_3O_4. This is easily explained by the fact that manganese oxides are more difficult to reduce (0.3 V [83]) than cobalt oxides (0.9–1.1 V [100, 125]). The same phenomenon has been observed for $Co_{3-x}Fe_xO_4$ (x = 0, 1, or 2) [125]. Upon the delithiation, the spinel structure is never recovered; instead, a mixture of single TMOs is observed, as demonstrated by Sharma for $ZnCo_2O_4$ [126]; two voltage plateaus are observed afterward.

5.4.1 *AMn_2O_4*

Manganese-based oxides are very interesting because of their low delithiation voltage plateau; it ranges from 1.0 to 1.4 V (see Fig. 5.1). However, they also show a large irreversibility due to the formation of an unstable gel-like layer on the nanoparticles of the Mn/Li_2O composite material. This manganite series has first been investigated by Pasero et al. via a brief study of the cycling performance of the off-stoichiometric $Co_{0.4}Mn_{2.6}O_4$. It provides a low but stable capacity of about 400 mAh g^{-1} after 10 cycles, and an irreversibility of about 60% [127]. Unfortunately, no information on the particle size or long-term cycling was provided. We recently reported on the performance of nanoparticles of $NiMn_2O_4$ (10 nm) and $CoMn_2O_4$ (200–500 nm) [73]. We showed that $NiMn_2O_4$ nanoparticles perform well for the first four cycles only. A predicted irreversible capacity of 210 mAh g^{-1} was observed that is due to the oxidation of Mn^0 to Mn^{2+} and not Mn^{3+}. Even though the capacity retention between the second and the fourth discharge is very good, it unfortunately drops drastically below 200 mAh g^{-1} after 20 cycles [73]. This drastic capacity fade could be explained by the adverse reaction of nickel nanoparticles with the electrolyte which leads to a thick ionically resistive organic SEI that does not allow lithium ions to travel through. According to cyclic voltammetry, the nickel reduction and oxidation peaks are drastically reduced in intensity after the tenth cycle. $CoMn_2O_4$ nanoparticles showed better capacity performance with a first discharge capacity of 975 mAh g^{-1} and a reversible capacity of 330 mAh g^{-1} after 50 cycles; however, capacity retention is still an issue [73].

An additional capacity is observed when the post-transition metal zinc is used. $ZnMn_2O_4$ is a good alternative when considering the low oxidation potential of zinc and manganese nanoparticles: 1.2 and 1.5 V, respectively [73]. These values will potentially increase the battery output voltage. In addition, their price and toxicity are lower compared to nickel and cobalt; they are abundant and environmentally friendly. Three groups reported the performance of $ZnMn_2O_4$ [73, 128]. In the first case it was synthesized via polymer-pyrolysis method, the obtained nanoparticles (30–60 nm) showed a capacity value of 626 mAh g^{-1} over 50 cycles [128, 129]. In the second case, $ZnMn_2O_4$ nanoflake-shaped particles prepared via calcination of an agglomerated Zn-Mn citrate complex precursor provided capacity values of 650 mAh g^{-1} at 100 mA g^{-1} [130]. In the third case, $ZnMn_2O_4$ particles were prepared via decomposition of a precipitate of zinc and manganese oxalate. We showed that it is necessary to investigate the effect of the size of the particles on the performance of the battery (see Fig. 5.5). An optimum particle size ranging from 75 to 150 nm provided a capacity of 690 mAh g^{-1} after 70 cycles, which represents 88% capacity retention. Good rate capability has also been observed for this material: 170 mAh g^{-1} at 3 C.

5.4.2 AFe_2O_4

Even though iron exhibits a higher oxidation potential than manganese, some members of the AFe_2O_4 series (A = Co, Ni, Cu, or Zn) have shown interesting performance. However, these compounds still exhibit large irreversible capacities ranging from 35% to 60% [125, 131–134] with capacity values of 600–800 mAh g^{-1} for $CoFe_2O_4$ [131], 600 mAh g^{-1} or less for $NiFe_2O_4$ [131, 133], and 550 mAh g^{-1}

for $CuFe_2O_4$ [134]. Even though an extra capacity is added by using zinc, $ZnFe_2O_4$ nanoparticles (100–300 nm) show a stable reversible capacity of only 615 mAh g^{-1} which still represents an irreversible capacity of 46% [135].

5.4.3 ACo_2O_4

Because of the cobalt content, compounds from the ACo_2O_4 (A = Mn, Fe, Ni, Cu, or Zn) series show higher delithiation potentials than spinels from the two other series [126]. The first member of the series, $MnCo_2O_4$ does not show good performance with a capacity of only 400 mAh g^{-1} after 50 cycles, which represents an irreversibility of 65% [124]. Depending on the preparation method, $FeCo_2O_4$ performed quite well with a good capacity retention with values ranging from 630 to 750 mAh g^{-1} after 50 cycles, which represents an irreversibility of 45% and 35%, respectively [125, 136]. Nanoparticles of $NiCo_2O_4$/C nanocomposite (10–20 nm) provided a capacity of 715 mAh g^{-1} after 50 cycles that decays significantly with cycling; an irreversibility of about 40% was measured [137]. Nanoparticles of $CuCo_2O_4$ (10–20 nm) showed a stable capacity of 745 mAh g^{-1} after 50 cycles, which represents an irreversibility of only 38% [138]. For the last member of the series, an extra capacity is added using zinc. $ZnCo_2O_{4r}$ nanoparticles (15–20 nm) showed a very stable capacity of 900 mAh g^{-1} after 60 cycles and an irreversible capacity that represents only 25% [126]. Other groups also reported very stable capacity values: 750 mAh g^{-1} for $ZnCo_2O_4$ nanoflakes [139] and up to 1,200 mAh g^{-1} for $ZnCo_2O_4$ nanowires (100–300 nm in diameter), but for 20 cycles only [140].

5.5 TMXs with X = S, N, P, and F

In the previous sections, we have shown that lithiation and delithiation potentials can be tuned by changing the transition metal. Another possibility is to use mixed oxides such as AB_2O_4 spinels with A and B being two different TMs. In this case, the potential can be tuned, but however occurs only during the first lithiation reaction, as shown by Fig. 5.8. For the subsequent delithiations, the active material is composed of a mixture of both oxides; the latter then reacts at two different potentials. Another solution for tuning lithiation and delithiation potentials is to change the anion. The previous sections focused on oxides, while this section will deal with fluorides, sulfides, nitrides, and phosphides, which will be referred to as TMFs, TMSs, TMNs, and TMPs respectively. They react with lithium the same way as TMOs and form a matrix of LiF, Li_2S, Li_3N, and Li_3P, respectively. For the same transition metal, the following trend is usually observed when looking at the lithiation potential of the different TMXs: TMFs > TMSs > TMOs > TMNs > TMPs. This observation is actually related to the decrease in the M-X bond polarization when going from M-F to M-P.

Due to the very high ionicity of the metal-fluorine chemical bond, very high oxidation and reduction potentials are observed for TMFs. The electronegativity difference between most transition metals and fluorine ranges from 2.5 to 2.1, which makes the TMFs more attractive as cathodes than anodes, especially FeF_3. The higher reduction potential related to the higher ionicity of the metal-fluorine chemical bond makes the conversion reaction possible even with titanium and vanadium [59], which was something impossible for TMOs. FeF_3 is one of the TMFs that shows the highest theoretical capacity (712 mAh g^{-1}); it also provides the best battery performance [141]. Other TMFs, such as NiF_2, CuF_2, CoF_2 and MnF_2, and MnF_3, do not show any good long-term capacity retention [45, 69].

TMSs were initially used as cathode material for primary cells (e.g.,TiS_2, FeS_2) [142–144]. Recently, it has been shown that TMSs can also go through a conversion reaction that is reversible to some extent. As shown in Tables 5.2 and 5.3, in general, TMSs show higher oxidation potentials than TMOs, which means that the reduction of the metal is easier but that the oxidation is more difficult. For

Table 5.3 Summary table of electromotive forces, reduction and oxidation potentials, and theoretical capacities of TMFs, TMSs, TMNs, and TMPs

TM	X	Phase	EMF[b] [45] (V)	E first reduction (V vs. Li/Li^+)	E second reduction (V vs. Li/Li^+)	E oxidation (V vs. Li/Li^+)	Number of Li per metal	Theoretical specific capacity [45] (mAh g^{-1})
Ti		TiF_3 [59]	1.396	1.5/1.0–0.5	1.2–1.0	1.7–2.2	3	767
		TiS_2	1.233	–	–	–	4	957
V		VF_3 [59]	1.863	0.3	1.8/1.0/ 0.1	0.1/2.5	3	745
		VN	–	–	–	–	3	1,239
Cr		CrF_3 [141]	–	1.7	1.9	2.6	3	738
		CrN [156]	–	a	a	1/2.7[a]	3	1,218
Mn		MnF_2	1.919	–	–	–	2	577
		MnF_3	2.647	–	–	–	3	719
		MnS [157]	1.144	–	0.7	1.2	2	616
		MnS_2	1.692	–	–	–	4	900
		Mn_4N	0.083	–	–	–	0.75	459
		Mn_5N_2	0.176	–	–	–	1.2	531
		MnP_4 [151]	–	0.65/0.1	0.78/0.1	1.2/1.56	12	1,798
Fe		FeF_2	2.664	–	–	–	2	571
		FeF_3 [141]	2.742	3.0/2.0	2.1/3.1	2.8/3.3	3	712
		FeS	1.747	–	–	–	2	610
		FeS_2[158]	1.861	1.5	2.0/1.5[a]	1.9/2.5[a]	4	893
		Fe_4N	0.432	–	–	–	0.75	339
		Fe_3N [148]	–	2.0/1.8	0.9	1.5–2.2[a]	1	443
		FeP [56]	–	0.1	0.5/0.2	1	3	926
		FeP_2 [56, 153]	–	0.25	0.7	1.1	6	1,365

(continued)

Table 5.3 (continued)

TM	X	Phase	EMF[b] [45] (V)	E first reduction (V vs. Li/Li$^+$)	E second reduction (V vs. Li/Li$^+$)	E oxidation (V vs. Li/Li$^+$)	Number of Li per metal	Theoretical specific capacity [45] (mAh g^{-1})
Co		CoF_2	2.854	–	–	–	2	553
		CoF_3	3.617	–	–	–	3	694
		$CoS_{0.9}$ [52]	–	1.1	1.4	2	1.8	545
		CoS_2 [159]	1.898	1.6/1.4[a]	1.9/1.4[a]	2/2.3	4	871
		Co_3S_4	1.644	–	–	–	2.7	703
		CoN [150]	–	0.6/0.25[a]	0.8/0.3[a]	0.7 and up[a]	3	1,103
		Co_3N [148]	0.326	1.1/0.8[a]	1.5–1.7[a]	2.1	1	421
		CoP [154]	–	From 0.9 and down[a]	0.6	1	3	894
		CoP_3 [155]	–	0.3	0.7	1.1	9	1,588
Ni		NiF_2 [45]	2.964	1.5	2.5[a]	1.4/3.2	2	554
		NiS [52]	–	1.2–0.8	1.8/1.4	2.8/2	2	591
		NiS_2 [160]	–	1.7–1.3[a]	1.8/1.4	2.1	4	873
		Ni_3N [161]	–	0.55	1.7–0.7[a]	1.5/2.1[a]	1	423
		NiP_2 [152]	–	0.5–0.2[a]	0.7	1.2	6	1,332
		NiP_3 [54]	–	0.6	0.75	1.2	9	1,590
Cu		CuF_2 [45]	3.553	2.1/1.2	3–0[a]	0–4[a]	2	528
		CuS [52]	1.998	2.1/1.6	2.1/1.6[a]	2.3	2	561
		Cu_2S	1.827	–	–	–	1	337
		Cu_3N [162]	–	0.5 and less[a]	–	1.7	1	393
		CuP_2 [163]	–	0.6	0.8	1.1	6	1,284
Zn		ZnF_2	2.404	–	–	–	3	778

[a]Slope
[b]EMF: Electromotive force

example, the first lithiation occurs at a voltage plateau of 0.7 V for MnS versus 0.2 V for MnO. As for lithium-sulfur batteries [145, 146], the main issue with TMSs (Li_2S) is the dissolution of sulfur species in the carbonate electrolytes, which means that the reversibility of the reaction is jeopardized at each discharge. Better performance is obtained when using either all solid-state batteries or a polymer electrolyte such as polyethylene oxide: lithium salt at 80°C [147].

Nitrides usually show lower potentials than oxides [148, 149]. It has been reported that TMNs can also go through the reversible conversion reaction process starting from vanadium. Nitrides show high first discharge capacity; however, they do not show very good capacity retention except for CoN that has a capacity over 700 mAh g^{-1} [150].

TMPs are probably the next most interesting TMX anode materials after TMOs since they exhibit the lowest delithiation potentials of all investigated TMXs. The metal-phosphorus bond shows a very strong covalent bond that is more difficult to

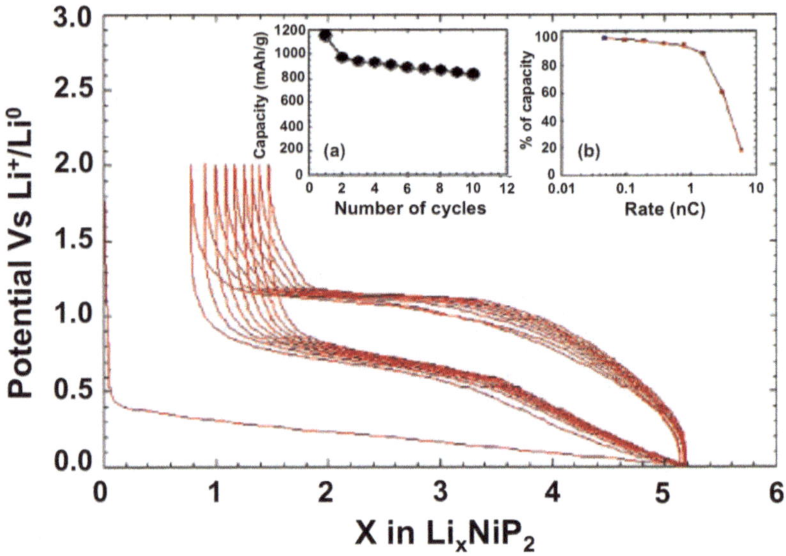

Fig. 5.9 Voltage profile for a foam-NiP_2/Li cell at C/10 from 2 to 0 V at a C/2 rate with, as insets, (**a**) its corresponding capacity retention and (**b**) rate capability performance of a foam-NiP_2 electrode measured, for reasons of current density associated to Li-metal, in a lithium-ion-type configuration using $LiFePO_4$ as the positive electrode (Reprinted with permission from [152]. Copyright (2011) American Chemical Society)

break than an ionic bond [151, 152]. As a matter of fact, only transition metals starting from manganese can be reduced. In addition, due to their high phosphorus content and the uptake of three lithium ions per phosphorus atom, they exhibit high theoretical capacities (see Table 5.3). As shown in Table 5.3, MnP_4 shows optimal potentials and a high theoretical capacity of 1,798 mAh g^{-1}. Unfortunately, the efficiency of the conversion reaction process is quite low as shown by Gillot et al. [145]. FeP and FeP_2 exhibit very low delithiation potential of about 1.1 V. During their first lithiation, they almost reach their theoretical capacity values, which are over 900 mAh g^{-1} [56, 153]. After 100 cycles at C/10, FeP still shows a capacity of 300 mAh g^{-1} [56]. Whereas CoO showed very high delithiation potential over 2 V, its phosphide counterparts CoP and CoP_3 show a much lower delithiation potential of about 1.1 V [154, 155]. CoP_3 almost reaches its theoretical capacity during the first lithiation; however, capacity retention is still an issue. As shown by Figs. 5.9 and 5.10, NiP_2 and NiP_3 are also of interest, especially because of their low delithiation potentials (1.2 V, see Table 5.3) and also their flat lithiation and delithiation plateaus [54, 152]. In addition, they have theoretical specific capacities over 1,300 mAh g^{-1}. During the first lithiation, NiP_2 reaches the theoretical value and still has a capacity of 850 mAh g^{-1} after 10 cycles and also shows good rate capability up to 1 C [152].

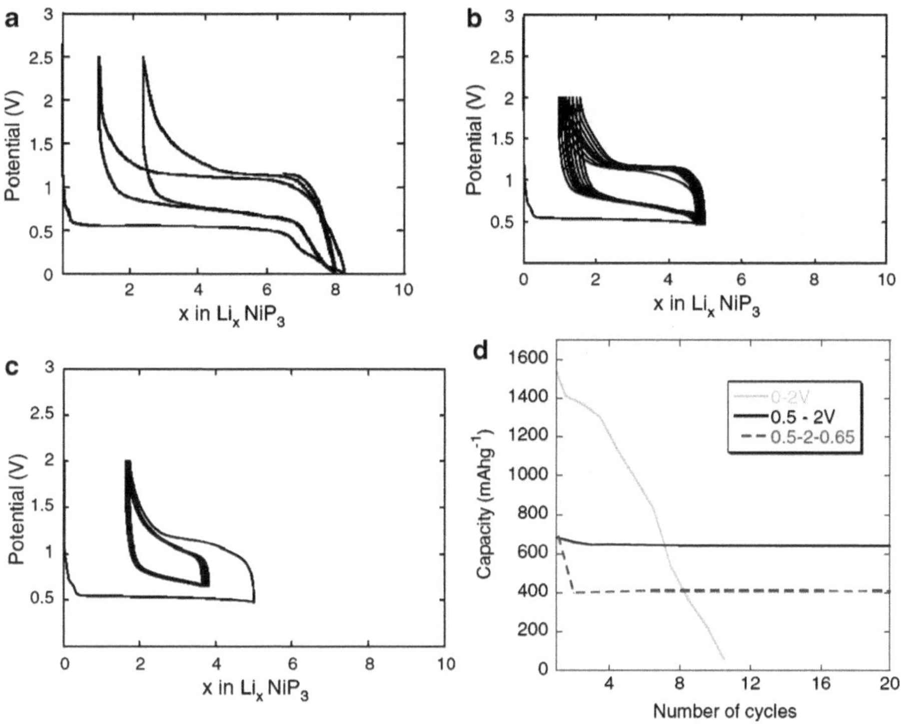

Fig. 5.10 Voltage profiles for a NiP_3/Li cell cycled at a C/10 rate between (**a**) 2.5 and 0 V, (**b**) 2 and 0.5 V, and (**c**) 0.5, 2, and 0.65 V. (**d**) The capacity retention at a C/10 rate for a NiP_3/Li cell cycled between 2 and 0 V (in *gray line*), 2 and 0.5 V (in *black line*), and between 2 and 0.5 V in the first discharge and 2 and 0.65 in the further cycles (in *dotted line*) (With kind permission from Springer Science + Business Media: Ionics [52] (2011))

5.6 Summary and Outlook

Due to the increasing demand for high energy density lithium-ion batteries, R&D laboratories around the world are actively looking for high specific capacity anode materials. In this regard TMOs, among others, have been of much interest. TMOs (and TMXs in general) having low oxidation number transition metals react in an unusual way with lithium. They undergo a reversible conversion reaction with at least two electrons or more exchanged per TM atom. This reaction also involves the reversible formation and decomposition of a Li_2O matrix. The conversion reaction is reversible only because TMOs are reduced into an intimate mixture of nanoparticles of transition metals and Li_2O. This reversible solid-state reaction does not occur at a macroscopic scale. This chapter also showed that the particle size and particle morphology of the starting material are crucial for the battery performance; each TMO has an optimum particle size and morphology. The large hysteresis observed between the charge and the discharge potential profile curve is one of the major issues with TMOs. However, it improves after the first lithiation, since the second lithiation potential is higher than the first. The delithiation potential of the anode is crucial for LIBs since it will set the output cell voltage; the lower the value, the higher the cell output voltage. TMOs generally show high delithiation voltages. One way to tune this voltage plateau is to change the transition metal, the lowest one being chromium and manganese around 1.4 V and the iron around 1.8 V. Another way would be to use TMOs made of a mix of two different transition metals, such as $CoMn_2O_4$, in order to have an average potential. This strategy works well for the first lithiation, but however does not work for the subsequent lithiation/delithiation since a mixture of both oxides is then obtained. The last strategy to change the voltage plateaus is to change the oxide counterions to fluorides, sulfides, nitrides, and phosphides. TMFs show the highest delithiation potentials and are more interesting as cathodes, while TMPs provide the lowest delithiation potentials, so potentially the highest output cell voltage of all TMXs.

References

1. Ravet N, Goodenough J, Besner S, Simoneau M, Hovington P, Armand M (1999) Improved iron based cathode material. In: Electrochemical society meeting abstracts, 99(2):Abstract #127
2. Huang H, Yin SC, Nazar LF (2001) Approaching theoretical capacity of $LiFePO_4$ at room temperature at high rates. Electrochem Solid State Lett 4(10):A170–A172
3. Armand M, Gauthier M, Magnan J-F, Ravet N (2002) Method for synthesis of carbon-coated redox materials with controlled size. Patent number: WO 02/27823, PCT/CA01/01349
4. Yabuuchi N, Ohzuku T (2003) Novel lithium insertion material of $LiCo_{1/3}Ni_{1/3}Mn_{1/3}O_2$ for advanced lithium-ion batteries. J Power Sources 119–121:171–174
5. Belharouak I, Lu W, Vissers D, Amine K (2006) Safety characteristics of Li $(Ni_{0.8}Co_{0.15}Al_{0.05})O_2$ and $Li(Ni_{1/3}Co_{1/3}Mn_{1/3})O_2$. Electrochem Commun 8(2):329–335
6. Weaving JS, Coowar F, Teagle DA, Cullen J, Dass V, Bindin P, Green R, Macklin WJ (2001) Development of high energy density Li-ion batteries based on $LiNi_{1-x-y}Co_xAl_yO_2$. J Power Sources 97–98:733–735
7. Primak W, Fuchs LH (1954) Electrical conductivities of natural graphite crystals. Phys Rev 95(1):22
8. Inoue H (2006) High capacity negative-electrode materials next to carbon; Nexelion. In: International meeting on lithium batteries, Biarritz
9. Mizutani S, Inoue H (2005) Negative active material and method for production thereof, non-aqueous electrolyte secondary cell using the same. Patent number: 2005-0208378
10. Kawakami S, Asao M (2005) Electrode material for anode of rechargeable lithium battery, electrode structural body using said electrode material, rechargeable lithium battery using said electrode structural body, process for producing said electrode structural body, and process for producing said rechargeable lithium battery. Patent number: 6,949,312
11. Dahn JR, Mar RE, Abouzeid A (2006) Combinatorial study of $Sn_{1-x}Co_x$ ($0 < x < 0.6$) and $[Sn_{0.55}Co_{0.45}]_{1-y}C_y$ ($0 < y < 0.5$) alloy negative electrode materials for Li-ion batteries. J Electrochem Soc 153(2):A361–A365

12. Hassoun J, Ochal P, Panero S, Mulas G, Bonatto Minella C, Scrosati B (2008) The effect of $CoSn/CoSn_2$ phase ratio on the electrochemical behaviour of $Sn_{40}Co_{40}C_{20}$ ternary alloy electrodes in lithium cells. J Power Sources 180(1):568–575
13. Hassoun J, Mulas G, Panero S, Scrosati B (2007) Ternary Sn-Co-C Li-ion battery electrode material prepared by high energy ball milling. Electrochem Commun 9(8):2075–2081
14. Ferguson PP, Todd ADW, Dahn JR (2008) Comparison of mechanically alloyed and sputtered tin-cobalt-carbon as an anode material for lithium-ion batteries. Electrochem Commun 10(1):25–31
15. Hassoun J, Panero S, Mulas G, Scrosati B (2007) An electrochemical investigation of a Sn-Co-C ternary alloy as a negative electrode in Li-ion batteries. J Power Sources 171(2):928–931
16. SonyCorporation (2011) Sony, the market for notebook PC; development of a tin-based amorphous anode, for high-capacity rechargeable lithium-ion battery 3.5Ah: the "Nexelion" (trans). Accessed January, 2012. Available from http://www.sony.co.jp/SonyInfo/News/Press/201107/11-078/
17. Ohzuku T, Ueda A, Yamamoto N (1995) Zero-strain insertion material of $Li[Li_{1/3}Ti_{5/3}]O_4$ for rechargeable lithium cells. J Electrochem Soc 142(5):1431–1435
18. Amatucci GG, Badway F, Pasquier AD, Zheng T (2001) An asymmetric hybrid nonaqueous energy storage cell. J Electrochem Soc 148(8):A930–A939
19. Thackeray MM, Goodenough JB (1985) Solid state cell wherein an anode, solid electrolyte and cathode each comprise a cubic-close-packed framework structure. Patent number: 4,504,371
20. Guerfi A, Sévigny S, Lagacé M, Hovington P, Kinoshita K, Zaghib K (2003) Nano-particle $Li_4Ti_5O_{12}$ spinel as electrode for electrochemical generators. J Power Sources 119–121:88–94
21. Plitz I, DuPasquier A, Badway F, Gural J, Pereira N, Gmitter A, Amatucci GG (2006) The design of alternative nonaqueous high power chemistries. Appl Phys Mater Sci Process 82(4):615–626
22. Du Pasquier A (2008) Nanosized titanium oxides for energy storage and conversion. In: Eftekhari A (ed) Nanostructured materials in electrochemistry. Wiley-VCH, Weinheim, pp 387–408
23. Nazri G-A, Pistoia G (2004) Lithium batteries: science and technology. Kluwer, Boston, 708 pages
24. Hamon Y, Brousse T, Jousse F, Topart P, Buvat P, Schleich DM (2001) Aluminum negative electrode in lithium ion batteries. J Power Sources 97–98:185–187
25. Rao BML, Francis RW, Christopher HA (1977) Lithium-aluminum electrode. J Electrochem Soc 124(10):1490–1492
26. Li J, Lewis RB, Dahn JR (2007) Sodium carboxymethyl cellulose. Electrochem Solid State Lett 10(2):A17–A20
27. Kasavajjula U, Wang C, Appleby AJ (2007) Nano- and bulk-silicon-based insertion anodes for lithium-ion secondary cells. J Power Sources 163(2):1003–1039
28. Chan CK, Peng H, Liu G, McIlwrath K, Zhang XF, Huggins RA, Cui Y (2008) High-performance lithium battery anodes using silicon nanowires. Nat Nanotechnol 3(1):31–35
29. Yang J, Winter M, Besenhard JO (1996) Small particle size multiphase Li-alloy anodes for lithium-ion batteries. Solid State Ion 90(1–4):281–287
30. Derrien G, Hassoun J, Panero S, Scrosati B (2007) Nanostructured Sn-C composite as an advanced anode material in high-performance lithium-ion batteries. Adv Mater 19(17):2336–2340
31. Nara H, Fukuhara Y, Takai A, Komatsu M, Mukaibo H, Yamauchi Y, Momma T, Kuroda K, Osaka T (2008) Cycle and rate properties of mesoporous tin anode for lithium ion secondary batteries. Chem Lett 37(2):142–143
32. Balbuena PB, Wang Y (2004) Lithium-ion batteries: solid-electrolyte interphase. Imperial College Press, London

33. Li H, Huang X, Chen L, Wu Z, Liang Y (1999) A high capacity nano-Si composite anode material for lithium rechargeable batteries. Electrochem Solid State Lett 2(11):547–549
34. Beattie SD, Larcher D, Morcrette M, Simon B, Tarascon JM (2008) Si electrodes for Li-ion batteries – a new way to look at an old problem. J Electrochem Soc 155(2):A158–A163
35. Buqa H, Holzapfel M, Krumeich F, Veit C, Novák P (2006) Study of styrene butadiene rubber and sodium methyl cellulose as binder for negative electrodes in lithium-ion batteries. J Power Sources 161(1):617–622
36. Kovalenko I, Zdyrko B, Magasinski A, Hertzberg B, Milicev Z, Burtovyy R, Luzinov I, Yushin G (2011) A major constituent of brown algae for use in high-capacity Li-ion batteries. Science 334(6052):75–79
37. Poizot P, Laruelle S, Grugeon S, Dupont L, Tarascon JM (2000) Nano-sized transition-metal oxides as negative-electrode materials for lithium-ion batteries. Nature 407:496–499
38. Cabana J, Monconduit L, Larcher D, Palacín MR (2010) Beyond intercalation-based Li-ion batteries: the state of the art and challenges of electrode materials reacting through conversion reactions. Adv Mater 22:E170–E192
39. Iijima T, Toyoguchi Y, Nishimura J, Ogawa H (1980) Button-type lithium battery using copper oxide as a cathode. J Power Sources 5(1):99–109
40. Novák P (1985) CuO cathode in lithium cells–II. Reduction mechanism of CuO. Electrochim Acta 30(12):1687–1692
41. Li T, Ai XP, Yang HX (2011) Reversible electrochemical conversion reaction of Li_2O/CuO nanocomposites and their application as high-capacity cathode materials for Li-ion batteries. J Phys Chem C 115(13):6167–6174
42. Thackeray MM, Coetzer J (1981) A preliminary investigation of the electrochemical performance of α-Fe_2O_3 and Fe_3O_4 cathodes in high-temperature cells. Mater Res Bull 16(5):591–597
43. Brousse T, Crosnier O, Santos-Peña J, Sandu I, Fragnaud P, Schleich DM (2002) Recent progress in the development of tin-based negative electrodes for Li-ion batteries. In: Kumagai N, Komaba S (eds) Materials chemistry in lithium batteries. Research Signpost, Kerala
44. Li H, Huang X, Chen L (1999) Anodes based on oxide materials for lithium rechargeable batteries. Solid State Ion 123(1–4):189–197
45. Li H, Balaya P, Maier J (2004) Li-storage via heterogeneous reaction in selected binary metal fluorides and oxides. J Electrochem Soc 151(11):A1878–A1885
46. Jamnik J, Maier J (2003) Nanocrystallinity effects in lithium battery materials: aspects of nano-ionics. Part IV. Phys Chem Chem Phys 5(23):5215–5220
47. Thackeray MM (1997) Manganese oxides for lithium batteries. Prog Solid State Chem 25(1–2):1–71
48. Binotto G, Larcher D, Prakash AS, Herrera Urbina R, Hegde MS, Tarascon JM (2007) Synthesis, characterization, and Li-electrochemical performance of highly porous Co_3O_4 659 powders. Chem Mater 19(12):3032–3040
49. Larcher D, Sudant G, Leriche J-B, Chabre Y, Tarascon J-M (2002) The electrochemical reduction of Co_3O_4 in a lithium cell. J Electrochem Soc 149(3):A234–A241
50. Fang X, Lu X, Guo X, Mao Y, Hu Y-S, Wang J, Wang Z, Wu F, Liu H, Chen L (2010) Electrode reactions of manganese oxides for secondary lithium batteries. Electrochem Commun 12(11):1520–1523
51. Debart A, Dupont L, Poizot P, Leriche J-B, Tarascon JM (2001) A transmission electron microscopy study of the reactivity mechanism of tailor-made CuO particles toward lithium. J Electrochem Soc 148(11):A1266–A1274
52. Débart A, Dupont L, Patrice R, Tarascon JM (2006) Reactivity of transition metal (Co, Ni, Cu) sulphides versus lithium: the intriguing case of the copper sulphide. Solid State Sci 8(6):640–651
53. Crosnier O, Nazar LF (2004) Facile reversible displacement reaction of Cu_3P with lithium at low potential. Electrochem Solid State Lett 7(7):A187–A189

54. Boyanov S, Gillot F, Monconduit L (2008) The electrochemical reactivity of the NiP_3 skutterudite-type phase with lithium. Ionics 14(2):125–130
55. Larcher D, Masquelier C, Bonnin D, Chabre Y, Masson V, Leriche J-B, Tarascon J-M (2003) Effect of particle size on lithium intercalation into α-Fe_2O_3. J Electrochem Soc 150(1): A133–A139
56. Boyanov S, Bernardi J, Gillot F, Dupont L, Womes M, Tarascon JM, Monconduit L, Doublet ML (2006) FeP: another attractive anode for the Li-ion battery enlisting a reversible two-step insertion/conversion process. Chem Mater 18(15):3531–3538
57. Zhang W-M, Wu X-L, Hu J-S, Guo Y-G, Wan L-J (2008) Carbon coated Fe_3O_4 nanospindles as a superior anode material for lithium-ion batteries. Adv Funct Mater 18(24):3941–3946
58. Dupont L, Laruelle S, Grugeon S, Dickinson C, Zhou W, Tarascon JM (2008) Mesoporous Cr_2O_3 as negative electrode in lithium batteries: TEM study of the texture effect on the polymeric layer formation. J Power Sources 175(1):502–509
59. Li H, Richter G, Maier J (2003) Reversible formation and decomposition of LiF clusters using transition metal fluorides as precursors and their application in rechargeable Li batteries. Adv Mater 15(9):736–739
60. Hu J, Li H, Huang X (2005) Cr_2O_3-based anode materials for Li-ion batteries. Electrochem Solid State Lett 8(1):A66–A69
61. Sun B, Chen Z, Kim H-S, Ahn H, Wang G (2010) MnO/C core-shell nanorods as high capacity anode materials for lithium-ion batteries. J Power Sources 196(6):3346–3349
62. Liu J, Li Y, Fan H, Zhu Z, Jiang J, Ding R, Hu Y, Huang X (2010) Iron oxide-based nanotube arrays derived from sacrificial template-accelerated hydrolysis: large-area design and reversible lithium storage. Chem Mater 22(1):212–217
63. Wang G, Liu T, Xie X, Ren Z, Bai J, Wang H (2011) Structure and electrochemical performance of Fe_3O_4/graphene nanocomposite as anode material for lithium-ion batteries. Mater Chem Phys 128(3):336–340
64. Varghese B, Reddy MV, Yanwu Z, Lit CS, Hoong TC, Subba Rao GV, Chowdari BVR, Wee ATS, Lim CT, Sow C-H (2008) Fabrication of NiO nanowall electrodes for high performance lithium ion battery. Chem Mater 20(10):3360–3367
65. Park JC, Kim J, Kwon H, Song H (2009) Gram-scale synthesis of Cu_2O nanocubes and subsequent oxidation to CuO hollow nanostructures for lithium-ion battery anode materials. Adv Mater 21(7):803–807
66. Gao XP, Bao JL, Pan GL, Zhu HY, Huang PX, Wu F, Song DY (2004) Preparation and electrochemical performance of polycrystalline and single crystalline CuO nanorods as anode materials for Li ion battery. J Phys Chem B 108(18):5547–5551
67. Xiang J, Tu J, Huang X, Yang Y (2008) A comparison of anodically grown CuO nanotube film and Cu_2O film as anodes for lithium ion batteries. J Solid State Electrochem 12(7):941–945
68. Xiang JY, Tu JP, Yuan YF, Huang XH, Zhou Y, Zhang L (2009) Improved electrochemical performances of core-shell Cu_2O/Cu composite prepared by a simple one-step method. Electrochem Commun 11(2):262–265
69. Liu J, Li Y, Ding R, Jiang J, Hu Y, Ji X, Chi Q, Zhu Z, Huang X (2009) Carbon/ZnO nanorod array electrode with significantly improved lithium storage capability. J Phys Chem C 113(13):5336–5339
70. Lee J-H, Hon M-H, Chung Y-W, Leu I-C (2010) The effect of TiO_2 coating on the electrochemical performance of ZnO nanorod as the anode material for lithium-ion battery. Appl Phys A Mater Sci Process 102(3):545–550
71. Hu J, Li H, Huang X, Chen L (2006) Improve the electrochemical performances of Cr_2O_3 anode for lithium ion batteries. Solid State Ion 177(26–32):2791–2799
72. Balaya P, Li H, Kienle L, Maier J (2003) Fully reversible homogeneous and heterogeneous Li storage in RuO_2 with high capacity. Adv Funct Mater 13:621–625

73. Courtel FM, Duncan H, Abu-Lebdeh Y, Davidson IJ (2011) High capacity anode materials for Li-ion batteries based on spinel metal oxides AMn_2O_4 (A = Co, Ni, and Zn). J Mater Chem 21(27):10206–10218
74. Yuan Z, Huang F, Feng C, Sun J, Zhou Y (2003) Synthesis and electrochemical performance of nanosized Co_3O_4. Mater Chem Phys 79(1):1–4
75. Oh SW, Bang HJ, Bae YC, Sun Y-K (2007) Effect of calcination temperature on morphology, crystallinity and electrochemical properties of nano-crystalline metal oxides (Co_3O_4, CuO, and NiO) prepared via ultrasonic spray pyrolysis. J Power Sources 173(1):502–509
76. Maier J (2007) Mass storage in space charge regions of nano-sized systems (nano-ionics. Part V). Faraday Discuss 134:51–56
77. Balaya P, Bhattacharyya AJ, Jamnik J, Zhukovskii YF, Kotomin EA, Maier J (2006) Nano-ionics in the context of lithium batteries. J Power Sources 159(1):171–178
78. Dupont L, Grugeon S, Laruelle S, Tarascon JM (2007) Structure, texture and reactivity versus lithium of chromium-based oxides films as revealed by TEM investigations. J Power Sources 164(2):839–848
79. Anisimov VI, Korotin MA, Kurmaev EZ (1990) Band-structure description of Mott insulators (NiO, MnO, FeO, CoO). J Phys Condens Matter 2(17):3973–3987
80. Jiang L-Y, Xin S, Wu X-L, Li H, Guo Y-G, Wan L-J (2010) Non-sacrificial template synthesis of Cr_2O_3-C hierarchical core/shell nanospheres and their application as anode materials in lithium-ion batteries. J Mater Chem 20(35):7565–7569
81. Grugeon S, Laruelle S, Dupont L, Chevallier F, Taberna PL, Simon P, Gireaud L, Lascaud S, Vidal E, Yrieix B, Tarascon M (2005) Combining electrochemistry and metallurgy for new electrode designs in Li-ion batteries. Chem Mater 17(20):5041–5047
82. Liu X, Yasuda H, Yamachi M (2005) Solid solution of nickel oxide and manganese oxide as negative active material for lithium secondary cells. J Power Sources 146(1–2):510–515
83. Cai Y, Liu S, Yin X, Hao Q, Zhang M, Wang T (2010) Facile preparation of porous one-dimensional Mn_2O_3 nanostructures and their application as anode materials for lithium-ion batteries. Physica E 43(1):70–75
84. Gao J, Lowe MA, Abruna HD (2011) Spongelike nanosized Mn_3O_4 as a high-capacity anode material for rechargeable lithium batteries. Chem Mater 23(13):3223–3227
85. Wang H, Cui L-F, Yang Y, Sanchez Casalongue H, Robinson JT, Liang Y, Cui Y, Dai H (2010) Mn_3O_4-graphene hybrid as a high-capacity anode material for lithium ion batteries. J Am Chem Soc 132(40):13978–13980
86. Yu XQ, He Y, Sun JP, Tang K, Li H, Chen LQ, Huang XJ (2009) Nanocrystalline MnO thin film anode for lithium ion batteries with low overpotential. Electrochem Commun 11(4):791–794
87. Zhong K, Xia X, Zhang B, Li H, Wang Z, Chen L (2010) MnO powder as anode active materials for lithium ion batteries. J Power Sources 195(10):3300–3308
88. Zhong K, Zhang B, Luo S, Wen W, Li H, Huang X, Chen L (2010) Investigation on porous MnO microsphere anode for lithium ion batteries. J Power Sources 196(16):6802–6808
89. Liu J, Pan Q (2010) MnO/C nanocomposites as high capacity anode materials for Li-ion batteries. Electrochem Solid State Lett 13(10):A139–A142
90. Ren Y, Armstrong AR, Jiao F, Bruce PG (2010) Influence of size on the rate of mesoporous electrodes for lithium batteries. J Am Chem Soc 132(3):996–1004
91. Zhong K, Xia X, Zhang B, Li H, Wang Z, Chen L (2010) MnO powder as anode active materials for lithium ion batteries. J Power Sources 195(10):3300–3308
92. Thackeray MM, David WIF, Goodenough JB (1982) Structural characterization of the lithiated iron oxides $Li_xFe_3O_4$ and $Li_xFe_2O_3$ ($0 < x < 2$). Mater Res Bull 17(6):785–793
93. Ban C, Wu Z, Gillaspie DT, Chen L, Yan Y, Blackburn JL, Dillon AC (2010) Nanostructured Fe_3O_4/SWNT electrode: binder-free and high-rate Li-ion anode. Adv Mater 22(20): E145–E149

94. Song Y, Qin S, Zhang Y, Gao W, Liu J (2010) Large-scale porous hematite nanorod arrays: direct growth on titanium foil and reversible lithium storage. J Phys Chem C 114(49):21158–21164
95. Tartaj P, Amarilla JM (2011) Iron oxide porous nanorods with different textural properties and surface composition: preparation, characterization and electrochemical lithium storage capabilities. J Power Sources 196(4):2164–2170
96. Chun L, Wu X, Lou X, Zhang Y (2010) Hematite nanoflakes as anode electrode materials for rechargeable lithium-ion batteries. Electrochim Acta 55(9):3089–3092
97. Muraliganth T, Vadivel Murugan A, Manthiram A (2009) Facile synthesis of carbon-decorated single-crystalline Fe_3O_4 nanowires and their application as high performance anode in lithium ion batteries. Chem Commun 47:7360–7362
98. Zhou G, Wang D-W, Li F, Zhang L, Li N, Wu Z-S, Wen L, Lu GQ, Cheng H-M (2010) Graphene-wrapped Fe_3O_4 anode material with improved reversible capacity and cyclic stability for lithium ion batteries. Chem Mater 22(18):5306–5313
99. Zhu Y, Murali S, Cai W, Li X, Suk JW, Potts JR, Ruoff RS (2010) Graphene and graphene oxide: synthesis, properties, and applications. Adv Mater 22:3906–3924
100. Badway F, Plitz I, Grugeon S, Laruelle S, Dolle M, Gozdz AS, Tarascon JM (2002) Metal oxides as negative electrode materials in Li-ion cells. Electrochem Solid State Lett 5(6): A115–A118
101. Li Y, Tan B, Wu Y (2008) Mesoporous Co_3O_4 nanowire arrays for lithium ion batteries with high capacity and rate capability. Nano Lett 8(1):265–270
102. Du N, Zhang H, Chen BD, Wu JB, Ma XY, Liu ZH, Zhang YQ, Yang DR, Huang XH, Tu JP (2007) Porous Co_3O_4 nanotubes derived from $Co_4(CO)_{12}$ clusters on carbon nanotube templates: a highly efficient material for Li-battery applications. Adv Mater 19(24):4505–4509
103. Lou XW, Deng D, Lee JY, Archer LA (2008) Thermal formation of mesoporous single-crystal Co_3O_4 nano-needles and their lithium storage properties. J Mater Chem 18(37):4397–4401
104. Wang X, Wu X-L, Guo Y-G, Zhong Y, Cao X, Ma Y, Yao J (2010) Synthesis and lithium storage properties of Co_3O_4 nanosheet-assembled multishelled hollow spheres. Adv Funct Mater 20(10):1680–1686
105. Jiang J, Liu J, Ding R, Ji X, Hu Y, Li X, Hu A, Wu F, Zhu Z, Huang X (2009) Direct synthesis of CoO porous nanowire arrays on Ti substrate and their application as lithium-ion battery electrodes. J Phys Chem C 114(2):929–932
106. Wang J, Du G, Zeng R, Niu B, Chen Z, Guo Z, Dou S (2010) Porous Co_3O_4 nanoplatelets by self-supported formation as electrode material for lithium-ion batteries. Electrochim Acta 55(16):4805–4811
107. Wu Z-S, Ren W, Wen L, Gao L, Zhao J, Chen Z, Zhou G, Li F, Cheng H-M (2010) Graphene anchored with Co_3O_4 nanoparticles as anode of lithium ion batteries with enhanced reversible capacity and cyclic performance. ACS Nano 4(6):3187–3194
108. Wang Y, Qin Q-Z (2002) A nanocrystalline NiO thin-film electrode prepared by pulsed laser ablation for Li-ion batteries. J Electrochem Soc 149(7):A873–A878
109. Hosono E, Fujihara S, Honma I, Zhou H (2006) The high power and high energy densities Li ion storage device by nanocrystalline and mesoporous Ni/NiO covered structure. Electrochem Commun 8(2):284–288
110. Needham SA, Wang GX, Liu HK (2006) Synthesis of NiO nanotubes for use as negative electrodes in lithium ion batteries. J Power Sources 159(1):254–257
111. Wang H, Pan Q, Wang X, Yin G, Zhao J (2009) Improving electrochemical performance of NiO films by electrodeposition on foam nickel substrates. J Appl Electrochem 39(9):1597–1602
112. Liu L, Li Y, Yuan S, Ge M, Ren M, Sun C, Zhou Z (2009) Nanosheet-based NiO microspheres: controlled solvothermal synthesis and lithium storage performances. J Phys Chem C 114(1):251–255

113. Nuli Y-N, Zhao S-L, Qin Q-Z (2003) Nanocrystalline tin oxides and nickel oxide film anodes for Li-ion batteries. J Power Sources 114(1):113–120
114. Pan Q, Liu J (2009) Facile fabrication of porous NiO films for lithium-ion batteries with high reversibility and rate capability. J Solid State Electrochem 13(10):1591–1597
115. Cheng M-Y, Hwang B-J (2010) Mesoporous carbon-encapsulated NiO nanocomposite negative electrode materials for high-rate Li-ion battery. J Power Sources 195(15):4977–4983
116. Grugeon S, Laruelle S, Herrera-Urbina R, Dupont L, Poizot P, Tarascon J-M (2001) Particle size effects on the electrochemical performance of copper oxides toward lithium. J Electrochem Soc 148(4):A285–A292
117. Chen LB, Lu N, Xu CM, Yu HC, Wang TH (2009) Electrochemical performance of polycrystalline CuO nanowires as anode material for Li ion batteries. Electrochim Acta 54(17):4198–4201
118. Ke F-S, Huang L, Wei G-Z, Xue L-J, Li J-T, Zhang B, Chen S-R, Fan X-Y, Sun S-G (2009) One-step fabrication of CuO nanoribbons array electrode and its excellent lithium storage performance. Electrochim Acta 54(24):5825–5829
119. Wang H, Pan Q, Cheng Y, Zhao J, Yin G (2009) Evaluation of ZnO nanorod arrays with dandelion-like morphology as negative electrodes for lithium-ion batteries. Electrochim Acta 54(10):2851–2855
120. Dey AN (1971) Electrochemical alloying of lithium in organic electrolytes. J Electrochem Soc 118(10):1547–1549
121. Xiao L, Yang Y, Yin J, Li Q, Zhang L (2009) Low temperature synthesis of flower-like $ZnMn_2O_4$ superstructures with enhanced electrochemical lithium storage. J Power Sources 194(2):1089–1093
122. Wang J, King P, Huggins RA (1986) Investigations of binary lithium-zinc, lithium-cadmium and lithium-lead alloys as negative electrodes in organic solvent-based electrolyte. Solid State Ion 20(3):185–189
123. Zhang CQ, Tu JP, Yuan YF, Huang XH, Chen XT, Mao F (2007) Electrochemical performances of Ni-coated ZnO as an anode material for lithium-ion batteries. J Electrochem Soc 154(2):A65–A69
124. Lavela P, Tirado JL, Vidal-Abarca C (2007) Sol-gel preparation of cobalt manganese mixed oxides for their use as electrode materials in lithium cells. Electrochim Acta 52(28):7986–7995
125. Lavela P, Ortiz GF, Tirado JL, Zhecheva E, Stoyanova R, Ivanova S (2007) High-performance transition metal mixed oxides in conversion electrodes: a combined spectroscopic and electrochemical study. J Phys Chem C 111(38):14238–14246
126. Sharma Y, Sharma N, Subba Rao GV, Chowdari BVR (2007) Nanophase $ZnCo_2O_4$ as a high performance anode material for Li-ion batteries. Adv Funct Mater 17(15):2855–2861
127. Pasero D, Reeves N, West AR (2005) Co-doped Mn_3O_4: a possible anode material for lithium batteries. J Power Sources 141(1):156–158
128. Xiao L, Yang Y, Yin J, Li Q, Zhang L (2009) Low temperature synthesis of flower-like $ZnMn_2O_4$ superstructures with enhanced electrochemical lithium storage. J Power Sources 194(2):1089–1093
129. Yang Y, Zhao Y, Xiao L, Zhang L (2008) Nanocrystalline $ZnMn_2O_4$ as a novel lithium-storage material. Electrochem Commun 10(8):1117–1120
130. Deng Y, Tang S, Zhang Q, Shi Z, Zhang L, Zhan S, Chen G (2011) Controllable synthesis of spinel nano-$ZnMn_2O_4$ via a single source precursor route and its high capacity retention as anode material for lithium ion batteries. J Mater Chem 21(32):11987–11995
131. Lavela P, Tirado JL (2007) $CoFe_2O_4$ and $NiFe_2O_4$ synthesized by sol-gel procedures for their use as anode materials for Li ion batteries. J Power Sources 172(1):379–387
132. Chu Y-Q, Fu Z-W, Qin Q-Z (2004) Cobalt ferrite thin films as anode material for lithium ion batteries. Electrochim Acta 49(27):4915–4921

133. Alcántara R, Jaraba M, Lavela P, Tirado JL, Jumas JC, Olivier-Fourcade J (2003) Changes in oxidation state and magnetic order of iron atoms during the electrochemical reaction of lithium with $NiFe_2O_4$. Electrochem Commun 5(1):16–21
134. Bomio M, Lavela P, Tirado JL (2007) Fe Mössbauer spectroscopy and electron microscopy study of metal extraction from $CuFe_2O_4$ electrodes in lithium cells. Chemphyschem 8:1999–2007
135. Sharma Y, Sharma N, Rao GVS, Chowdari BVR (2008) Li-storage and cyclability of urea combustion derived $ZnFe_2O_4$ as anode for Li-ion batteries. Electrochim Acta 53(5):2380–2385
136. Sharma Y, Sharma N, Subba Rao GV, Chowdari BVR (2008) Studies on spinel cobaltites, $FeCo_2O_4$ and $MgCo_2O_4$ as anodes for Li-ion batteries. Solid State Ion 179(15–16):587–597
137. NuLi Y, Zhang P, Guo Z, Liu H, Yang J (2008) $NiCo_2O_4$/C nanocomposite as a highly reversible anode material for lithium-ion batteries. Electrochem Solid State Lett 11(5): A64–A67
138. Sharma Y, Sharma N, Rao GVS, Chowdari BVR (2007) Lithium recycling behaviour of nano-phase-$CuCo_2O_4$ as anode for lithium-ion batteries. J Power Sources 173(1):495–501
139. Qiu Y, Yang S, Deng H, Jin L, Li W (2010) A novel nanostructured spinel $ZnCo_2O_4$ electrode material: morphology conserved transformation from a hexagonal shaped nanodisk precursor and application in lithium ion batteries. J Mater Chem 20(21):4439–4444
140. Du N, Xu Y, Zhang H, Yu J, Zhai C, Yang D (2011) Porous $ZnCo_2O_4$ nanowires synthesis via sacrificial templates: high-performance anode materials of Li-ion batteries. Inorg Chem 50(8):3320–3324
141. Plitz I, Badway F, Al-Sharab J, DuPasquier A, Cosandey F, Amatucci GG (2005) Structure and electrochemistry of carbon-metal fluoride nanocomposites fabricated by solid-state redox conversion reaction. J Electrochem Soc 152(2):A307–A315
142. Gabano JP, Dechenaux V, Gerbier G, Jammet J (1972) D-size lithium cupric sulfide cells. J Electrochem Soc 119(4):459–461
143. Whittingham MS (1978) Chemistry of intercalation compounds: metal guests in chalcogenide hosts. Prog Solid State Chem 12(1):41–99
144. Henriksen GL, Jansen AN (2002) Lithium batteries. In: Linden D, Reddy TB (eds) Handbook of batteries. McGraw-Hill, New York
145. Ellis BL, Lee KT, Nazar LF (2010) Positive electrode materials for Li-ion and Li-batteries. Chem Mater 22(3):691–714
146. Ji X, Lee KT, Nazar LF (2009) A highly ordered nanostructured carbon-sulphur cathode for lithium-sulphur batteries. Nat Mater 8(6):500–506
147. Han S-C, Kim K-W, Ahn H-J, Ahn J-H, Lee J-Y (2003) Charge-discharge mechanism of mechanically alloyed NiS used as a cathode in rechargeable lithium batteries. J Alloys Compd 361(1–2):247–251
148. Fu Z-W, Wang Y, Yue X-L, Zhao S-L, Qin Q-Z (2004) Electrochemical reactions of lithium with transition metal nitride electrodes. J Phys Chem B 108(7):2236–2244
149. Huggins RA (2002) Alternative materials for negative electrodes in lithium systems. Solid State Ion 152–153:61–68
150. Das B, Reddy MV, Malar P, Osipowicz T, Subba Rao GV, Chowdari BVR (2009) Nanoflake CoN as a high capacity anode for Li-ion batteries. Solid State Ion 180(17–19):1061–1068
151. Gillot F, Monconduit L, Doublet ML (2005) Electrochemical behaviors of binary and ternary manganese phosphides. Chem Mater 17(23):5817–5823
152. Gillot F, Boyanov S, Dupont L, Doublet ML, Morcrette M, Monconduit L, Tarascon JM (2005) Electrochemical reactivity and design of NiP_2 negative electrodes for secondary Li-ion batteries. Chem Mater 17(25):6327–6337
153. Silva DCC, Crosnier O, Ouvrard G, Greedan J, Safa-Sefat A, Nazar LF (2003) Reversible lithium uptake by FeP_2. Electrochem Solid State Lett 6(8):A162–A165
154. Zhang Z, Yang J, Nuli Y, Wang B, Xu J (2005) CoP_x synthesis and lithiation by ball-milling for anode materials of lithium ion cells. Solid State Ion 176(7–8):693–697

155. Pralong V, Souza DCS, Leung KT, Nazar LF (2002) Reversible lithium uptake by CoP_3 at low potential: role of the anion. Electrochem Commun 4(6):516–520
156. Sun Q, Fu Z-W (2007) An anode material of CrN for lithium-ion batteries. Electrochem Solid State Lett 10(8):A189–A193
157. Zhang N, Yi R, Wang Z, Shi R, Wang H, Qiu G, Liu X (2008) Hydrothermal synthesis and electrochemical properties of alpha-manganese sulfide submicrocrystals as an attractive electrode material for lithium-ion batteries. Mater Chem Phys 111(1):13–16
158. Montoro LA, Rosolen JM (2003) Gelatin/DMSO: a new approach to enhancing the performance of a pyrite electrode in a lithium battery. Solid State Ion 159(3–4):233–240
159. Yan JM, Huang HZ, Zhang J, Liu ZJ, Yang Y (2005) A study of novel anode material CoS_2 for lithium ion battery. J Power Sources 146(1–2):264–269
160. Takeuchi T, Sakaebe H, Kageyama H, Sakai T, Tatsumi K (2008) Preparation of NiS_2 using spark-plasma-sintering process and its electrochemical properties. J Electrochem Soc 155(9): A679–A684
161. Wang Y, Fu Z-W, Yue X-L, Qin Q-Z (2004) Electrochemical reactivity mechanism of Ni_3N with lithium. J Electrochem Soc 151(4):E162–E167
162. Pereira N, Dupont L, Tarascon JM, Klein LC, Amatucci GG (2003) Electrochemistry of Cu_3N with lithium. J Electrochem Soc 150(9):A1273–A1280
163. Wang K, Yang J, Xie J, Wang B, Wen Z (2003) Electrochemical reactions of lithium with CuP_2 and $Li_{1.75}Cu_{1.25}P_2$ synthesized by ballmilling. Electrochem Commun 5(6):480–483

Chapter 6
Graphene-Based Composite Anodes for Lithium-Ion Batteries

Nathalie Lavoie, Fabrice M. Courtel, Patrick R.L. Malenfant, and Yaser Abu-Lebdeh

Abstract Graphene has emerged as a novel, highly promising material with exceptional properties and potential application in a wide range of technologies. As an anode material for lithium-ion batteries, it was shown that it cannot be used in the pure form due to its large irreversible capacity but as part of a composite with other active materials. Transition metal oxides, silicon, and tin have been explored as active anode materials to replace graphite because of their high theoretical capacities. However, these materials have large volume changes during cycling that leads to the failure of the batteries. To resolve this problem, additives have been added to these materials to mitigate this volume change. In recent years, graphene has been employed as an encapsulating agent for these materials. In this chapter, an overview of the work exploring composites made of graphene as a novel support for nanoscale materials that react with lithium and provide high capacities will be presented.

6.1 Introduction

The world is moving toward low-carbon "clean energy," driven by increasing concern over climate change and energy security. New technologies are required to address clean energy generation, distribution, storage, and system management. Implementation of the smart grid, renewable integration, and sustainable transportation will require innovation across many scientific disciplines. Energy storage, and specifically electrochemical energy storage, will play an integral role in the introduction of electric vehicles and the stabilization of the electrical power grid that will increasingly include intermittent renewable energy sources. Many technical

N. Lavoie • P.R.L. Malenfant • Y. Abu-Lebdeh (✉)
National Research Council of Canada, Ottawa, ON K1A 0R6, Canada
e-mail: Yaser.Abu-Lebdeh@nrc-cnrc.gc.ca

F.M. Courtel
Atomic Energy of Canada Limited, Chalk River, ON K0J 1J0, Canada

Y. Abu-Lebdeh and I. Davidson (eds.), *Nanotechnology for Lithium-Ion Batteries*, Nanostructure Science and Technology, DOI 10.1007/978-1-4614-4605-7_6,

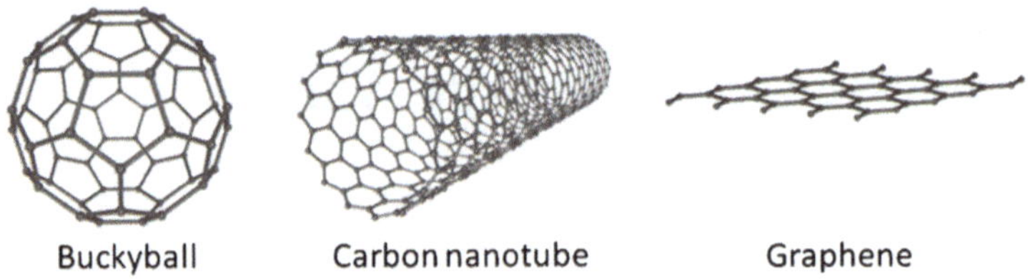

Fig. 6.1 Forms of carbon [3]

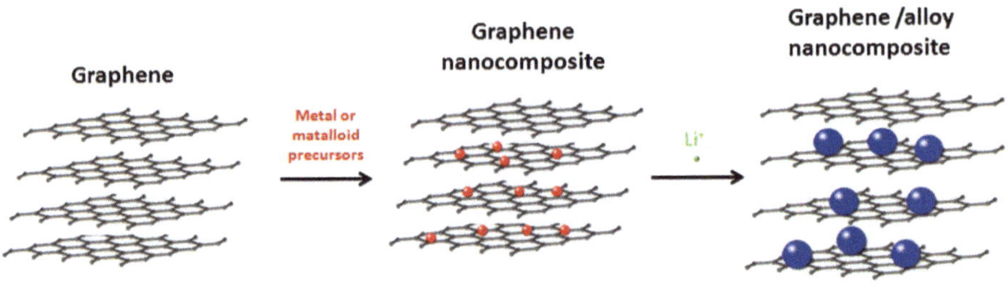

Fig. 6.2 Graphene-metal/oxide nanocomposite electrodes

challenges remain with regard to lithium-ion battery technology such as energy density, cost, reliability, and cycle life. One approach to address energy density is the use of composite anodes capable of reacting with lithium and thus providing two- to threefold enhancements in capacity compared to graphite. This chapter will focus on recent work exploring the use of graphene as a novel support for nanoscale materials that react with lithium and provide high capacities.

In 2010, the Noble prize in physics was awarded to Andre Geim and Konstantin Novoselov for their groundbreaking work on graphene [1]. Graphene is the name used to describe a single 2D sheet of sp^2-hybridized carbon [2]. Stacks of graphene in 3D will provide graphite, whereas the 2D rolled form provides nanotubes; it can also be wrapped up into buckyballs (0D) (see Fig. 6.1).

Graphite has very impressive electrical (~10^4 S cm^{-1}) and thermal (~3,000 W mK^{-1}) conductivity values [4–9] and a good lithium-ion diffusion coefficient (10^{-7} to 10^{-10} cm^2 s^{-1} [10–12]), which makes it a potentially excellent material for negative electrodes in lithium-ion batteries. Graphene is equally impressive in terms of mechanical integrity. Furthermore, stacked sheets of graphene derived from exfoliated graphite provide a modular approach to exploring lithium storage in layered carbon as well as layered carbon/metal nanocomposites (see Fig. 6.2). For example, composite graphene electrodes can be made with nanometals or nanometal oxides (0D, 1D, 2D) capable of reacting with lithium while dispersed between the graphene sheets, thus providing a novel route to significantly enhance gravimetric and volumetric capacities.

Thin stacks of graphene sheets, often referred to as graphene platelets, can be prepared using scalable processes starting from abundantly available graphite. Alternatively, many avenues can be pursued in order to oxidize and exfoliate graphite to provide graphene oxide (GO) [4–9]. From there, the material can be

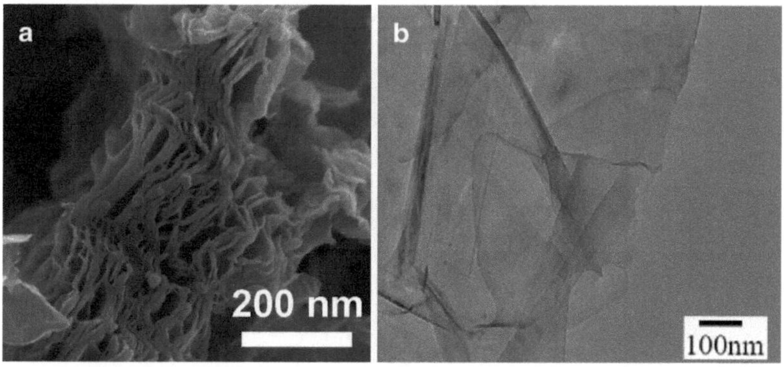

Fig. 6.3 Graphene paper (**a**) (Reprinted with permission from [13]. Copyright (2009) American Chemical Society) and graphene single sheet (**b**) (Reprinted from [14]. Copyright (2011), with permission from Elsevier)

processed in different ways to obtain graphene paper, or graphene sheets (see Fig. 6.3), that can be used to prepare graphene/metal nanocomposites. With high electrical conductivity (~10^x S cm^{-1} [15]) and good mechanical properties, graphene paper will provide a layered carbon electrode material that may not require the need for binders or other additives. This will maximize the content of active material, eliminate components from the formulation, simplify the fabrication process, and reduce the cost of the final product. With a more open structure, the lithium-ion mobility may also be enhanced versus graphite, yet higher surface areas are prone to high irreversible capacities, and this issue must be addressed for practical devices to emerge [16]. While multi-walled nanotubes/single-walled nanotubes (MWNTs/SWNTs) have provided encouraging results as an alternative to graphite in lithium-ion batteries, there are many challenges that limit their use in commercial processes that need to be addressed [17, 18]. In particular, a consistent supply of adequate scale and purity of SWNTs (metal content, tube diameter, tube chirality) must be provided, thus making their implementation, at this juncture, prohibitively expensive. In contrast, the established large-scale supply of graphite coupled with scalable processing methods toward graphene makes the graphene paper approach conceivable, yet the economics still need to be demonstrated. To date, a limited number of results have been published using graphene as the active anode material, and several reports have been published on graphene/metal or graphene/metal oxide nanocomposites. This chapter will review progress in both areas.

6.2 Graphene Anodes

Carbon-based anodes have been studied extensively, and great progress has been made in understanding lithium intercalation mechanisms since the pioneering work of Dahn [19]. Several reviews have examined work on ordered and disordered

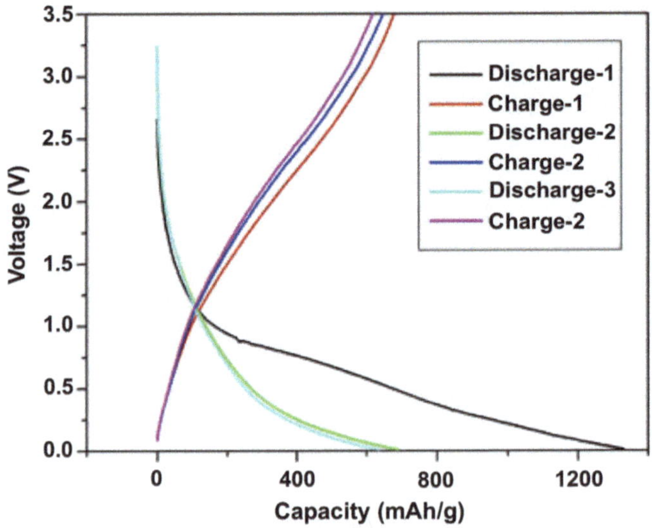

Fig. 6.4 Three first charge–discharge curves for graphene nanosheets (Reprinted from [22]. Copyright (2011), with permission from Elsevier)

carbon-based anodes in lithium batteries; hence, this area will not be discussed here [20, 21]. The following discussion will focus on graphene-based anodes in which graphene is used as the active material. In the following examples, graphene is either isolated from exfoliated graphite, from reduced GO (RGO) or via unzipping carbon nanotubes to provide graphene nanoribbons (GNRs).

Guo et al. prepared graphene from artificial graphite by oxidation, rapid expansion, and ultrasonic treatment. They tested the graphene sample in half-cells and obtained an irreversible capacity of approximately 1,250 mAh g^{-1} for the first cycle and 672 mAh g^{-1} of reversible capacity up to 30 cycles [22].

Figure 6.4 shows the charge-discharge curves for the three first cycles. In the first cycle, the graphene materials have a plateau at 0.7 V that is due to the formation of the solid electrolyte interface (SEI) which is a layer that forms on the surface of the electrode due to the electrolyte decomposition. The formation of the SEI contributes to the large irreversible capacity.

With an inherently high surface area, graphene paper has been shown to be prone to high irreversible lithium insertion upon the first cycle. As shown by Wang et al., graphene paper made from reduced GO dispersions can yield a discharge capacity of 680 mAh g^{-1}, yet upon the second cycle, the capacity drops significantly to 84 mAh g^{-1} [23]. While annealing the graphene paper at 800°C under N_2 for 1 h further removes oxygen functional groups and provides 301 mAh g^{-1} of reversible discharge capacities after 10 cycles. This performance is still significantly lower than that of commercial graphite.

Similar observations have been made by Abouimrane et al. in which non-annealed graphene paper, prepared by preparing GO using a modified Hummers

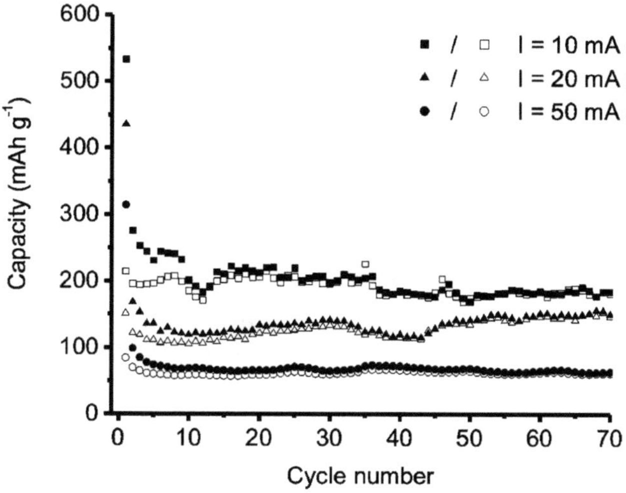

Fig. 6.5 Cycle performance of a Li/graphene paper half-cell at three different current densities. Charge and discharge capacities under each current density are represented by *open* and *solid shapes*, respectively (Reprinted from [24]. Copyright (2011) with permission from American Chemical Society)

method followed by hydrazine reduction, was used to evaluate its performance in a half-cell [24]. As shown in Fig. 6.5, it exhibited an initial reversible capacity of 84 mAh g^{-1} (at 50 mA g^{-1}), whereas an anode made of graphene powder using polyvinylidene fluoride (PVDF) as a binder had a reversible capacity of 288 mAh g^{-1}. Various current rates were explored with the binder-free graphene paper devices, in which case initial reversible capacities of 214, 151, and 84 mAh g^{-1} were obtained at 10, 20, and 50 mA g^{-1}, respectively. After 70 cycles, the capacities were approximately 190, 150, and 55 mA g^{-1}. A large part of the capacity loss is due to the formation of the SEI that is more prominent in graphene than in graphite due to the higher surface area.

Recent work by Bhardwaj et al. explored the use of GNRs derived from MWNTs in which the tubes are unzipped to yield narrow strips of GNRs [25]. The GNRs exhibit higher first charge and discharge capacities compared to MWNTs. However, high irreversible charge capacity is reported (1,400 mAh g^{-1}) with a discharge capacity of 820 mAh g^{-1} for the oxidized graphene nanoribbons (ox-GNRs). Only 14 cycles were shown, and a capacity loss of 3% per cycle indicates that these cells degrade rapidly and are not likely to provide extended cycling capability. The ox-GNR can be annealed (900°C; H_2/Ar; 15 min) to give GNRs. These ribbons yielded cells with an irreversible capacity of approximately 775 mAh g^{-1} and a reversible capacity of approximately 200 mAh g^{-1} after 14 cycles, still much lower than graphite.

Yoo et al. took an interesting approach at investigating graphene as an anode active material for lithium storage in that fullerenes and carbon nanotubes (CNTs)

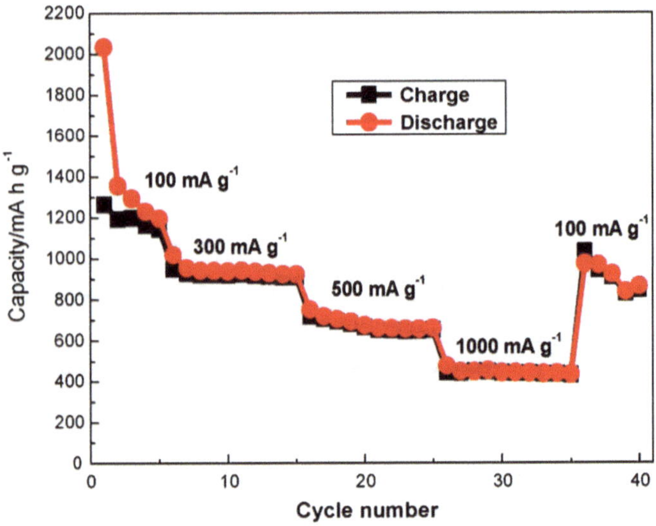

Fig. 6.6 Cycle performance of graphene sheets at current densities from 100 to 1,000 mAg^{-1} (Reprinted from [28]. Copyright (2010), with permission from Elsevier)

were used as spacers to increase the distance between sheets to enhance lithium-ion storage capacity [26]. Graphene nanosheets used in these experiments were made of 10–20 layers (3–7 nm thick). Voltage curves clearly indicate that lithium insertion into the graphene/CNT or graphene/fullerene devices is different. Furthermore, reversible capacities of 540, 730, and 784 mAh g^{-1} are obtained for graphene, graphene/CNTs, and graphene/fullerene, respectively. These are high values, but as observed previously, the rate at which these devices degrade is significant, and data is only shown for 20 cycles. It is unclear at this juncture, if the increased d-spacing obtained with CNTs and fullerenes can yield enhanced lithium accommodation as observed with polyacenic semiconductors (PAS) since Yoo's results are very similar to what was observed with GNRs [27].

High capacities were also observed for graphene sheets prepared by oxidation of graphite, followed by rapid thermal expansion [28]. After 40 cycles, at current densities from 100 to 1,000 mA g^{-1}, the reversible capacity was maintained at 848 mAh g^{-1} as shown in Fig. 6.6. In addition, this graphene also provided good rate capability with a reversible specific capacity that remained at 718 mAh g^{-1} at high current density of 500 mA g^{-1}. The influence of the graphene oxide reduction temperature was studied by Wan et al. [29]. The irreversible capacities of the graphene nanosheet in the first cycle decrease with increasing annealing temperatures (2,137 mAh g^{-1} for 300°C, 1,523 mAh g^{-1} for 600°C, and 1,167 mAh g^{-1} for 800°C). This is potentially due to the larger number of lithium insertion active sites in larger surface area graphene obtained with lower annealing temperature. After 100 cycles, the capacity of the three cells is much closer, varying from 478 mAh g^{-1} for the graphene prepared at 300°C to approximately 350 mAh g^{-1} for the sample prepared at 800°C.

Table 6.1 Graphene employed as anode materials in lithium-ion batteries

Graphene source	Irr. cap. (mAh g^{-1})	Rev. cap. (mAh g^{-1})	Number of cycles	Current density (mA g^{-1})	Voltage range (V vs. Li/Li^+)	Ref.
Hydrazine reduction of GO	680	84	50	50	3.0–0	[23]
Hydrazine reduction of GO	310	50	70	50	3.0–0	[24]
Unzipping of MWCNTs for ?ox-GNR? followed by H_2 reduction for GNR	ox-GNR 1400 GNR 775	ox-GNR 500 GNR 200	14	0.1 mA cm^{-2} (~C/10)	3.5–0	[25] [14]
Reduction of OGS with hydrazine		290	20	50	3.5–0.01	[26]
GO exfoliated at 1,050°C for several minutes	2,035	848	40	100	3.5–0.01	[28]
GO rapid heat at 1,050°C and ultrasonic agitation at 400 W for 4 h	1,233	502	30	0.2 mA cm^{-2}	3.5–0	[22]
Thermal reduction of GO under argon atmosphere	2,137	478	100	200	3.0–0.01	[29]
Sonication of natural graphite, ammonium peroxydisulfate, and hydrogen peroxide, followed by microwave irradiation	580	420	50	0.1 C	3.0–0.01	[30]

It is important to note that these high capacities are at odds with several reports that suggest that the capacity of anodes where graphene is the active material is typically <200 mAh g^{-1}. Graphene characteristics are highly dependent on the processing method, resulting defect density and surface area, among other critical properties such as electrical conductivity that will strongly impact the capacity for lithium-ion storage. At this juncture, there is much to be learned about the mechanism of lithium-ion storage in graphene, and this should be a fruitful area for further research (Table 6.1).

6.3 Graphene Composite Anodes

Carbon/metal composites made of metals that are able to alloy with lithium hold great promise at enhancing the capacity of lithium-ion batteries, yet reports in the literature are mixed [20, 31–34]. In fact, next-generation products are already hitting the market in portable electronics applications whereby manufacturers are claiming 30% improvements in battery capacities where alloying is used, yet these batteries do not fully exploit the potential of alloys [Miyaki, Y. US Patent, Patent

No: US 6,908,709 B2 (2005).]. While it may be impractical to attain the maximum theoretical capacities that can be potentially achieved with alloy-based anodes (e.g., due to capacity limitations of cathode materials), there is great utility in having new anode materials that can reliably yield 600–1,000 mAh g^{-1} of reversible capacity for long cycle use. Hence, improved materials that can withstand the enormous volume expansion (up to 400% [35]) and mechanical stresses that come with the alloying/de-alloying process must still be developed. The literature points to binders [36], nanostructured materials [37–41], and graphite/metal composites [42] as a possible solution to the mechanical instability caused by the repeated alloying/de-alloying process.

Graphene, as a support material for active nanostructured metal or metal oxide species distributed onto the surface of graphene or between the graphene layers, is a very compelling approach given graphene's large surface area (2,600 m^2 g^{-1}) [43] and both its mechanical [44] and electrical [45] conductivity. This section will discuss graphene/metal or graphene/metal oxide nanocomposites made either by mechanical mixing or fabricated by stacking graphene sheets in the presence of the corresponding organometallic precursor. The rationale is that this material nano-architecture will provide the necessary means of limiting the aggregation of the nanoparticles and accommodating large volume expansions that occur upon cycling while maintaining the mechanical integrity and electrical conductivity needed to provide long-term cyclability.

6.3.1 Graphene/Silicon-Based Materials

Silicon has been explored as anode material for lithium-ion batteries because of its high theoretical capacity (4,200 mAh g^{-1} or 9,805 mAh mL^{-1}) and its natural abundance [46–48]. Silicon reacts with lithium via the following alloying/de-alloying reactions:

$$\mathrm{Si} + \mathrm{xLi}^{+} + \mathrm{xe}^{-} \leftrightarrow \mathrm{SiLi_x} \text{ with } 0 \le \mathrm{x} \le 4.4(4,200\ \mathrm{mAh\ g^{-1}}) \qquad (6.1)$$

Up to 4.4, lithium can alloy with silicon leading to a maximum capacity of 4,200 mAh g^{-1}. However, this large capacity is associated with a large volume change of 300–400% upon cycling [35, 49, 50]. The latter gives rise to mechanical stresses that lead to cracks, eventual disintegration of the electrode, and a failure of the battery [51]. All strategies to fix this problem have shown their limitations: use of nanoparticles, cellulose-based binder, composites, and nanostructures [37–41]. Graphene/silicon nanocomposite materials, where silicon nanoparticles are trapped between graphene sheets, might be a way to better mitigate the effect of the large volume change. Several research groups worked on making these nanocomposites using mixing procedures of either graphene or GO with commercial silicon nanoparticles. Table 6.2 summarizes the preparation procedure, ratio, and different capacity values of the composites prepared by different research groups around the world.

Table 6.2 Graphene/silicon nanocomposites preparation methods and capacity values

Graphene/GO	Preparation of composite material	C/Si ratio	Particle size (nm)	Irr. cap. Rev. cap. (mAh g^{-1})	Number of cycles	Current density (mA g^{-1})	Voltage range (V vs. Li/Li^+)	Ref.
Graphene Kovtyukhova's method	– GO and nano-Si suspension (sonication) – Annealing at 700°C, Ar/H_2	~40/60	20–25	1,750 800	300	1,000–80	1.5–0.005	[52]
GO modified Hummers' method	– GO and nano-Si suspension (stirring) – Annealing at 500°C in Ar	~40/60	50	~2,500 ~750	30	300	1.2–0.005	[14]
Graphene heat treatment of graphite	– Graphene sheets and nano Si mechanically mixed	~33/67	50	3,500 2,300	30	300	1.2–0.005	[14]
Graphene solvothermal method	– Graphene and nano Si mechanically mixed	50/50	~40	2,158 1,168	30	100	1.2–0.02	[53]
Graphene prepared by liquid-phase exfoliation technique	– Preparation of Si nanowires via an electroless etching process – Graphene and Si nanowires mechanically are mixed	50/50	Length: 10 μm Diameter: 100	3,646 2,470	20	105	0.8–0.005	[54]
GO modified Hummers method Exfoliated by ultrasonication	– Silicon nanoparticles and GO suspended in water (sonication) – Suspension filtered to produce a self-supporting film – Film is thermally reduced under Ar at 700°C	33/67	100–200 nm	1,650 1,040	30	50	1.5–0.05	[55]

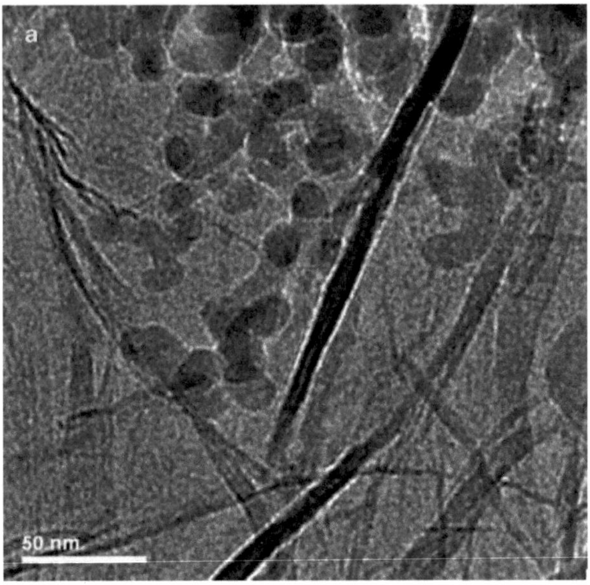

Fig. 6.7 TEM micrograph of a graphene/Si nanocomposite paper [52] (Reproduced by permission of the Royal Society of Chemistry)

Two groups worked on making composites of GO and silicon nanoparticles by mechanical mixing [14, 52]. As shown by the TEM micrograph in Fig. 6.7, Lee et al. obtained graphene/silicon nanocomposite paper [52]. The latter was prepared by annealing (700°C, Ar/H_2, 1 h) a nanocomposite paper of GO and silicon nanocomposite (20–25 nm) [52]. An initial discharge capacity of about 1,750 mAh g^{-1} was obtained when cycled at 50 mA g^{-1} with almost no first irreversible capacity. With only 0.5% capacity decrease per cycle, this nanocomposite provided good capacity retention with a discharge capacity of 1,000 mAh g^{-1} after 200 cycles [52].

Xiang et al. also used a composite made of GO and silicon nanoparticles (50 nm) in order to prepare their graphene/silicon nanocomposite, obtained after annealing at 500°C in argon for 1 h [14]. Three different graphene to silicon ratios were experimented, 1:1, 1:2, and 1:3 and named SG1, SG2, and SG3, respectively. The TEM micrographs showed in Fig. 6.8 confirm that the silicon nanoparticles bonded to graphene are well dispersed between the graphene sheets.

Figure 6.9 shows the cycling performance of pristine silicon nanoparticles and the three different nanocomposites (SG1, SG2, and SG3). The nanocomposites show better performance than the pristine silicon nanoparticles. Having the silicon nanoparticles trapped between graphene sheets helps to mitigate the effect of the volume change by keeping an electrical contact between the silicon particles and the electrical path. In addition, upon inspection the pure silicon anode was brittle, whereas the nanocomposite anodes were not. Nonetheless, large irreversible capacities ranging from 600 to 750 mAh g^{-1} are observed with these samples.

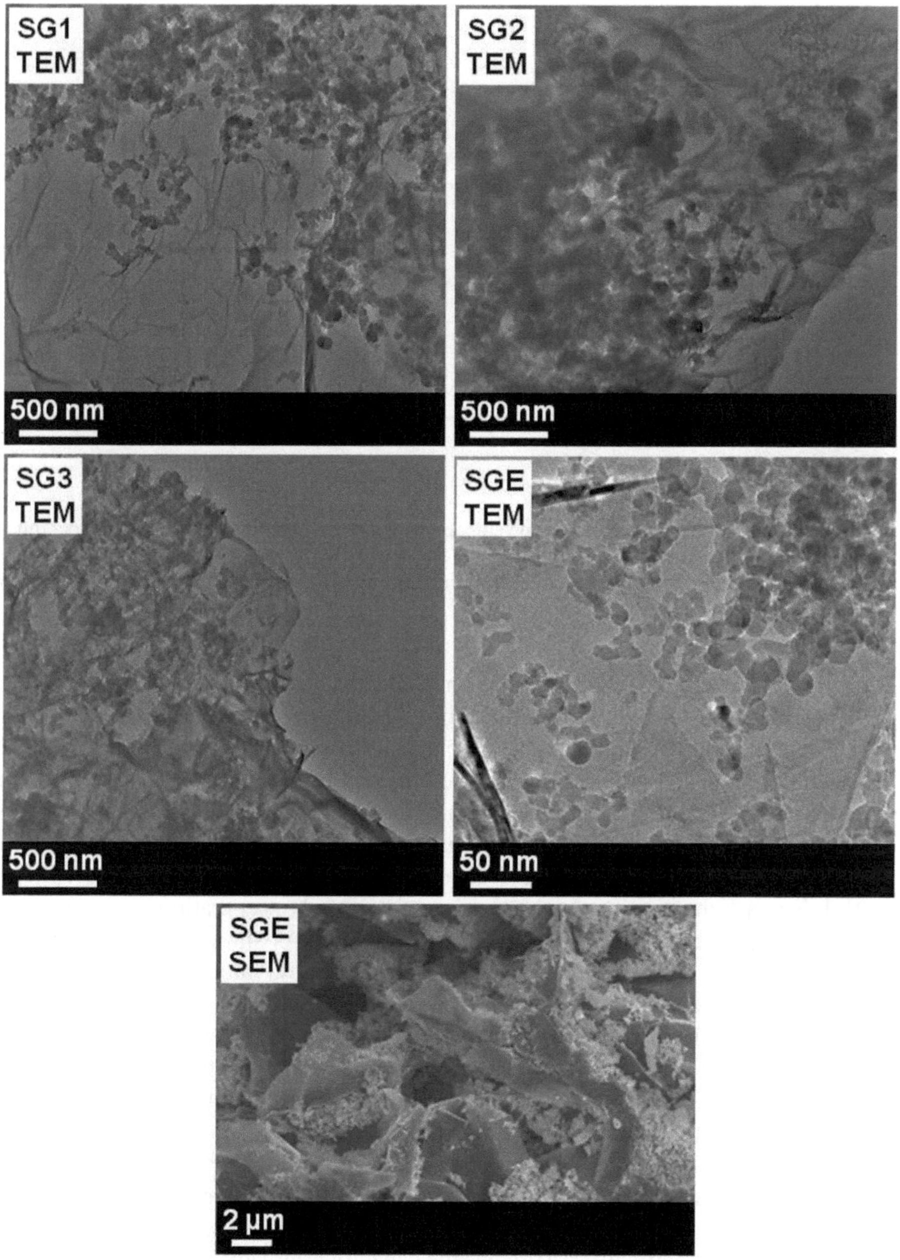

Fig. 6.8 TEM micrographs of graphene/silicon nanocomposites SG1, SG2, SG3, and SGE and SEM micrographs of SGE (see text for details) (Reprinted from [14]. Copyright (2011) with permission from Elsevier)

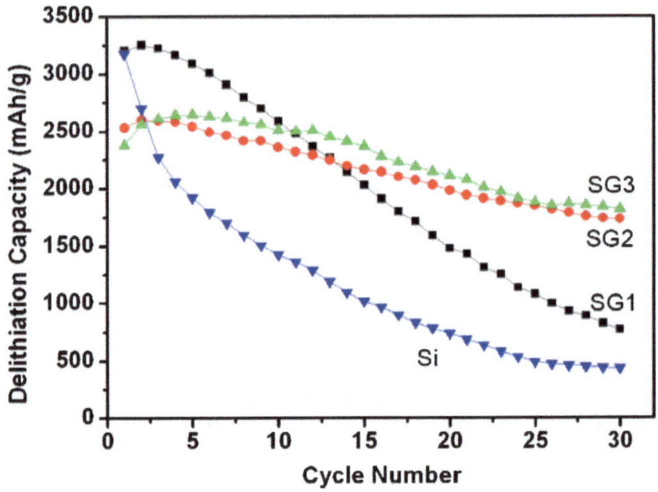

Fig. 6.9 Cycling performance of pristine silicon nanoparticles and graphene/silicon nanocomposites having graphene to silicon ratios of 1:1 (SG1), 1:2 (SG2), and 1:3 (SG3). Capacities are calculated using the weight of composite (Reprinted from [14], Copyright (2011), with permission from Elsevier)

After 30 cycles, samples SG2 and SG3 showed capacities of 700–800 mAh g^{-1} at a rate of 300 mA g^{-1} which represents about 70–80% of the initial capacity.

Using expandable graphite, a last graphene/silicon nanocomposite was prepared by Xiang et al. [14] Expandable graphite was rapidly heated to 1,050°C and then blended with silicon nanoparticles using mechanical blending in a 1:2 ration (SGE sample). TEM and SEM micrographs of an SGE sample (see Fig. 6.8) show that the silicon nanoparticles occupy the pores of the expanded graphene [14]. It is worth mentioning that sonication of SGE led to separation of some of the silicon nanoparticles from the graphene sheets [14]. As shown in Fig. 6.10, this modified method led to a material with higher capacity values (1,800 mAh g^{-1}) than the ones observed with SG2 and SG3.

Another useful study was performed by Chou et al. where they studied the performance of a graphene/silicon nanocomposite versus neat graphene and neat silicon nanoparticles [53]. The graphene used was synthesized via a solvothermal method [56], and the nanocomposite was prepared via mechanical mixing of the as-prepared graphene with silicon nanoparticles (~40 nm). The comparative study performed is shown in Fig. 6.11. The cycling data obtained clearly show that due to the large volume change that occurs upon cycling, silicon nanoparticles cannot sustain a stable capacity; after 20 cycles the capacity is already below 500 mAh g^{-1}. When the nanoparticles are dispersed within a graphene matrix, much better capacity retention is observed; the nanocomposite yields a lithium cell with a capacity of about 1,400 mAh g^{-1} after 20 cycles at 100 mA g^{-1} [53].

Nanostructured silicon arrays, such as silicon nanowires, have recently been investigated by Cui et al. [57]. Due to their better tolerance to strain and also the

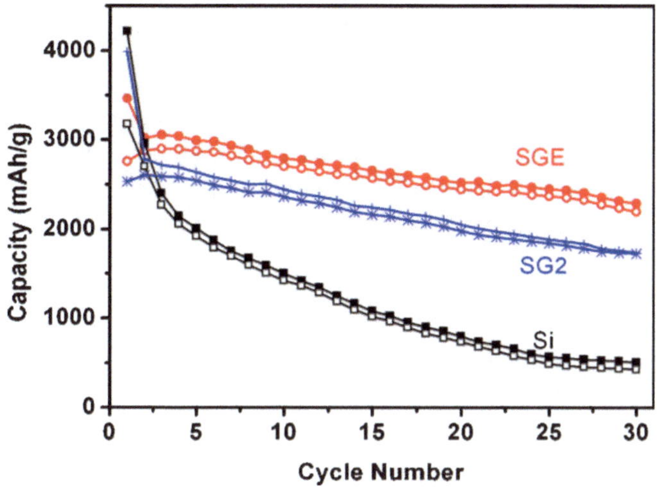

Fig. 6.10 Cycling performance of pristine silicon nanoparticles and graphene/silicon nanocomposites having a graphene to silicon ratio of 1:2 (SG2 and SGE) (see text for details). Capacities are calculated using the weight of composite (Reprinted from [14], Copyright (2011), with permission from Elsevier)

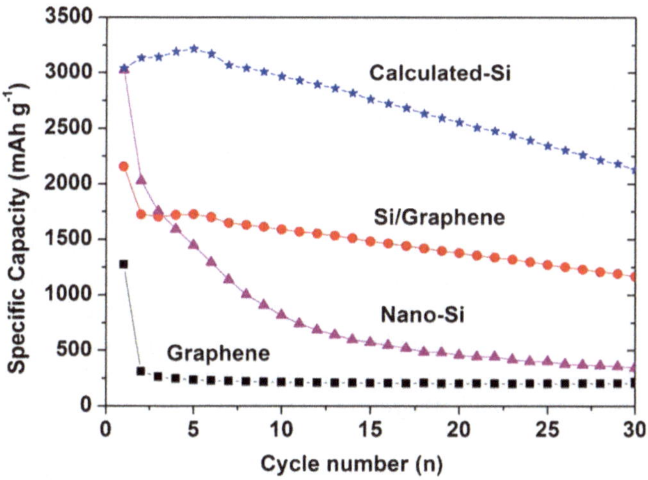

Fig. 6.11 Cycling performance of pristine graphene (*squares*), silicon nanoparticles (*triangles*), graphene/silicon nanocomposite (*circles*) electrodes, and the calculated pure silicon contribution from the graphene/silicon composite (*stars*) (Reprinted from [53], Copyright (2011), with permission from Elsevier)

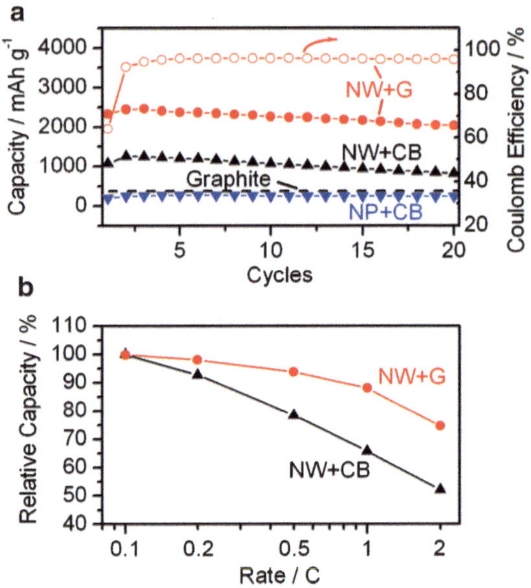

Fig. 6.12 (**a**) Charge capacities and Coulomb efficiency of cells for 20 cycles. *NW* silicon nanowires, *G* graphene, *CB* carbon black, *NP* silicon nanoparticles. (**b**) Rate capacities from 0.1 C to 2 C, with the rate for discharge fixed to 0.1 C (Reprinted with permission from [54]. Copyright (2011) American Chemical Society)

empty space between the wires, nanowires can better accommodate the volume change that occurs upon cycling. Wang et al. recently prepared single-crystalline silicon nanowires, and they showed that adding graphene significantly enhances the reversible capacity of the silicon nanowires [54]. As shown in Fig. 6.12a, the graphene/silicon-nanowire nanocomposite showed a charge capacity of 2,470 mAh g^{-1} after 20 cycles, which is higher than the carbon-black/silicon-nanowire nanocomposite (1,256 mAh g^{-1}). The rate capability of the graphene nanocomposite is also enhanced; for instance, at 2C, a discharge capacity representing 75% of the capacity obtained at C/10 was observed (see Fig. 6.12b). This is due to the favorable charge-transportation properties of the graphene additive. Before making any conclusions, longer cycling data would be required since it is often the case that these silicon-based types of anodes tend to catastrophically fail between 30 and 50 cycles.

6.3.2 Graphene/Tin-Based Materials

Like silicon, tin is a metalloid that reacts reversibly with lithium via an alloying/de-alloying reaction depicted by the following equation:

$$Sn + xLi^+ + xe^- \leftrightarrow SnLi_x \text{ with } 0 \le x \le 4.4 (993 \text{ mAh g}^{-1}) \qquad (6.2)$$

With a maximum of 4.4 lithium per tin, a capacity up to 993 mAh g^{-1} (7,313 mAh mL^{-1}) can be reached. However, like silicon, tin suffers from large volume changes upon cycling, about 260% [35]. Mechanical stresses usually lead to cracks and disintegration of the electrode. To prevent pulverization during the battery cycling, it has been suggested that minimizing the particle size along with the use of binders that can better accommodate volume changes, such as sodium carboxymethylcellulose (NaCMC) [42] or styrene butadiene rubber [58]. Other approaches would be to use an inert matrix such as a transition metal that alloys with tin (Fe, Cu, Co, Mn, etc.) [59], Li_2O as in the case of SnO_2 [60], or even a carboneous matrix such as nanotubes or graphene [61].

Tin oxide (SnO_2) reacts with lithium via a two-step reaction: First it is irreversibly reduced to tin embedded in a Li_2O matrix, and then tin reversibly alloys with lithium [62], as shown by the following equations:

$$4Li^+ + SnO_2 + 4e^- \rightarrow Sn + 2Li_2O(707\ \text{mAh g}^{-1}) \quad (6.3)$$

$$Sn + xLi^+ + xe^- \leftrightarrow SnLi_x \text{ with } 0 \leq x \leq 4.4(783\ \text{mAh g}^{-1} \text{ of } SnO_2) \quad (6.4)$$

As a result, the Li_2O matrix prevents severe volume change and keeps tin metal in nano-sized form. Even though building up the Li_2O helps to mitigate the volume expansion associated to the alloying reaction of Sn with lithium, it is usually not enough to prevent the degradation of the electrode. In addition, the reduction of SnO_2 to tin metal leads to an irreversible capacity of 707 mAh g^{-1}, which represents about half of the electrode capacity. Preparations of graphene/SnO_2 and graphene/tin nanocomposite materials have been explored to reduce the negative effects of the volume change that occurs during cycling (Table 6.3).

It is mostly tin oxide that has been studied in a composite with graphene [13, 61, 63, 64, 66]; up to now, only one publication reports the use of elemental tin and graphene nanocomposite. In the case of SnO_2, usually large first irreversible capacities are observed due to the irreversible reduction of SnO_2 to Sn and also the irreversible capacity associated with graphene. This is illustrated by Fig. 6.13b where an initial discharge capacity of 1,800 mAh g^{-1} and reversible discharge capacity of 665 mAh g^{-1} (at 50 mA g^{-1}) after 50 cycles were observed for the graphene/SnO_2 nanocomposite [63]. The as-prepared graphene gave a capacity of 380 mAh g^{-1} after 100 cycles. Figure 6.13a shows a TEM micrograph of the nanocomposite made of SnO_2 nanoparticles of 2–4 nm evenly distributed on the graphene sheets [63]. The nanocomposite was prepared via reflux of $SnCl_2$ and GO at 190°C in ethylene glycol which was then annealed (500°C, Ar, 2 h).

An interesting study was performed by Yao et al. where they demonstrated the advantage of a graphene/SnO_2 nanocomposite over pristine SnO_2 nanoparticles and pristine graphene. Figure 6.14a–d shows four TEM micrographs of the graphene/SnO_2 nanocomposite prepared using $SnCl_2$ as the precursor and $NaBH_4$ as a reducing agent [64]. The nanocomposite is made of SnO_2 nanoparticles of 4–6 nm uniformly distributed on the graphene sheets. Figure 6.14c–d shows HRTEM

Table 6.3 Graphene/SnO_2 and graphene-/Sn-based nanocomposites preparation methods and capacity values

Active material Graphene/GO source	Preparation of composite material	C/SnO_2 C/Sn ratio	Particle size (nm)	Irr. cap. Rev. cap. (mAh g^{-1})	Number of cycles	Current density (mA g^{-1})	Voltage range (V vs. Li/Li^+)	Ref.
SnO_2 GO prepared by modified Hummers method	– GO and $SnCl_2$ in ethylene glycol and H_2O. – Reflux 500°C, Ar	22/78	2–4	~1,800 665	50	50	2.0–0.005	[63]
SnO_2 GO prepared by modified Hummers method	– $SnCl_2$ and GO – addition of citric acid and $NaBH_4$ – Heated at 120°C	60/40	4–6	1,420 520	100	55	3.0–0.1	[64]
SnO_2 Graphene produce by arc-discharge evaporation of graphite in a NH_3-He mixed atmosphere [65]	– NaOH solution is added to an aqueous solution of $SnCl_4$ – The solution is then mixed with a graphene dispersion in ethylene glycol – Addition of H_2SO_4 – Precipitate is to 400°C under argon		~4	~675 403	60	130, 450, 1,400, 6,000, 8,000, 450	3.0–0.05	[61]
SnO_2 GO prepared by modified Hummers method	– Hydrolysis of $SnCl_4$ with NaOH – Mixing of graphene and SnO_2 in ethylene glycol – Heat treatment at 400°C in Ar	40/60	~5.4	~1,900 570	30	50	2.0–0.05	[13]
SnO_2 Chemical exfoliation of graphite	– GO and HCl in H_2O – Addition of $SnCl_2$ – Heat treatment at 400°C in Ar	2.4/97.6	3–5	~875 ~590	50	400	2.0–0.01	[66]

Sn@C Thermal exfoliation of graphite oxide	– Graphene in ethylene glycol – $SnCl_4$ and NaOH in H_2O – Two solutions mixed – Hydrothermal carbon coating (glucose), 180°C – Annealing 800°C in N_2/H_2	46.9/ 53.1	50–200	846 660	100	100	3.0–0.01	[67]
SnO_2 GO prepared by modified Hummers method	– GO and $SnCl_2$ in isopropanol – Ultrasonification for 30 min – Annealing at 500°C in N_2	25/75	20	1,700 775	50	100	3.0–0.005	[68]
SnO_2 GO prepared by modified Hummers method	– $SnCl_4$, urea and GO in H_2O – Ultrasonication for 30 min – Microwaved 60 s – Addition of hydrazine – Microwaved 60 s – Annealing 500°C in N_2	20/80	100–200	2,140 650	30	50	2.0–0.005	[69]
SnS_2 GO prepared using modified Hummers method	– L-cysteine and $SnCl_4{\cdot}5H_2O$ dissolved in water – Addition of GO – Heated in autoclave 180°C – Product filtered and washed		Curved nanosheets ~2 μm diameter	1,664 920	50	100	1.5–0.01	[70]

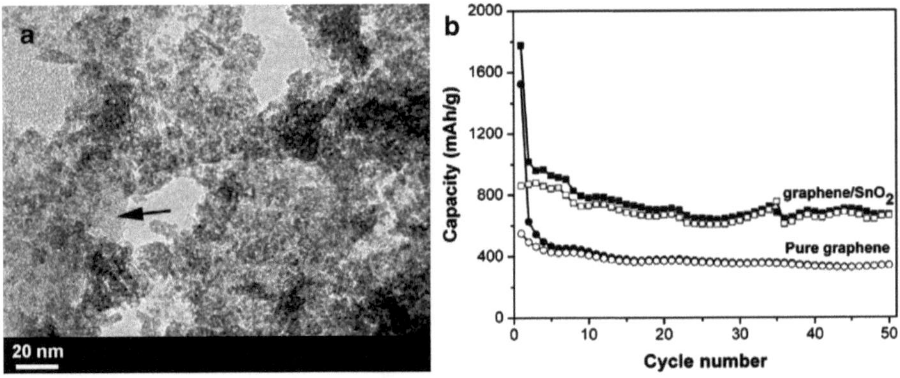

Fig. 6.13 (**a**) TEM micrograph of the graphene/SnO_2 nanocomposite [63]. (**b**) Discharge (*solid*) and charge (*hollow*) capacity versus cycle number for a graphene/SnO_2 nanocomposite and pure graphene nanosheets (Reprinted from [63], Copyright (2011), with permission from Elsevier)

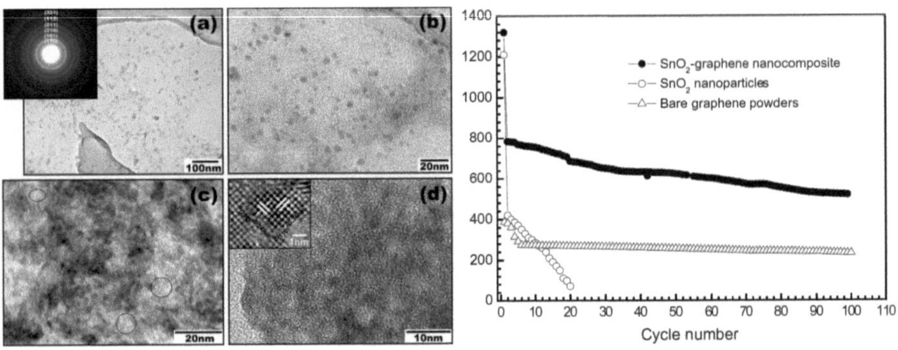

Fig. 6.14 (**a**) Low-magnification TEM micrograph of the graphene/SnO_2 nanocomposite. (**b**) High-magnification TEM micrograph of graphene/SnO_2 nanocomposite (**c**) HRTEM micrograph of graphene/SnO_2 nanocomposite showing. (**d**) Lattice-resolved HRTEM image of graphene/SnO_2 nanocomposite. (**e**) Cycling performance of SnO_2/graphene nanocomposite electrode, bare SnO_2 nanoparticle electrode, and bare graphene electrode (Reprinted from [64], Copyright (2011), with permission from Elsevier)

micrograph of graphene/SnO_2 nanocomposite and lattice-resolved HRTEM image of graphene/SnO_2 nanocomposite, in which the lattices of SnO_2 nanoparticles and graphene nanosheets are clearly visible. The inset is an atomically resolved lattice image of a SnO_2 nanoparticle, from which two perpendicular crystal planes, (110) and (200), can be distinguished. As previously mentioned, for all SnO_2 containing electrodes, a large first irreversible capacity ranging from 500 to 900 mAh g^{-1} is observed. After 100 cycles, the nanocomposite shows a capacity of 520 mAh g^{-1}, whereas the capacity of the pristine SnO_2 nanoparticles declines very rapidly with a capacity of less than 50 mAh g^{-1} after 20 cycles. The controlled experiment

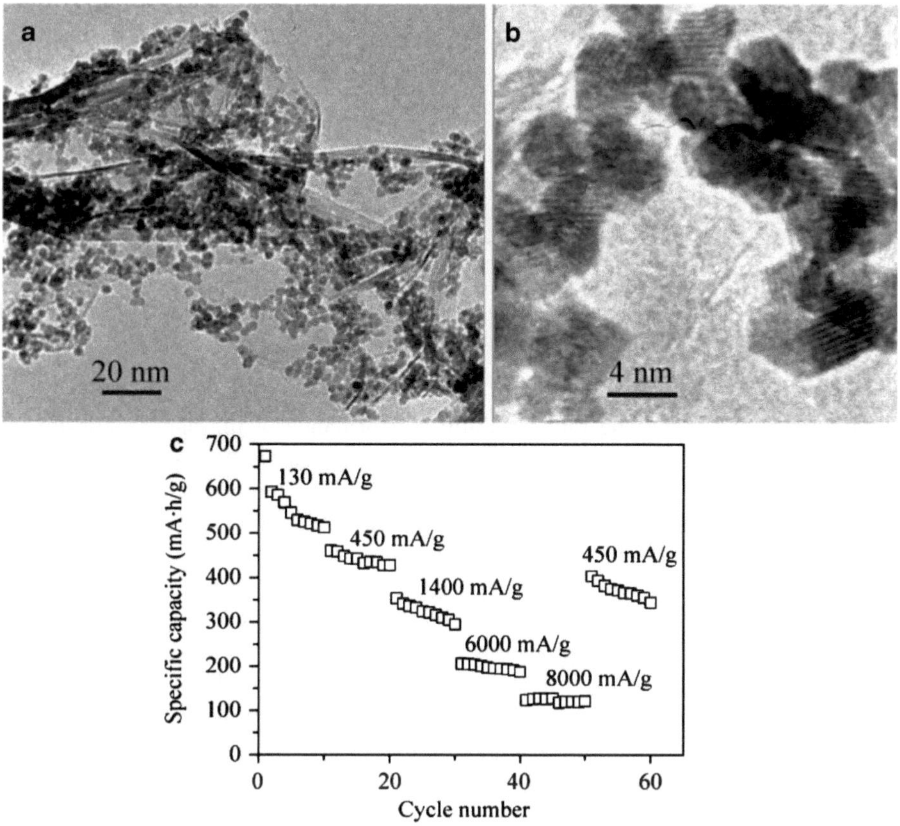

Fig. 6.15 (**a**) Low-magnification (*top left*) and (**b**) high-magnification TEM micrographs of laterally confined graphene/SnO_2 composites (*top right*). (**c**) Cycle performance of the graphene/SnO_2 composites at various charge–discharge current densities (*bottom*) (Reprinted from [61]. Copyright (2010), with permission from Springer)

showed that the graphene alone provides a capacity less than 300 mAh g^{-1} after 100 cycles.

Graphene is known for its high electronic conductivity which is a good add-on for obtaining high rate capability with active materials that are poor electronic conductors. Wang et al. performed these rate capability measurements on a graphene/SnO_2 nanocomposite [61]. They prepared the GNS via arc-discharge evaporation of graphite in a NH_3/He atmosphere. The SnO_2 nanoparticles were prepared by the hydrolysis of $SnCl_4$, and the resulting hydrosol was combined with an ethylene glycol dispersion of GNS. The isolated powder was then annealed (400°C, Ar, 2 h) [61].

Figure 6.15a–b shows TEM micrographs of the nanocomposite which are quite similar to the previous examples with well-distributed 4 nm SnO_2 nanoparticles on GNS.

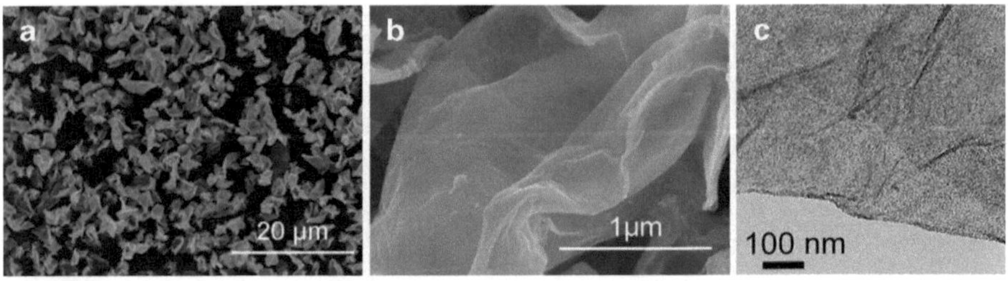

Fig. 6.16 SEM and TEM micrographs of the spray-dried graphene/SnO_2 nanocomposite, (**a**) low-magnification SEM, (**b**) high-magnification SEM, and (**c**) high-magnification TEM (Reprinted from [66], Copyright (2011), with permission from Elsevier)

Figure 6.15c shows the rate capability performance of the nanocomposite for various current densities. The performance is pretty impressive: At 130 mA g^{-1}, a stable capacity of about 515 mAh g^{-1} was observed, whereas at current density as high as 8,000 mA g^{-1}, a capacity value of about 100 mAh g^{-1} was measured.

Two other papers reported the synthesis and the use of a graphene/SnO_2 nanocomposite as anode materials [13, 66]. One was prepared from GO and $SnCl_2$, and upon cycling, it was found that an optimal graphene to SnO_2 ratio of 1:3.2 provided the best performance with a charge capacity of about 590 mAh g^{-1} after 50 cycles at 400 mA g^{-1} [66]. Surprisingly enough, a relatively low irreversible capacity is obtained. Figure 6.16 shows SEM and TEM micrographs of this composite. The other one was reported by Paek et al. who combined chemically reduced GO and in situ synthesized SnO_2 nanoparticles (ca. 5 nm) made from $SnCl_4$, in ethylene glycol, and then annealed (400°C, Ar, 2 h) [13]. TEM images of this material show SnO_2 nanoparticles well dispersed between graphene sheets (see Fig. 6.17). A comparative study on the performance of bare SnO_2 nanoparticle, graphite, and graphene was also performed by Paek et al. [13] The graphite and graphene exhibit capacities slightly above 250 mAh g^{-1} after 30 cycles, while the bare SnO_2 nanoparticles have a capacity of almost zero after only 15 cycles. The composite made of a 1:1.5 ratio of graphene to SnO_2 yields a first irreversible capacity of ca. 1,900 mAh g^{-1}, whereas a reversible capacity of 810 mAh g^{-1} on the second cycle and 570 mAh g^{-1} after 30 cycles is observed at a cycling rate of 50 mA g^{-1} (see Fig. 6.18).

The major issue that prevents the use of SnO_2 for commercial lithium-ion batteries is its unavoidable irreversible capacity associated with the first lithiation reaction that represents about half of SnO_2 theoretical capacity. In this sense, elemental tin is more interesting in this regard with a theoretical capacity of 993 mAh g^{-1} or 7,313 mAh mL^{-1}. A graphene/Sn nanocomposite material was prepared from a dispersion of graphene and carbon-coated SnO_2 which was then reduced (800°C, N_2/10% H_2, 12 h). As shown in Fig. 6.19a, a graphene and carbon-coated Sn nanocomposite was obtained. It was tested in half-cells and showed a capacity of 660 mAh g^{-1} (100 mA g^{-1}) after 100 cycles and very good rate capability (see Fig. 6.19b) [67]. A similar nanocomposite made of GNS and Sn

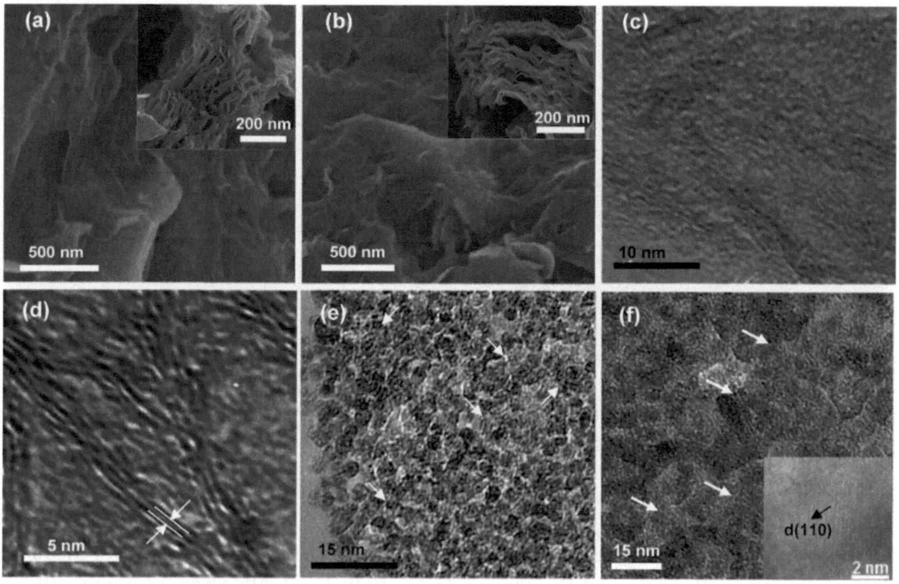

Fig. 6.17 SEM and TEM micrographs of graphene and graphene/SnO_2 nanocomposite, SEM micrographs for (**a**) graphene and (**b**) graphene/SnO_2 nanocomposite cross-sectional TEM micrographs for (**c**) graphene, (**d**) graphene (high magnification), (**e**) as-prepared graphene/SnO_2 nanocomposite, and (**f**) heat-treated graphene/SnO_2 nanocomposite. The *white arrows* denote the graphene nanosheets (Reprinted with permission from [13], Copyright (2009) American Chemical Society)

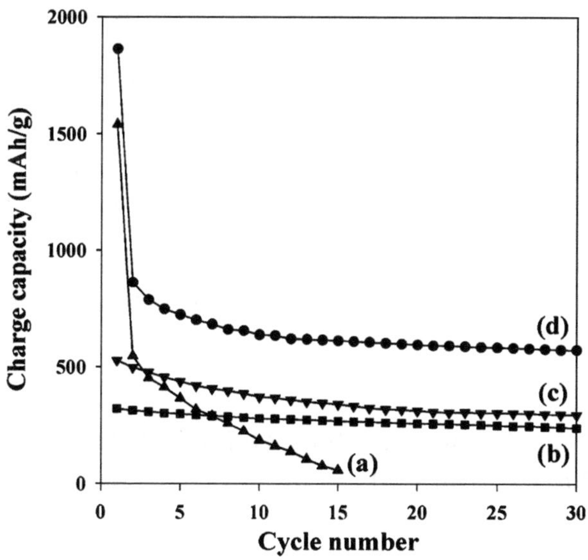

Fig 6.18 Cyclic performances for (**a**) bare SnO_2 nanoparticle, (**b**) graphite, (**c**) graphene, and (**d**) graphene/SnO_2 nanocomposite (Reprinted with permission from [13], Copyright (2009) American Chemical Society)

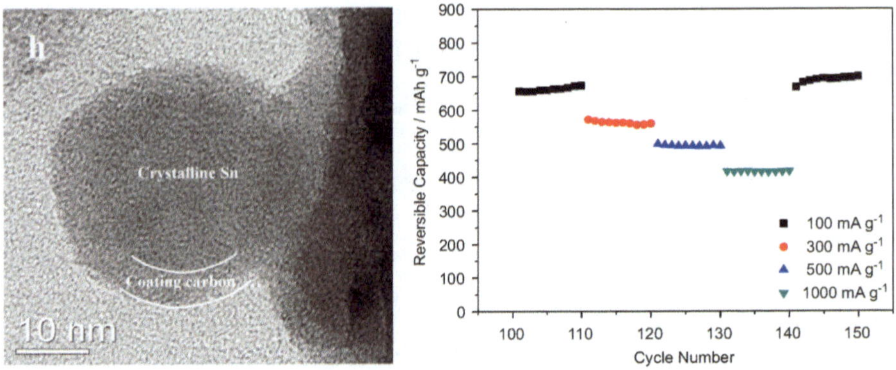

Fig. 6.19 TEM micrograph of the Sn-coated nanoparticle anchored on the graphene (*left*). Rate capability of the nanocomposite of graphene/Sn carbon-coated nanoparticles (*right*) (Reprinted from [67], Copyright (2011), with permission from Elsevier)

nanoparticles chemically reduced with $NaBH_4$ was prepared by Wang et al. [71]. The observed well-dispersed Sn nanoparticles between GNS showed a capacity of 508 mAh g^{-1} (55 mA g^{-1}) after 100 cycles. The performance of the nanocomposite was better than the bare graphene and pristine Sn powder. However, a large first irreversible capacity of 450 mAh g^{-1} was also observed. Finally a composite made of GNS (10–20 layers) and carbon-coated Sn-Sb nanoparticles (50–150 nm) provided a very impressive capacity retention with a capacity of about 700 mAh g^{-1} at 2C (1,600 mA g^{-1}) [72].

SnS_2/graphene hybrid materials were prepared and tested in lithium-ion batteries by Chant et al. [70]. SnS_2 reacts with Li^+ in the same way SnO_2 does, (Eqs. 6.3 and 6.4) producing Li_2S and $SnLi_x$. The authors believe that since SnS_2 is structurally and morphologically analogous to graphene, it would be more compatible for the preparation of nanocomposite than SnO_2 particles, leading to high capacities with good cycle stability. SEM images of the SnS_2/graphene are presented in Fig. 6.20. The images show a three-dimensional architecture made of curved nanosheets. The performance of the composite in lithium-ion batteries was evaluated. The first discharge capacity for the SnS_2/graphene composite was 1,664 mAh g^{-1} and dropped down to 920 mAh g^{-1} after 50 cycles and appears to be stable (Fig. 6.20).

6.4 Graphene/Transition Metal Oxide (TMO) Materials (Mn, Fe, Co, and Cu)

In 2000, Poizot et al. were the first to introduce the concept of using TMOs as anode materials for lithium-ion batteries [73]. TMOs have the advantage of delivering large theoretical capacities, usually ranging from 600 to 1,000 mAh g^{-1} [35] which

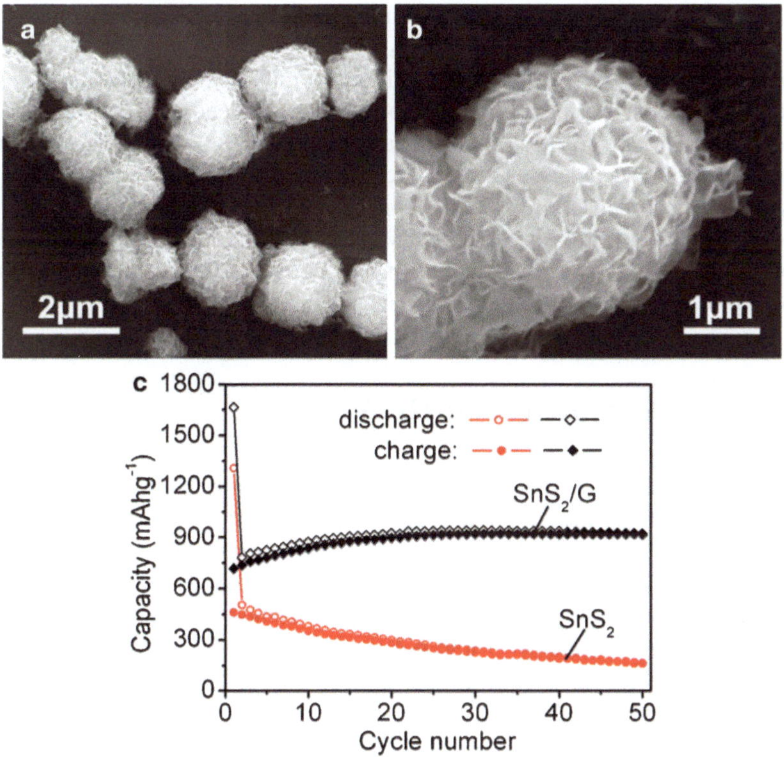

Fig. 6.20 (**a**) and (**b**) SEM images of SnS_2/graphene composite and (**c**) cycling performance of SnS_2/graphene composite at 100 mA g^{-1} (Reprinted with permission from Journal of Power Sources, doi:10.1016/j.jpowsour.2011.10 Copyright (2011) Springer)

translates to volumetric capacities ranging from 3,000 to 5,000 mAh mL^{-1} (assuming an average density of ~5 g mL^{-1} [74]). They react with lithium via an unusual way. They undergo a conversion reaction in which the TMO is converted into metallic nanoparticles embedded in a matrix of Li_2O as in the following example:

$$NiO + 2Li^{+} + 2e^{-} \leftrightarrow Ni + Li_2O\left(718 \text{ mAh g}^{-1}\right) \tag{6.5}$$

However, unlike in the case of the SnO_2, the Li_2O matrix formed during the first lithiation is in this case decomposable upon delithiation, and the metallic nanoparticles are oxidized back into TMO nanoparticles. This solid-state oxidation occurs only at nanometric scale. It is believed that the size confinement of the metallic nanoparticles enhances their electrochemical activity toward the decomposition of the Li_2O matrix [73]; a similar phenomenon has also been observed with LiF for transition metal fluorides (TMFs) [75].

However, the use of TMOs lithium-ion batteries is limited by their low intrinsic electrical conductivity (i.e., Mn_3O_4: 2×10^{-7} S cm^{-1} [76], Cr_2O_3: 1.8×10^{-7} S cm^{-1} [77]). This observation generally applies for all TMOs which exhibit typical insulator or semiconductor behavior; indeed they exhibit band gaps ranging from 3 to 4 eV [78]. Graphene can potentially serve as an excellent matrix due to its tunable surface area, mechanical flexibility, and high electrical conductivity that can compensate for the low conductivity of TMOs and lead to improved capacity retention [79–82]. Several TMOs ranging from manganese oxides to copper oxides have been investigated in nanocomposites with graphene, and the following section focuses on their batteries performance.

6.4.1 Graphene/Manganese Oxide

Manganese oxides are attractive materials for lithium-ion batteries due to the high abundance of manganese [48] and its low cost but more importantly because of its very low oxidation potential (1.2–1.3 V vs. Li/Li^+) compared to other TMOs [83]. The anode oxidation potential (or delithiation potential) is very important for a battery since it will dictate the battery output voltage; the lower the potential, the larger the output voltage. Table 6.4 summarizes the preparation method and the performance of graphene/manganese oxide nanocomposites prepared by several research groups around the world.

Two different reversible processes with lithium have been reported in the literature for Mn_3O_4 [84, 85]. According to Fang et al., first insertion of lithium occurs to form $LiMn_3O_4$ as shown by Eq. 6.6. A second reaction with lithium then takes place to obtain MnO, as shown by Eq. 6.7. It has been shown that once MnO is obtained, reactions (6.6) and (6.7) are not reversible anymore. From MnO, the reversible conversion reaction occurs, as shown by Eq. 6.8; the latter observation has been demonstrated by high-resolution TEM and selected area electron diffraction (SAED) measurements [84]. A second path has also been proposed by Gao et al. where they state that the reaction shown by Eq. 6.9 is totally reversible [85].

$$Mn_3O_4 + Li^+ + e^- \rightarrow LiMn_3O_4 \tag{6.6}$$

$$LiMn_3O_4 + Li^+ + e^- \rightarrow 3MnO + Li_2O \tag{6.7}$$

$$MnO + 2Li^+ + 2\,e^- \leftrightarrow Mn + Li_2O \tag{6.8}$$

$$Mn_3O_4 + 8Li^+ + 8e^- \leftrightarrow 3Mn + 4Li_2O \tag{6.9}$$

The use of manganese oxides in lithium-ion batteries has been limited by the low intrinsic electrical conductivity (2×10^{-7} S cm^{-1}) [76]. Wang et al. [79] and Lavoie et al. [80] improved this issue by preparing graphene/Mn_3O_4 nanocomposites whereas Xing et al. [81] and Yu et al. [82] investigated graphene/MnO_2 nanocomposites.

Table 6.4 Graphene-/manganese oxide-based nanocomposites preparation methods and capacity values

TMO and graphene/GO source	Preparation of composite material	C/TMO wt. ratio	Particle size (nm)	Irr. cap. Rev. cap. (mAh g^{-1})	Number of cycles	Current density (mA g^{-1})	Voltage range (V vs. Li/Li^+)	Ref.
Mn_3O_4	– GO and MnAc. in MF/H_2O (10:1) at 80°C	10/90	10–20	1,300 810	5	40	3.0–0.1	[79]
GO prepared by modified Hummers method	– Hydrothermal 180°C							
Mn_3O_4	– Graphene, MnOOH, and sodium dodecyl sulfate are suspended in water	35/65	200–450	*(A)* ~ 1,180 ~680 *(B)* ~ 1,370 ~720	100	75	3.0–0.01	[80]
(A) GO prepared by modified Hummers method and reduced under H_2 atmosphere or *(B)* commercial graphene platelets	– Mixture is sonicated for 20 min, filtered, and washed with water – Annealed at 450°C for 1 h under argon atmosphere							
MnO_2	– Graphene and sodium dodecyl sulfate in H_2O	N/A	N/A	1,400 635.5	30	C/10	3.0–0.005	[81]
Graphene prepared via solvothermal reaction with Na and EtOH	– Addition of Na_2SO_4 and $KMnO_4$ – Precipitate washed and dried							
MnO_2	– MnO_2 NT prepared via hydrothermal reaction using $KMnO_4$	50/50	Length: 1,000	1,250 495	40	100	3.0–0.01	[82]
GO prepared by modified Hummers method	– Assembly of the electrode layer-by-layer		Diameter: 70–80 Wall thickness: 30					

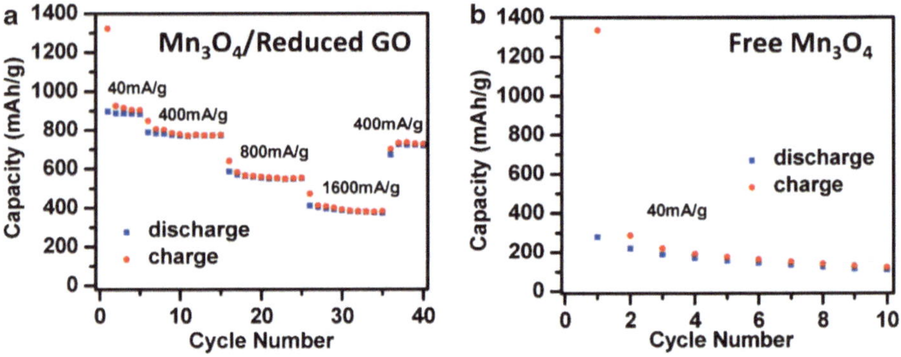

Fig. 6.21 Electrochemical characterizations of a half-cell composed of graphene/Mn_3O_4 and Li. The specific capacities are based on the mass of Mn_3O_4 in the graphene/Mn_3O_4 nanocomposite. (**a**) Capacity retention of the graphene/Mn_3O_4 nanocomposite at various current densities. (**b**) Capacity retention of free Mn_3O_4 nanoparticles without graphene at a current density of 40 mA g^{-1} (Reprinted with permission from [79]. Copyright (2010) American Chemical Society)

Wang et al. prepared a nanocomposite made of Mn_3O_4 nanoparticles (ca. 10–15 nm) grown directly on RGO. This nanocomposite showed a reversible capacity of 810 mAh g^{-1} when cycled at 40 mA g^{-1}, and at a higher cycling rate, a capacity of 730 mAh g^{-1} was retained after 40 cycles at 400 mA g^{-1} (see Fig. 6.21a). An initial irreversible capacity of ca. 400 mAh g^{-1} was observed, which represents about 30% capacity loss. It is interesting to note that the electrical conductivity of Mn_3O_4 is much lower than that of cobalt or iron oxides [76], yet the graphene nanocomposite architecture seems to address this limitation, as shown by Fig. 6.21b, which may in part be due to the nanoscale size of the Mn_3O_4 particles and the high electrical conductivity of graphene [86]. We prepared similar composites made of graphene platelets or RGO and Mn_3O_4 needles [80]. We demonstrated the advantage of using a composite with graphene over pristine Mn_3O_4 needles. A capacity of 720 mAh g^{-1} was obtained when using graphene platelets (XGM-5) and 675 mAh g^{-1} when using RGO; half-cells were cycled at 75 mA g^{-1} for 100 cycles (see Fig. 6.22).

As depicted by Ref. [84], MnO_2 is known for having a large irreversible capacity due to the thick SEI that forms on manganese and also due to the fact that the starting oxidation state of manganese is +4. Xing et al. reported a capacity of 718 mAh g^{-1} at C/20 after 30 cycles [81]. An irreversible capacity of about 50% is observed. Yu et al. made a comparative study on the performance of freestanding MnO_2 nanotubes and a free-standing porous nanocomposite made of an intimate mixture of MnO_2 nanotubes and graphene [82]. They clearly showed the advantage of having graphene as a conductive matrix. The nanocomposite had a capacity of 500 mAh g^{-1} after 40 cycles when cycled at 100 mA g^{-1}, while pristine MnO_2 nanotubes showed a capacity of about 150 mAh g^{-1}.

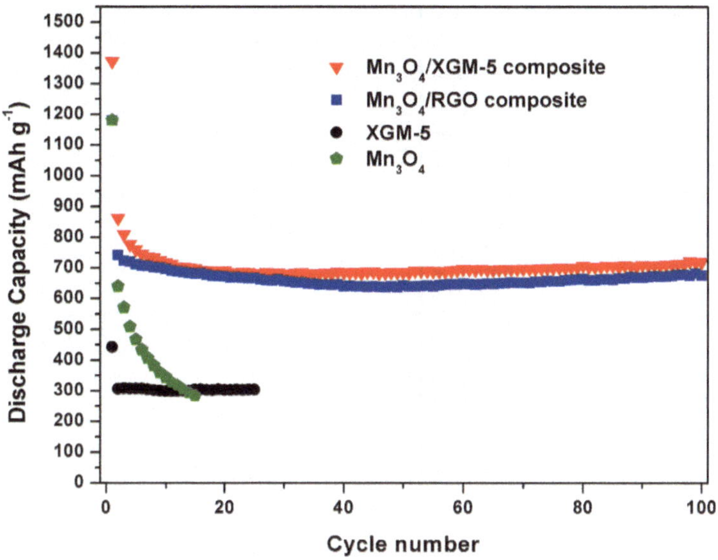

Fig. 6.22 Cycling performance of Mn_3O_4/graphene composites (0.01–3.0 V), XGM-5 (0.005–2 V), and Mn_3O_4 (0.01–3 V) cycled at 75 mAh g^{-1} with LiCMC binder [80] (Reprinted with permission from Elsevier)

6.4.2 *Graphene/Iron Oxide*

Iron oxide is one of the most interesting TMOs. Iron oxides have an oxidation potential of about 1.8 V versus Li/Li^+; the latter is a bit higher than manganese oxide but lower than cobalt oxide. Iron is naturally very abundant [48] and is quite inexpensive. In addition, it shows high theoretical capacities, about 1,000 mAh g^{-1}, and lower irreversible capacity than other TMOs [87], but like many other materials, one of the problems with iron oxide is the pulverization due to volume change upon cycling. No graphene/FeO nanocomposite has been prepared up to now; however, several graphene/Fe_2O_3 [88, 89] and graphene/Fe_3O_4 [90–95] have been reported in the literature. Table 6.5 summarizes the preparation method and the performance of these graphene/iron oxide nanocomposites. The electrochemical reaction of iron oxides with lithium is presented by Eqs. 6.10 and 6.11:

$$Fe_3O_4 + 8e^- + 8Li^+ \leftrightarrow 3Fe^0 + 4Li_2O \qquad 926\ \text{mAh g}^{-1} \tag{6.10}$$

$$Fe_2O_3 + 6e^- + 6Li^+ \leftrightarrow 2Fe^0 + 3Li_2O \qquad 1{,}007\ \text{mAh g}^{-1} \tag{6.11}$$

Two graphene/Fe_2O_3 nanocomposites have been reported. The first one used GO reduced by hydrazine and Fe_2O_3 nanoparticles (60 nm) made via a microwave-assisted method [88]. As shown in Fig. 6.23, Zhu et al. demonstrated the advantage

Table 6.5 Graphene-/iron oxide-based nanocomposites preparation methods and capacity values

TMO and graphene/GO source	Preparation of composite material	C/TMO wt. ratio	Particle size (nm)	Irr. cap. Rev. cap. (mAh g^{-1})	Number of cycles	Current density (mA g^{-1})	Voltage range (V vs. Li/Li^+)	Ref.
Fe_2O_3 GO prepared by modified Hummers method	– $FeCl_3$ and urea in H_2O – Addition of GO and ultrasonication – Addition of hydrazine – Microwaved for 2 min.	20/80	60	1,693 1,027	50	100	3.0–0.005	[88]
Fe_2O_3 Graphene prepared via thermal reduction of GO under H_2	– $FeCl_3$,ascorbic acid, PEG-6000, and urea in H_2O – Addition of graphene and ultrasonication – Hydrothermal 120°C	30/70	2,000–5,000	1,800 660	100	160	3.0–0.005	[89]
Fe_3O_4 GO prepared by modified Hummers method	– Graphene, $(NH_4)_2Fe(SO_4)$ •$6H_2O$, and $NH_4Fe(SO_4)_2$• $12H_2O$ in H_2O – Addition of NH_4OH – Annealing 600°C in Ar	5.5/94.5	15–30	1,237 857	20	46.3	2.8–0.002	[90]
Fe_3O_4 GO prepared by modified Hummers method	– $FeCl_2$ and GO in H_2O – Addition of NaOH, pyrenebutyric acid, and hydrazine	38/62	3–15	1,100 650	100	100	3.0–0.0	[91]

	– Hydrothermal 150°C							
Fe_3O_4 Graphene prepared via by rapid thermal expansion in N_2	– $Fe(NO_3)_3$, ethylene glycol, and graphene – Ultrasonication for 6 h – Addition of NH_4OH – Hydrothermal 180°C	22.7/77.3	12.5	1,400 1,048	90	100	3.0–0.01	[92]
Fe_3O_4 GO prepared by modified Hummers method	– GO, $Fe(NO_3)_3$, urea, and ascorbic acid in H_2O – Refluxed for 1 h with a microwave heater – Annealing 600°C in Ar	N/A	20–70	1,350 650	50	C/10	3.0–0.0	[93]
Fe_3O_4 Graphene prepared via thermal reduction of GO under Ar and then H_2	– Graphene and $FeCl_3$ in H_2O – Sonication for 1 h – Annealing 600°C in Ar	13.3/86.7	100–500	~1,400 1,026	30	35	3.0–0	[94]
Fe_3O_4 Graphene prepared via thermal reduction of GO under H_2	– $Fe(NO_3)_3$, NaAc, PEG-20000 in ethylene glycol – Addition of graphene – Hydrothermal 180°C	30/70	100–200	2,315 771	50	200	3.0–0.005	[95]

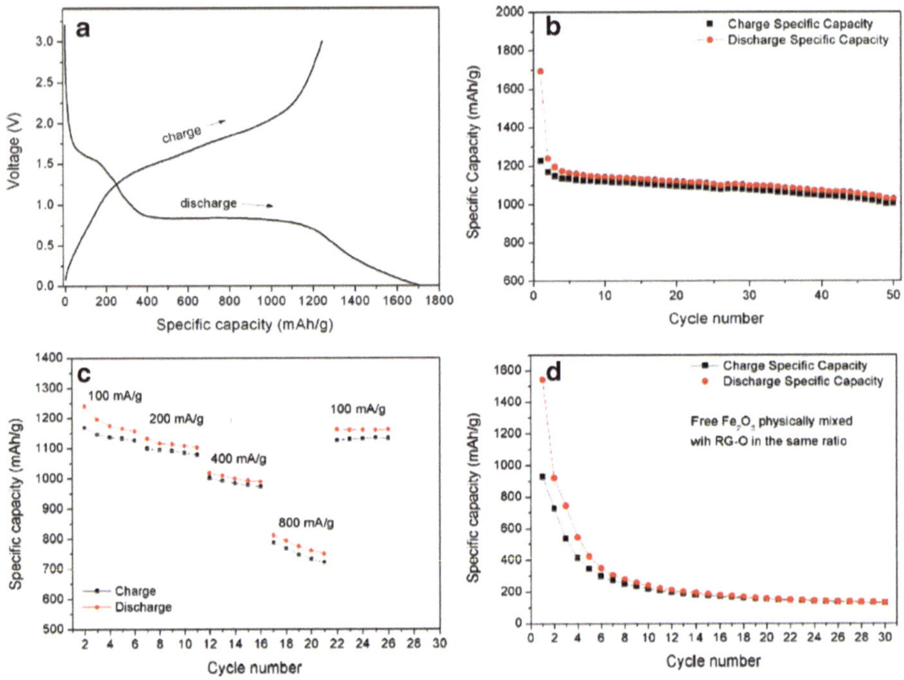

Fig. 6.23 Electrochemical performance of the RGO/Fe_2O_3 composite. (**a**) Voltage profile (discharge/charge) of RGO/Fe_2O_3 composite for the first cycle at the current density of 100 mA g^{-1}. (**b**) Cycling performance of RGO/Fe_2O_3 composite at the current density of 100 mA g^{-1}. (**c**) Rate capacity of RGO/Fe_2O_3 composite between 0.005 and 3.0 V with increasing current density. (**d**) Capacity retention of free Fe_2O_3 nanoparticles physically mixed with RGO platelets at a current density of 100 mA g^{-1} (Reprinted with permission from [88]. Copyright 2011 American Chemical Society)

of using an in situ–prepared graphene/Fe_2O_3 nanocomposite over pristine Fe_2O_3 particles or mechanical mixing of graphene and Fe_2O_3 particles. They reported a first discharge capacity of 1,693 mAh g^{-1} with an irreversible capacity of about 30%. After 50 cycles the nanocomposite showed a discharge capacity of 1,027 mAh g^{-1}, whereas the mechanically mixed composite showed a capacity below 150 mAh g^{-1}. The second composite prepared by Wang et al. [89] was made using GO reduced via H_2 and Fe_2O_3 agglomerates (2–5 μm) prepared via a hydrothermal method and gave similar behavior to the first composite. The specific capacities are based on the mass of Fe_2O_3 in the RGO/Fe_2O_3 composite.

An interesting study on graphene/Fe_3O_4 nanocomposites was performed by Zhou et al., and a very high reversible capacities were obtained [94]. In this example, FeOOH spindles (ca. 200 nm) were used to prepare a graphene/FeOOH precursor composite. The latter was annealed at 600°C in Argon and gave graphene/Fe_3O_4 nanocomposite (see Fig. 6.24a–d). The composite has a reversible capacity of 1,026 mAh g^{-1} after 30 cycles (at 35 mAh g^{-1}) and 580 mAh g^{-1} after

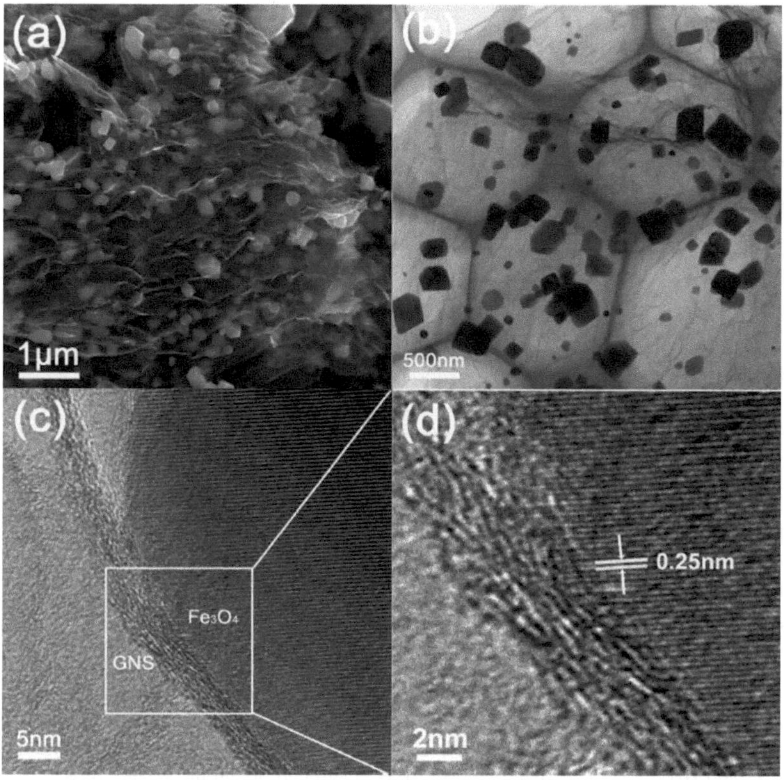

Fig. 6.24 (**a**) SEM micrograph of the cross-section of graphene/Fe_3O_4 nanocomposite, (**b**) TEM and (**c**) high-resolution TEM micrographs of graphene/Fe_3O_4 nanocomposite, and (**d**) detailed interface structure of the square area in (**c**) (Reprinted with permission from [94]. Copyright 2010 American Chemical Society)

100 cycles at 700 mA g^{-1} with an irreversible capacity of ca. 30% in the first cycle. Analysis of the iron oxide particle size after cycling reveals that in the graphene composite, the size of the particles is almost unchanged, whereas in the case graphene-free iron oxide particles, pulverization is obvious based on the fact that the particle size decreased by 50%.

Other groups observed similar behavior [90–93, 95]; for example, Wang et al. reported a nanocomposite having a reversible capacity of 771 mAh g^{-1} after 50 cycles at a current density of 200 mA g^{-1} [95]. Li et al. went further than just making half-cells; they actually made a full cell with the graphene/Fe_3O_4 nanocomposite as anode and a cathode of $LiNi_{1/3}Mn_{1/3}Co_{1/3}O_2$ [90]. The full cell was cycled between 1.2 and 3.2 V. As shown in Fig. 6.25a and b, the cell shows a working potential ranging from 2.8 to 1.6 V. Based on the cathode weight, after 10 cycles at C/10 and C/5 a capacity of 80 mAh g^{-1} was measured for this battery (see Fig. 6.25c and d).

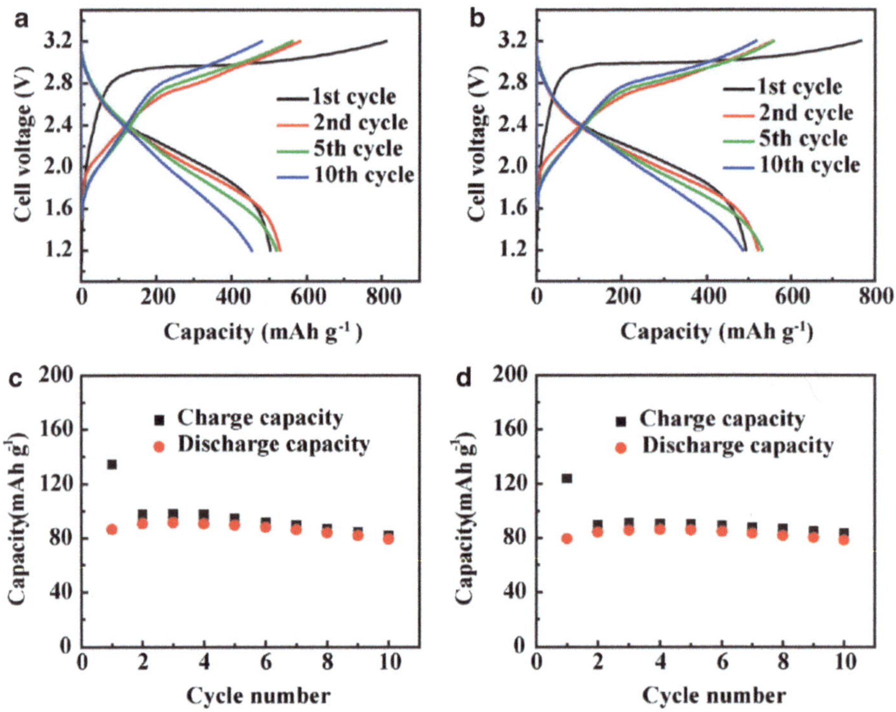

Fig. 6.25 Voltage profiles of graphene/Fe_3O_4 – $LiNi_{1/3}Mn_{1/3}Co_{1/3}O_2$/full cells at cycling rate of (**a**) 0.1 C and (**b**) 0.2 C. The specific capacity is calculated according to the mass of graphene/ Fe3O4 nanocomposites. The corresponding cycling performance curves in cycling rate of (**c**) 0.1 C and (**d**) 0.2 C, according to the $LiNi_{1/3}Mn_{1/3}Co_{1/3}O_2$ cathode weight (Reprinted from [90]. Copyright 2011, reproduced by permission of the PCCP Owner Societies)

6.4.3 Graphene/Co_3O_4

Co_3O_4 has been one of the most studied TMOs due to its high lithium-storage capacities and impressive performance [73]. Its electrochemical conversion reaction with lithium is $Co_3O_4 + 8Li^+ \leftrightarrow 4Li_2O + 3Co$. However, the large volume changes and aggregation of particles upon lithiation lead to limited cycling [96, 97]. Encapsulating or trapping the Co_3O_4 in a carbon material such as graphene was explored as a method of reducing the degradation of the electrodes [98–105]. Table 6.6 summarizes the preparation method and the performance of these graphene/Co_3O_4 nanocomposites as reported by several research groups around the world.

One example of a graphene/Co_3O_4 nanocomposite was reported by Wu et al. By annealing of a graphene/$Co(OH)_2$ precursor at 450°C in air, they made a nanocomposite of well-dispersed Co_3O_4 nanoparticles (10–30 nm) between graphene sheets made of three layers or less [98]. As shown by the TEM micrograph in Fig. 6.26a, Co_3O_4 nanoparticles are homogeneously attached to the

Table 6.6 Graphene-/Co_3O_4-based nanocomposites preparation methods and capacity values

Graphene/GO source	Preparation of composite material	C/TMO wt. ratio	Particle size (nm)	Irr. cap. Rev. cap. (mAh g^{-1})	Number of cycles	Current density (mA g^{-1})	Voltage range (V vs. Li/Li^+)	Ref.
Graphene prepared by H_2 reduction	– Graphene, $Co(NO_3)_2$, NH_4OH – Stirring in Ar – Annealing at 450°C in air	24.6/75.4	10–30	1,097 935	30	50	3.0–0.01	[98]
GO prepared by modified Hummers method	– Dispersion of GO and CoPc in H_2O by ultrasonication – Reduction GO by hydrazine – Annealing 800°C in Ar – Oxidation at 400°C in Air	34/66	20–30	1,200 754	30	74	3.0–0.01	[99]
GO	– GO and $CoCl_2$ in H_2O ultrasonicated – Addition of NaOH and H_2O_2 – Hydrothermal at 100°C – Addition of $NaBH_4$ – Hydrothermal at 120°C	10/90	10–50	941 740	60	200	3.0–0.001	[100]
GO prepared by modified Hummers method	– In EtOH: $Co(NO_3)_2$ and hexamethylenetetramine – Dispersion of GO – Microwaved at 180°C for 5 min in a sealed in glass tube	46.2/53.8	Sheets 100 × 1,000	~2,250 ~1,065	30	89	3.0–0.005	[101]
GO prepared by modified Hummers method	– Preparation of aminopropyltrimethoxysilane (APS)-modified Co_3O_4 nanoparticles – Addition of GO in the suspension – Reduction of with hydrazine	8.5/91.5	>500	1,700 1,000	130	74	3.0–0.01	[102]
GO prepared by modified Hummers method	– GO and CoAc in H_2O – Addition of NH_4OH and hydrazine – Annealing 200°C in Ar	30/70	5	1,550 800	42	200	3.0–0.001	[103]

(continued)

Table 6.6 (continued)

Graphene/GO source	Preparation of composite material	C/TMO wt. ratio	Particle size (nm)	Irr. cap. Rev. cap. (mAh g^{-1})	Number of cycles	Current density (mA g^{-1})	Voltage range (V vs. Li/Li^+)	Ref.
GO prepared by modified Hummers method	– GO and $Co(NO_3)_2$ in H_2O – Hydrothermal at 170°C – Annealing 350°C in air	10.7/89.3	Thickness 3–5	1,250 700	100	180	3.0–0.1	[104]
GO prepared by modified Hummers method	– GO and $CoCl_2$ in H_2O – Addition of $NaBH_4$ – Refluxed at 100°C – Annealing 200°C in air	45.4/54.6	15–25	1433 650	50	55	3.0–0.1	[105]
GO prepared by modified Hummers method	– GNS obtained via ultrasonication/exfoliation – $CoSO_4 \cdot 7H_2O$, urea dissolved in alcohol-water – GNS and 25% $NH_3 \cdot H_2O$ solution are added – Heated at 120°C for 12 h – Filtered and dried – Composite calcinated at 450°C under N_2 for 3 h	20/80	Rod shape 30 nm diameter 1–2 μm length	1,303 ~1,310	40	100	3.0–0.01	[106]

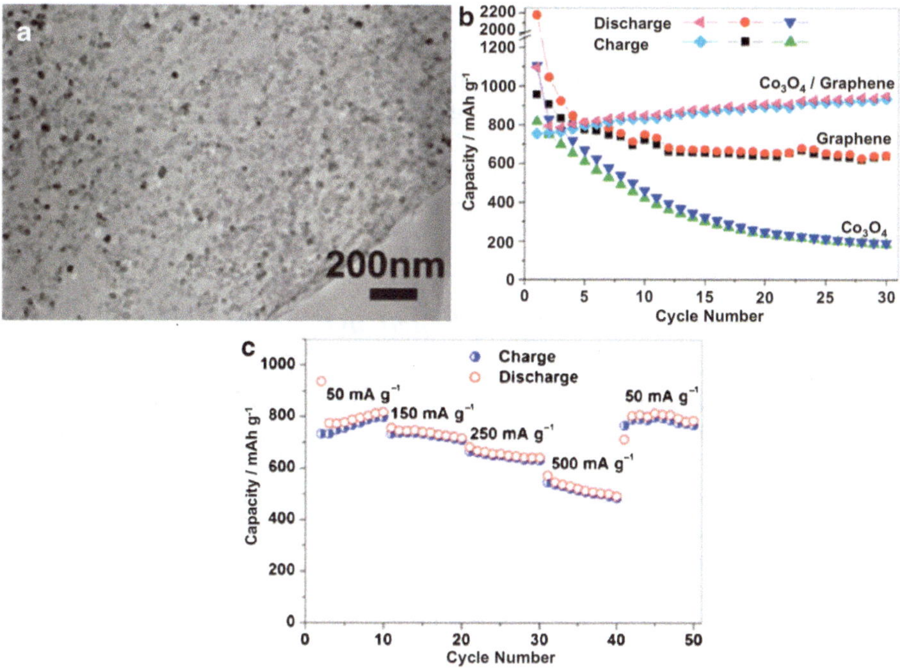

Fig. 6.26 (**a**) TEM micrograph of the graphene/Co_3O_4 nanocomposite; (**b**) cycling performance of graphene, Co_3O_4 nanoparticle, and the graphene/Co_3O_4 nanocomposite; (**c**) rate capability of the graphene/Co_3O_4 nanocomposite at various current densities between 50 and 500 mA g^{-1} (Reprinted with permission from [98]. Copyright (2011) American Chemical Society)

graphene sheets. Figure 6.26b shows a comparison of the performance of the nanocomposite versus bare graphene and bare Co_3O_4 nanoparticles. It is undeniable that graphene improves the performances of Co_3O_4 by reducing the agglomeration of particles upon cycling, by better accommodation of the volume changes, and also by improving the electrical and mechanical properties of the electrode. Here a reversible capacity of 935 mAh g^{-1} was obtained after 30 cycles for a battery cycled at 50 mA g^{-1}. A columbic efficiency of 98% was calculated. The performance of the graphene-free Co_3O_4 electrode showed a capacity of 184 mAh g^{-1} after 30 cycles. It is interesting to note that a graphene electrode provides a capacity of 638 mAh g^{-1} but with a large irreversible capacity. High-rate studies were also undertaken in which a capacity >500 mAh g^{-1} was obtained at 500 mA g^{-1} (see Fig. 6.26c). Because Co_3O_4 nanoparticles were not as well dispersed as those of Wu et al. [98], Yang et al. obtained slightly lower performance [99]; similar results were obtained by four other research groups [100, 103–105].

As shown in Table 6.6, better results were obtained by Chen and Wang when using a composite made of Co_3O_4 (0.1 × 1 μm) sheets. However this good performance has been associated with a large irreversible capacity of about 1,200 mAh g^{-1} [101]. Astonishing performance was reported by Yang et al. when using graphene-encapsulated Co_3O_4 particles [102]. A stable capacity of about 1,000 mAh g^{-1} was observed after 130 cycles at 74 mAg^{-1}. Even though a first irreversible capacity of about 500 mAh g^{-1} has been measured, the performance is impressive. It is believed that confining the Co_3O_4 particles in a graphene shell restrains the aggregation of nanoparticles and better accommodates the volume change that occurs upon cycling. In addition, it reduces the carbon content (increased capacity) and maintains a high electrical conductivity of the overall electrode [102].

6.4.4 Graphene/NiO and Graphene/CuO

Graphene/NiO and graphene/CuO are the two last graphene/TMO nanocomposites investigated. NiO reacts with lithium via the conventional reversible conversion reaction, as shown by Eq. 6.12. CuO interacts a bit differently with lithium, as shown by Eqs. 6.13 and 6.14. An insertion reaction followed by the reversible conversion reaction occurs during cycling (see Chap. 5 for more details). Table 6.7 summarizes the preparation method and the performance of these composites.

$$NiO + 2Li^+ + 2e^- \rightarrow Ni^0 + Li_2O \qquad 718\ \text{mAh g}^{-1} \tag{6.12}$$

$$Cu^{II}O + xLi^+ + xe^- \rightarrow \left[Cu^{II}{}_{1-x}Cu^{I}{}_{x}\right]O_{1-x/2} + x/2\,Li_2O_{(0\leq x\leq 1)} \qquad 674\ \text{mAh g}^{-1} \tag{6.13}$$

$$Cu^{I}{}_2O + 2Li^+ + 2e^- \rightarrow 2Cu^0 + Li_2O \qquad 674\ \text{mAh g}^{-1} \tag{6.14}$$

Only one graphene/NiO nanocomposite has been reported in the literature. It is made of graphene nanosheets (GNS) and NiO nanosheets [107]. Zou et al. made a sandwich of GNS and NiO nanosheets, as shown in Fig. 6.27b and c. They showed improved capacity retention when using this composite compared to bare NiO nanosheet or bare NiO nanoparticles, mostly due to the improved electrical conductivity and also the shorted path length for Li^+ transport. Conductivity values of 2.14×10^{-5} S cm^{-1} for the NiO nanosheets versus 1.36×10^{-3} S cm^{-1} for the nanocomposite were reported. The capacity performance is impressive; however, one should be careful with these values since the C to NiO ratio is quite high: 77.2:22.8. As shown by Fig. 6.27g, the reported capacity is about 1,050 mAh g^{-1} at 71.8 mA g^{-1} (C/10) after 40 cycles. This nanocomposite also showed good rate capability with capacities of 870, 660, and 500 at C, 2C, and 5C, respectively (see Fig. 6.27h).

CuO is a bit less interesting than other TMOs because of its lower capacity, 674 mAh g^{-1}, and like NiO it has a higher delithiation potential (over 2 V vs. Li/Li^+). However, it has a low cost and is environmentally benign and safe. Up to now, the four following references reported battery performance of graphene/CuO nanocomposites [108–111]. One of them has been reported by Wang et al. who prepared their graphene via an arc discharge method [65, 112] prior to the

Table 6.7 Graphene/NiO and graphene-/CuO-based nanocomposites preparation methods and capacity values

TMO and graphene/GO source	Preparation of composite material	C/TMO wt. ratio	Particle size (nm)	Irr. cap. Rev. cap. ($mAh\ g^{-1}$)	Cycles #	Current density ($mA\ g^{-1}$)	Voltage range (V vs. Li/Li^+)	Ref.
NiO Graphene prepared by modified Hummers method	– In H_2O: NiAc., hexamethylenetetramine, and cetyltrimethylammonium bromide – Addition of graphene – Hydrothermal 120°C – Annealing 360°C in Air	77.2/22.8	Thickness (500–2,000) × (30–50)	1,750 1,050	40	71.8	3.0–0.005	[107]
CuO Graphene prepared by arc discharge method	– Graphene and CuAc. dispersed in H_2O/EtOH – Addition of NH_4OH – Reflux for 2 h	26/74	1,500 Nanoplates 25 thick	850 650	100	65	3.0–0.01	[108]
CuO Graphene prepared by using modified Staudenmaier's method [9]	– GNS prepared by rapid heat treatment and ultrasonication – GNS and $Cu(NO_3)_2{\cdot}3H_2O$ dispersed in EtOH and ultrasonicated for 1 h – Reduction to Cu/graphene in $NaBH_4$ solution – Oxidation to CuO-HNPs/graphene in air at 300°C	35/65	50 Wall thickness 10	1,248 743	50	50	3.0–0.01	[109]

(continued)

Table 6.7 (continued)

TMO and graphene/GO source	Preparation of composite material	C/TMO wt. ratio	Particle size (nm)	Irr. cap. Rev. cap. (mAh g^{-1})	Cycles #	Current density (mA g^{-1})	Voltage range (V vs. Li/Li^+)	Ref.
CuO	– GO in H_2O and DMF	12/88	30	817	50	67	3.0–0.02	[110]
GO prepared by modified Hummers method	– Heating at 90°C							
	– Addition of CuAc.							
	– Heating at 90°C			423				
	– Hydrothermal 180°C							
	– GO reduction by hydrazine vapors at 50°C							
CuO	– GNS dispersed in H_2O	46/54	20–30	~2,150	40	70	3.0–0.01	[111]
GO modified Hummers method	– $CuCl_2$ is and solution heated to 60°C			~750				
	– Addition of NaOH							
	– Filtration							

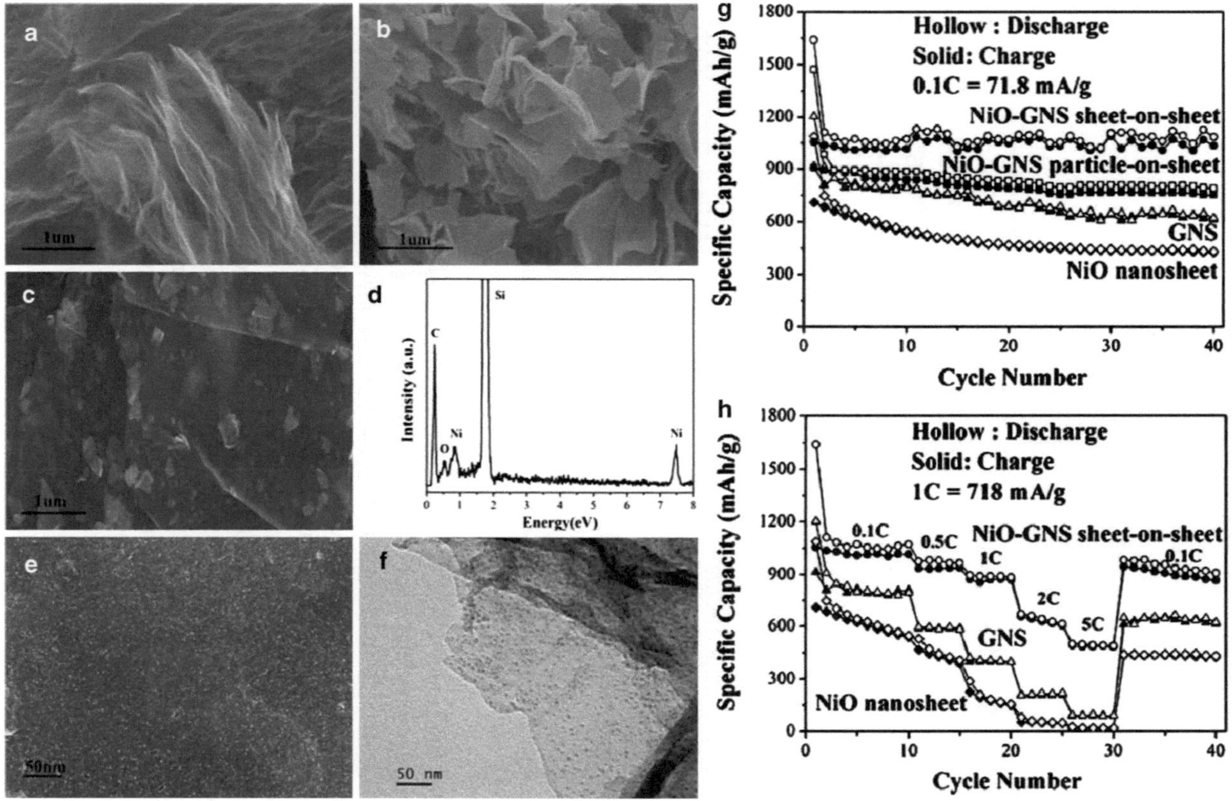

Fig. 6.27 SEM micrographs of (**a**) GNS; (**b**) NiO nanosheets; (**c**) GNS-/NiO-nanosheet nanocomposite; (**d**) the EDS spectrum of GNS-/NiO-nanosheet nanocomposite; (**e**) SEM image of GNS-/NiO-nanoparticle nanocomposite; (**f**) GNS-/NiO-nanoparticle nanocomposite. Electrochemical performances of GNS, NiO nanosheets, and GNS-/NiO-nanosheets and GNS-/NiO-nanoparticle nanocomposites; (**g**) cycling performances at 0.1 C; (**h**) cycling performances at stepwise increased current rates (Reprinted from [107]. Reproduced by permission of The Royal Society of Chemistry)

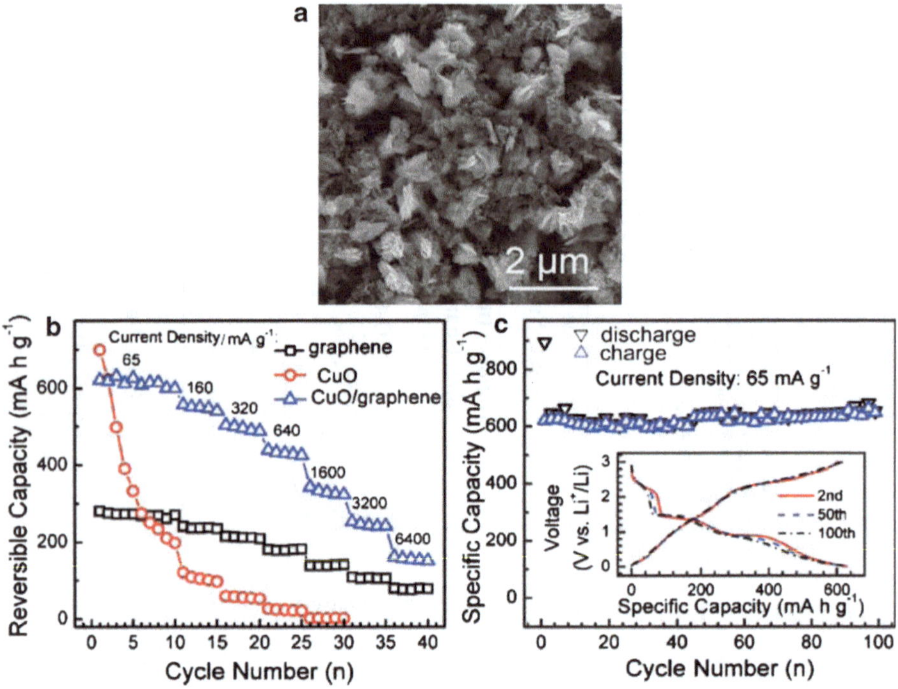

Fig. 6.28 (**a**) SEM micrograph of graphene/CuO nanocomposite. (**b**) Specific capacity of graphene/CuO nanocomposite, graphene, and CuO under different current density. (**c**) Discharge-charge capacity of the CuO/graphene up to 100 cycles under 65 mA g^{-1} and the corresponding 2nd, 50th, and 100th discharge-charge voltage profiles (Reprinted from [108]. Reproduced by permission of The Royal Society of Chemistry)

preparation of the graphene/CuO nanocomposite materials [108]. Figure 6.28 shows the SEM micrograph of the composite material made of graphene sheets and 1.5 μm urchin-like clusters themselves made of 25 nm nanoplates. Figure 6.28 shows cycling data for the CuO/graphene composite, bare CuO, and graphene. The results clearly show the advantage of the nanocomposite over bare CuO urchin-like clusters. A stable capacity of 650 mAh g^{-1} after 100 cycles at 65 mA g^{-1} was measured. Lu et al. also show the superiority of the bare CuO with capacities ~750 mAh g^{-1} of after 40 cycles at 70 mA g^{-1} [111]. Similar performance and observation were obtained by Mai et al. by using a composite of 30 nm CuO nanoparticles [110] and Zhou et al. who prepared a composite with 50 nm CuO hollow nanoparticles [109].

6.5 Summary and Outlook

Transition metal oxide, silicon, tin, and tin oxide have been studied as a replacement for graphite as anode materials in lithium-ion batteries because of their high theoretical capacities. Cycling of these materials however is often associated with large volume changes, giving rise to mechanical stresses that lead to cracks, eventual disintegration of the electrode, and a failure of the battery. Graphene can be utilized to encapsulate nanoparticles of these materials to mitigate the effect of the large volume expansion. Low intrinsic electrical conductivity has also limited the usefulness of some of these materials. The mechanical flexibility and high electrical conductivity of graphene makes it a good matrix to help compensate for the low conductivity of some TMOs, leading to improved capacity retention.

This chapter gives an overview of the work exploring graphene as a novel support for nanoscale materials that react with lithium and provide high capacities. In general, anodes made of graphene only do not look to be promising materials to replace graphite due to their high irreversible capacity and low capacities. When graphene is employed as a support for materials such as silicon, tin, or transition metal oxide, it does help improve the capacity retention, but long-term stability still needs improvement for these materials to be applicable for commercial applications. The graphene preparation method and the source of particles greatly influence the capacity of the composites. Further research will be needed to find the best combination that will give satisfactory capacity and capacity retention.

References

1. Novoselov KS, Geim AK, Morozov SV, Jiang D, Zhang Y, Dubonos SV, Grigorieva IV, Firsov AA (2004) Electric field effect in atomically thin carbon films. Science 306:666–669
2. Allen MJ, Tung VC, Kaner RB (2010) Honeycomb Carbon: A Review of Graphene. Chem Rev 110:132–145
3. Ströck M (2006) WikimediaCommons. http://commons.wikimedia.org/wiki/File:Eight_Allotropes_of_Carbon.png
4. Brodie BC (1859) On the atomic weight of graphite. Philos Trans Roy Soc Lond 149:249–259
5. Dreyer DR, Park S, Bielawski CW, Ruoff RS (2010) The chemistry of graphene oxide. Chem Soc Rev 39:228–240
6. Park S, Ruoff RS (2009) Chemical methods for the production of graphenes. Nat Nanotechnol 4:217–224
7. Hummers WS, Offeman RE (1958) Preparation of graphitic oxide. J Am Chem Soc 80:1339–1339
8. Marcano DC, Kosynkin DV, Berlin JM, Sinitskii A, Sun Z, Slesarev A, Alemany LB, Lu W, Tour JM (2010) Improved synthesis of graphene oxide. ACS Nano 4:4806–4814
9. Staudenmaier L (1898) Verfahren zur Darstellung der Graphitsäure. Berichte der deutschen chemischen Gesellschaft 31:1481–1487
10. Takami N, Satoh A, Hara M, Ohsaki T (1995) Structural and kinetic characterization of lithium intercalation into carbon anodes for secondary lithium batteries. J Electrochem Soc 142:371–379
11. Funabiki A, Inaba M, Ogumi Z, Yuasa S-i, Otsuji J, Tasaka A (1998) Impedance study on the electrochemical lithium intercalation into natural graphite powder. J Electrochem Soc 145:172–178
12. Guo H-j, Li X-h, Zhang X-m, Wang H-q, Wang Z-x, Peng W-j (2007) Diffusion coefficient of lithium in artificial graphite, mesocarbon microbeads, and disordered carbon. New Carbon Mater 22:7–10

13. Paek S-M, Yoo E, Honma I (2009) Enhanced cyclic performance and lithium storage capacity of SnO_2/graphene nanoporous electrodes with three-dimensionally delaminated flexible structure. Nano Lett 9:72–75
14. Xiang H, Zhang K, Ji G, Lee JY, Zou C, Chen X, Wu J (2011) Graphene/nanosized silicon composites for lithium battery anodes with improved cycling stability. Carbon 49:1787–1796
15. Fuhrer MS, Lau CN, MacDonald AH (2010) Graphene: materially better carbon. MRS Bull 35:289–295
16. Hu Y-S, Adelhelm P, Smarsly BM, Hore S, Antonietti M, Maier J (2007) Synthesis of hierarchically porous carbon monoliths with highly ordered microstructure and their application in rechargeable lithium batteries with high-rate capability. Adv Funct Mater 17:1873–1878
17. Wallace GG, Chen J, Li D, Moulton SE, Razal JM (2010) Nanostructured carbon electrodes. J Mater Chem 20:3553–3562
18. Landi BJ, Ganter MJ, Cress CD, DiLeo RA, Raffaelle RP (2009) Carbon nanotubes for lithium ion batteries. Energy Environ Sci 2:638–654
19. Dahn JR, Zheng T, Liu Y, Xue JS (1995) Mechanisms for lithium insertion in carbonaceous materials. Science 270:590–593
20. Liang M, Zhi L (2009) Graphene-based electrode materials for rechargeable lithium batteries. J Mater Chem 19:5871–5878
21. Kaskhedikar NA, Maier J (2009) Lithium storage in carbon nanostructures. Adv Mater 21:2664–2680
22. Guo P, Song H, Chen X (2009) Electrochemical performance of graphene nanosheets as anode material for lithium-ion batteries. Electrochem Commun 11:1320–1324
23. Wang C, Li D, Too CO, Wallace GG (2009) Electrochemical properties of graphene paper electrodes used in lithium batteries. Chem Mater 21:2604–2606
24. Abouimrane A, Compton OC, Amine K, Nguyen ST (2010) Non-annealed graphene paper as a binder-free anode for lithium-ion batteries. J Phys Chem C 114:12800–12804
25. Bhardwaj T, Antic A, Pavan B, Barone V, Fahlman BD (2010) Enhanced electrochemical lithium storage by graphene nanoribbons. J Am Chem Soc 132:12556–12558
26. Yoo E, Kim J, Hosono E, Zhou H-s, Kudo T, Honma I (2008) Large reversible Li storage of graphene nanosheet families for use in rechargeable lithium ion batteries. Nano Lett 8:2277–2282
27. Yata S, Kinoshita H, Komori M, Ando N, Kashiwamura T, Harada T, Tanaka K, Yamabe T (1994) Structure and properties of deeply Li-doped polyacenic semiconductor materials beyond C_6Li stage. Synth Met 62:153–158
28. Lian P, Zhu X, Liang S, Li Z, Yang W, Wang H (2010) Large reversible capacity of high quality graphene sheets as an anode material for lithium-ion batteries. Electrochim Acta 55:3909–3914
29. Wan L, Ren Z, Wang H, Wang G, Tong X, Gao S, Bai J (2011) Graphene nanosheets based on controlled exfoliation process for enhanced lithium storage in lithium-ion battery. Diamond Relat Mater 20:756–761
30. Shanmugharaj AM, Choi WS, Lee CW, Ryu SH (2011) Electrochemical performances of graphene nanosheets prepared through microwave radiation. J Power Sources 196:10249–10253
31. Magasinski A, Dixon P, Hertzberg B, Kvit A, Ayala J, Yushin G (2010) High-performance lithium-ion anodes using a hierarchical bottom-up approach. Nat Mater 9:353–358
32. Derrien G, Hassoun J, Panero S, Scrosati B (2007) Nanostructured Sn-C Composite as an advanced anode material in high-performance lithium-ion batteries. Adv Mater 19:2336–2340
33. Kim H, Han B, Choo J, Cho J (2008) Three-dimensional porous silicon particles for use in high-performance lithium secondary batteries. Angew Chem Int Ed 47:10151–10154
34. Zhang T, Fu L, Gao J, Yang L, Wu Y, Wu H (2006) Core-shell Si/C nanocomposite as anode material for lithium ion batteries. Pure Appl Chem 78:1889–1896

35. Nazri G-A, Pistoia G (2004) Lithium batteries: science and technology. Kluwer, Boston/Dordrecht/New York/London, p 708
36. Inoue H (2006) High capacity negative-electrode materials next to carbon; Nexelion. In: International meeting on lithium batteries, Biarritz
37. Kim H, Cho J (2008) Superior lithium electroactive mesoporous Si@Carbon core-shell nanowires for lithium battery anode material. Nano Lett 8:3688–3691
38. Park M-H, Kim MG, Joo J, Kim K, Kim J, Ahn S, Cui Y, Cho J (2009) Silicon nanotube battery anodes. Nano Lett 9:3844–3847
39. Cui L-F, Yang Y, Hsu C-M, Cui Y (2009) Carbon – silicon core – shell nanowires as high capacity electrode for lithium ion batteries. Nano Lett 9:3370–3374
40. Chan CK, Patel RN, O'Connell MJ, Korgel BA, Cui Y (2010) Solution-grown silicon nanowires for lithium-ion battery anodes. ACS Nano 4:1443–1450
41. Cui L-F, Ruffo R, Chan CK, Peng H, Cui Y (2009) Crystalline-amorphous core – shell silicon nanowires for high capacity and high current battery electrodes. Nano Lett 9:491–495
42. Beattie SD, Larcher D, Morcrette M, Simon B, Tarascon JM (2008) Si electrodes for Li-ion batteries—a new way to look at an old problem. J Electrochem Soc 155:A158–A163
43. Stoller MD, Park S, Zhu Y, An J, Ruoff RS (2008) Graphene-based ultracapacitors. Nano Lett 8:3498–3502
44. Fasolino A, Los JH, Katsnelson MI (2007) Intrinsic ripples in graphene. Nat Mater 6:858–861
45. Novoselov KS, Geim AK, Morozov SV, Jiang D, Katsnelson MI, Grigorieva IV, Dubonos SV, Firsov AA (2005) Two-dimensional gas of massless Dirac fermions in graphene. Nature 438:197–200
46. Larcher D, Beattie S, Morcrette M, Edstrom K, Jumas J-C, Tarascon J-M (2007) Recent findings and prospects in the field of pure metals as negative electrodes for Li-ion batteries. J Mater Chem 17:3759–3772
47. Obrovac MN, Christensen L (2004) Structural changes in silicon anodes during lithium insertion/extraction. Electrochem Solid State Lett 7:A93–A96
48. CRC Handbook of Chemistry and Physics, 91st ed. CRC Press (2011–2012) http://www.hbcpnetbase.com. Accessed December 2011
49. Li J, Christensen L, Obrovac MN, Hewitt KC, Dahn JR (2008) Effect of Heat Treatment on Si Electrodes Using Polyvinylidene Fluoride Binder. J Electrochem Soc 155:A234–A238
50. Kasavajjula U, Wang C, Appleby AJ (2007) Nano- and bulk-silicon-based insertion anodes for lithium-ion secondary cells. J Power Sources 163:1003–1039
51. Balbuena PB, Wang Y (2004) Lithium-Ion Batteries: Solid-Electrolyte Interphase. Imperial College Press, London
52. Lee JK, Smith KB, Hayner CM, Kung HH (2010) Silicon nanoparticles-graphene paper composites for Li ion battery anodes. Chem Commun 46:2025–2027
53. Chou S-L, Wang J-Z, Choucair M, Liu H-K, Stride JA, Dou S-X (2010) Enhanced reversible lithium storage in a nanosize silicon/graphene composite. Electrochem Commun 12:303–306
54. Wang X-L, Han W-Q (2011) Graphene Enhances Li Storage Capacity of Porous Single-Crystalline Silicon Nanowires. ACS Appl Mater Interfaces 2:3709–3713
55. Tao H-C, Fan L-Z, Mei Y, Qu X (2011) Self-supporting Si/reduced graphene oxide nanocomposite films as anode for lithium ion batteries. Electrochem Commun 13(12):1332–1335
56. Choucair M, Thordarson P, Stride JA (2009) Gram-scale production of graphene based on solvothermal synthesis and sonication. Nat Nano 4:30–33
57. Chan CK, Peng H, Liu G, McIlwrath K, Zhang XF, Huggins RA, Cui Y (2008) High-performance lithium battery anodes using silicon nanowires. Nat Nano 3:31–35
58. Buqa H, Holzapfel M, Krumeich F, Veit C, Novák P (2006) Study of styrene butadiene rubber and sodium methyl cellulose as binder for negative electrodes in lithium-ion batteries. J Power Sources 161:617–622
59. Besenhard JO, Yang J, Winter M (1997) Will advanced lithium-alloy anodes have a chance in lithium-ion batteries? J Power Sources 68:87–90

60. Brousse T, Crosnier O, Santos-Peña J, Sandu I, Fragnaud P, Schleich DM (2002) Recent progress in the development of tin-based negative electrodes for Li-ion batteries. In: Kumagai N, Komaba S (eds) Materials chemistry in lithium batteries. Research Signpost, Kerala
61. Wang Z, Zhang H, Li N, Shi Z, Gu Z, Cao G (2010) Laterally confined graphene nanosheets and graphene/SnO_2 composites as high-rate anode materials for lithium-ion batteries. Nano Res 3:748–756
62. Courtney IA, Dahn JR (1997) Electrochemical and in situ x-ray diffraction studies of the reaction of lithium with tin oxide composites. J Electrochem Soc 144:2045–2052
63. Du Z, Yin X, Zhang M, Hao Q, Wang Y, Wang T (2010) In situ synthesis of SnO_2/graphene nanocomposite and their application as anode material for lithium ion battery. Mater Lett 64:2076–2079
64. Yao J, Shen X, Wang B, Liu H, Wang G (2009) In situ chemical synthesis of SnO_2-graphene nanocomposite as anode materials for lithium-ion batteries. Electrochem Commun 11:1849–1852
65. Li N, Wang Z, Zhao K, Shi Z, Gu Z, Xu S (2010) Large scale synthesis of N-doped multi-layered graphene sheets by simple arc-discharge method. Carbon 48:255–259
66. Wang X, Zhou X, Yao K, Zhang J, Liu Z (2011) A SnO_2/graphene composite as a high stability electrode for lithium ion batteries. Carbon 49:133–139
67. Liang S, Zhu X, Lian P, Yang W, Wang H (2011) Superior cycle performance of Sn@C/graphene nanocomposite as an anode material for lithium-ion batteries. J Solid State Chem 184:1400–1404
68. Zhao B, Zhang G, Song J, Jiang Y, Zhuang H, Liu P, Fang T (2011) Bivalent tin ion assisted reduction for preparing graphene/SnO_2 composite with good cyclic performance and lithium storage capacity. Electrochim Acta 56:7340–7346
69. Zhu X, Zhu Y, Murali S, Stoller MD, Ruoff RS (2011) Reduced graphene oxide/tin oxide composite as an enhanced anode material for lithium ion batteries prepared by homogenous coprecipitation. J Power Sources 196:6473–6477
70. Chang K, Wang Z, Huang G, Li H, Chen W, Lee JY (2012) Few-layer SnS2/graphene hybrid with exceptional electrochemical performance as lithium-ion battery anode. J Power Sources 201:259–266
71. Wang G, Wang B, Wang X, Park J, Dou S, Ahn H, Kim K (2009) Sn/graphene nanocomposite with 3D architecture for enhanced reversible lithium storage in lithium ion batteries. J Mater Chem 19:8378–8384
72. Chen S, Chen P, Wu M, Pan D, Wang Y (2010) Graphene supported Sn-Sb@carbon core-shell particles as a superior anode for lithium ion batteries. Electrochem Commun 12:1302–1306
73. Poizot P, Laruelle S, Grugeon S, Dupont L, Tarascon JM (2000) Nano-sized transition-metal oxides as negative-electrode materials for lithium-ion batteries. Nature 407:496–499
74. Banerjee B, Lahiry S (1983) Superparamagnetism in γ-Mn_2O_3–α-Fe_2O_3–α-Mn_2O_3 system. Phys Status Solidi 76:683–694
75. Li H, Richter G, Maier J (2003) Reversible formation and decomposition of LiF clusters using transition metal fluorides as precursors and their application in rechargeable Li batteries. Adv Mater 15:736–739
76. Dhaouadi H, Madani A, Touati F (2010) Synthesis and spectroscopic investigations of Mn_3O_4 nanoparticles. Mater Lett 64:2395–2398
77. Hu J, Li H, Huang X, Chen L (2006) Improve the electrochemical performances of Cr_2O_3 anode for lithium ion batteries. Solid State Ion 177:2791–2799
78. Anisimov VI, Korotin MA, Kurmaev EZ (1990) Band-structure description of Mott insulators (NiO, MnO, FeO, CoO). J Phys Condens Matter 2:3973–3987
79. Wang H, Cui L-F, Yang Y, Sanchez Casalongue H, Robinson JT, Liang Y, Cui Y, Dai H (2010) Mn_3O_4-graphene hybrid as a high-capacity anode material for lithium ion batteries. J Am Chem Soc 132:13978–13980

80. Lavoie N, Malenfant PRL, Courtel FM, Abu-Lebdeh Y, Davidson IJ (2012) High gravimetric capacity and long cycle life in Mn_3O_4/graphene platelet/LiCMC composite lithium ion batteries anodes. J Power Sources 213:249–254
81. Xing L, Cui C, Ma C, Xue X (2011) Facile synthesis of α-MnO_2/graphene nanocomposites and their high performance as lithium-ion battery anode. Mater Lett 65:2104–2106
82. Yu A, Park HW, Davies A, Higgins DC, Chen Z, Xiao X (2011) Free-standing layer-by-layer hybrid thin film of graphene-MnO_2 nanotube as anode for lithium ion batteries. J Phys Chem Lett 2(15):1855–1860
83. Courtel FM, Duncan H, Abu-Lebdeh Y, Davidson IJ (2011) High capacity anode materials for Li-ion batteries based on spinel metal oxides AMn_2O_4 (A = Co, Ni, and Zn). J Mater Chem 21:10206–10218
84. Fang X, Lu X, Guo X, Mao Y, Hu Y-S, Wang J, Wang Z, Wu F, Liu H, Chen L (2010) Electrode reactions of manganese oxides for secondary lithium batteries. Electrochem Commun 12:1520–1523
85. Gao J, Lowe MA, Abruna HD (2011) Spongelike nanosized Mn_3O_4 as a high-capacity anode material for rechargeable lithium batteries. Chem Mater 23:3223–3227
86. Zhu Y, Stoller MD, Cai W, Velamakanni A, Piner RD, Chen D, Ruoff RS (2010) Exfoliation of graphite oxide in propylene carbonate and thermal reduction of the resulting graphene oxide platelets. ACS Nano 4:1227–1233
87. Ban C, Wu Z, Gillaspie DT, Chen L, Yan Y, Blackburn JL, Dillon AC (2010) Nanostructured Fe_3O_4/SWNT electrode: binder-free and high-rate Li-ion anode. Adv Mater 22:E145–E149
88. Zhu X, Zhu Y, Murali S, Stoller MD, Ruoff RS (2011) Nanostructured reduced graphene oxide/Fe_2O_3 composite as a high-performance anode material for lithium ion batteries. ACS Nano 5:3333–3338
89. Wang G, Liu T, Luo Y, Zhao Y, Ren Z, Bai J, Wang H (2011) Preparation of Fe_2O_3/graphene composite and its electrochemical performance as an anode material for lithium ion batteries. J Alloys Compd 509:L216–L220
90. Ji L, Tan Z, Kuykendall TR, Aloni S, Xun S, Lin E, Battaglia V, Zhang Y (2011) Fe_3O_4 nanoparticle-integrated graphene sheets for high-performance half and full lithium ion cells. Phys Chem Chem Phys 13:7170–7177
91. Wang J-Z, Zhong C, Wexler D, Idris NH, Wang Z-X, Chen L-Q, Liu H-K (2011) Graphene-encapsulated Fe_3O_4 nanoparticles with 3D laminated structure as superior anode in lithium ion batteries. Chem Eur J 17:661–667
92. Lian P, Zhu X, Xiang H, Li Z, Yang W, Wang H (2010) Enhanced cycling performance of Fe_3O_4-graphene nanocomposite as an anode material for lithium-ion batteries. Electrochim Acta 56:834–840
93. Zhang M, Lei D, Yin X, Chen L, Li Q, Wang Y, Wang T (2010) Magnetite/graphene composites: microwave irradiation synthesis and enhanced cycling and rate performances for lithium ion batteries. J Mater Chem 20:5538–5543
94. Zhou G, Wang D-W, Li F, Zhang L, Li N, Wu Z-S, Wen L, Lu GQ, Cheng H-M (2010) Graphene-wrapped Fe_3O_4 anode material with improved reversible capacity and cyclic stability for lithium ion batteries. Chem Mater 22:5306–5313
95. Wang G, Liu T, Xie X, Ren Z, Bai J, Wang H (2011) Structure and electrochemical performance of Fe_3O_4/graphene nanocomposite as anode material for lithium-ion batteries. Mater Chem Phys 128:336–340
96. Li Y, Tan B, Wu Y (2008) Mesoporous Co_3O_4 nanowire arrays for lithium ion batteries with high capacity and rate capability. Nano Lett 8:265–270
97. Yao W-L, Wang J-L, Yang J, Du G-D (2008) Novel carbon nanofiber-cobalt oxide composites for lithium storage with large capacity and high reversibility. J Power Sources 176:369–372
98. Wu Z-S, Ren W, Wen L, Gao L, Zhao J, Chen Z, Zhou G, Li F, Cheng H-M (2010) Graphene anchored with Co_3O_4 nanoparticles as anode of lithium ion batteries with enhanced reversible capacity and cyclic performance. ACS Nano 4:3187–3194

99. Yang S, Cui G, Pang S, Cao Q, Kolb U, Feng X, Maier J, Müllen K (2010) Fabrication of cobalt and cobalt oxide/graphene composites: towards high-performance anode materials for lithium ion batteries. ChemSusChem 3:236–239
100. Li B, Cao H, Shao J, Li G, Qu M, Yin G (2011) Co_3O_4@graphene composites as anode materials for high-performance lithium ion batteries. Inorg Chem 50:1628–1632
101. Chen SQ, Wang Y (2010) Microwave-assisted synthesis of a Co_3O_4-graphene sheet-on-sheet nanocomposite as a superior anode material for Li-ion batteries. J Mater Chem 20:9735–9739
102. Yang S, Feng X, Ivanovici S, Müllen K (2010) Fabrication of graphene-encapsulated oxide nanoparticles: towards high-performance anode materials for lithium storage. Angew Chem Int Ed 49:8408–8411
103. Kim H, Seo D-H, Kim S-W, Kim J, Kang K (2010) Highly reversible Co_3O_4/graphene hybrid anode for lithium rechargeable batteries. Carbon 49:326–332
104. Zhu J, Sharma YK, Zeng Z, Zhang X, Srinivasan M, Mhaisalkar S, Zhang H, Hng HH, Yan Q (2011) Cobalt oxide nanowall arrays on reduced graphene oxide sheets with controlled phase, grain size, and porosity for Li-ion battery electrodes. J Phys Chem C 115:8400–8406
105. Wang B, Wang Y, Park J, Ahn H, Wang G (2011) In situ synthesis of Co_3O_4/graphene nanocomposite material for lithium-ion batteries and supercapacitors with high capacity and supercapacitance. J Alloys Compd 509:7778–7783
106. Tao L, Zai J, Wang K, Zhang H, Xu M, Shen J, Su Y, Qian X (2012) Co_3O_4 nanorods/ graphene nanosheets nanocomposites for lithium ion batteries with improved reversible capacity and cycle stability. J Power Sources 202:230–235
107. Zou Y, Wang Y (2011) NiO nanosheets grown on graphene nanosheets as superior anode materials for Li-ion batteries. Nanoscale 3:2615–2620
108. Wang B, Wu X-L, Shu C-Y, Guo Y-G, Wang C-R (2010) Synthesis of CuO/graphene nanocomposite as a high-performance anode material for lithium-ion batteries. J Mater Chem 20:10661–10664
109. Zhou J, Ma L, Song H, Wu B, Chen X (2011) Durable high-rate performance of CuO hollow nanoparticles/graphene-nanosheet composite anode material for lithium-ion batteries. Electrochem Commun 13(12):1357–1360
110. Mai YJ, Wang XL, Xiang JY, Qiao YQ, Zhang D, Gu CD, Tu JP (2011) CuO/graphene composite as anode materials for lithium-ion batteries. Electrochim Acta 56:2306–2311
111. Lu LQ, Wang Y (2012) Facile synthesis of graphene supported shuttle- and urchin-like CuO for high and fast Li-ion storage. Electrochem Commun 14(1):82–85
112. Hashimoto A, Suenaga K, Urita K, Shimada T, Sugai T, Bandow S, Shinohara H, Iijima S (2005) Atomic correlation between adjacent graphene layers in double-wall carbon nanotubes. Phys Rev Lett 94:045504

Chapter 7
Nanosized and Nanostructured Cathode Materials for Lithium-Ion Batteries

Hugues Duncan, Ali Abouimrane, and Yaser Abu-Lebdeh

Abstract There is a great deal of interest in improving the properties of cathode materials for lithium-ion batteries to meet the energy and power demand of many applications including most consumer electronics, the electric vehicle and large-format energy storage. Traditionally, the rate capability of most cathode materials is intrinsically limited to the slow ionic diffusion within the crystalline structure, and for a few materials, the problem is exacerbated by poor bulk electronic conductivity. Nanostructuring and nanosizing cathode materials have proven to be a very useful method to overcome the problem and not only enhance the rate performance of the batteries but also render some materials electrochemically active. In this chapter, we review the most recent advances in the subject by summarizing new design and synthetic methods for nanomaterials, their characterization and performance in lithium-ion batteries with great emphasis on olivines ($LiFePO_4$) and spinel $LiMn_2O_4$ and its Ni substituent nanomaterials.

7.1 Introduction

Layered oxide materials are micron-sized particles that have been intensively studied and commercially used as cathode materials for lithium-ion batteries. These materials show some limitation in terms of rate capability and need structural or morphological modifications in order to enable them useful for high-power

H. Duncan
Lawrence Berkeley National Laboratory, Berkeley, CA 94720, USA

A. Abouimrane
Argonne National Laboratory, 9700 S. Cass Avenue, Lemont, IL 60439, USA

Y. Abu-Lebdeh (✉)
National Research Council of Canada, Ottawa, ON K1A 0R6, Canada
e-mail: Yaser.Abu-Lebdeh@nrc.gc.ca

Y. Abu-Lebdeh and I. Davidson (eds.), *Nanotechnology for Lithium-Ion Batteries*, Nanostructure Science and Technology, DOI 10.1007/978-1-4614-4605-7_7,

applications. Reducing the size of the particles or designing the architecture of the material to sub-micron level is one of these options and can lead to improvement in the performance of electrode materials [1] for the following reasons:

1. The reduced diffusion path length for Li^+ ions and electron transport compared to micron-sized particles can increase the rate of charge and discharge, particularly for materials with 1D diffusion such as olivines where structural defects might hinder the diffusion process as demonstrated by Malik et al. [2] for $LiFePO_4$.
2. The short path length permits the utilization of materials with low electronic conductivity.
3. There is an increased contact (surface) area between the electrode and the electrolyte that leads to better charge and discharge rates.
4. The volume changes caused by Li^+ insertion/de-insertion will be better accommodated by nanomaterials compared to bulk materials due to faster stress relaxation and that prolongs the battery cycle life.

However, there are a few drawbacks associated to their use as cathode materials for lithium-ion battery:

1. The enhanced surface area increases the rate of reactions between the electrolyte and the (charged) cathode material which can lead to a more rapid deterioration of the battery performance. In some cases, the high surface area could induce poor crystallinity, resulting in fewer storage sites for Li^+ ions. For example, in the case of nanosized $LiCoO_2$, it was found that the coulombic efficiencies and reversible capacities decreased when the $LiCoO_2$ particle size is decreased from bulk to 8 nm [3].
2. The nanoparticles lower the density of the materials which puts more challenges on the electrode and battery design and manufacture.

This chapter will explore nanosized and nanostructured cathode materials with emphasis on synthetic methods that can be used to obtain defined morphologies as well as optimized performance. Two classes of materials that benefit from the use of nanoparticles, olivines and manganese spinels, will be examined in great details.

7.2 Olivines ($LiMPO_4$) Cathode Materials

The use of olivines $LiMPO_4$ (M = Fe, Mn, Co) as cathode materials for lithium batteries have been reported for the first time by Pahdi et al. [4]. $LiFePO_4$ has been by far the most studied and promising. It offers a high theoretical capacity of 170 mAh g^{-1} and a flat voltage curve at 3.4 V on discharge and 3.6 V on charge indicative of a two-phase process. The mixed lithium manganese-iron phosphate ($Li(Mn,Fe)PO_4$) is a naturally occurring mineral, triphylite [5]. Initially, low capacities were observed due to the low electronic conductivity of this material,

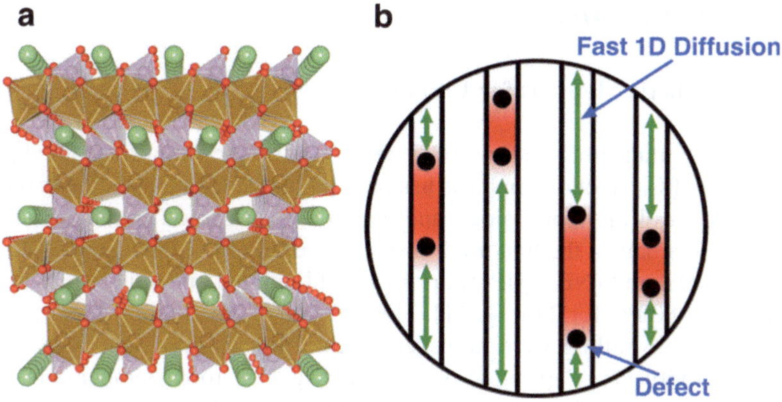

Fig. 7.1 (**a**) Crystal structure of $LiFePO_4$ illustrating 1D Li^+ diffusion channels oriented along the [010] direction. (**b**) Schematic illustration of Li^+ diffusion impeded by immobile point defects in 1D channels (Reprinted with permission from Ref. [2]. Copyright 2009 American Chemical Society)

in the range of 10^{-9} S cm^{-1} [6]. Padhi's first report of $LiFePO_4$ showed a discharge capacity of around 125 mAh g^{-1} at a charging rate of 2.1 mA g^{-1}, that is, C/80. However, a reduction of the particle size by ball milling and introduction of a carbon coating [7] resulted in a dramatic increase in the performance. Capacities approaching the theoretical capacity of 170 mAh g^{-1} were obtained at 1C rate (175 mA g^{-1}) using PEO-Li salt electrolyte at 80°C. It has been shown that nanoparticles of 140 nm can exhibit good electrochemical activity (147 mAh g^{-1} at 5C rate) without adding a carbon coating to the particles [8].

A detailed review of nanosized $LiFePO_4$ for high-power applications is the subject of Chap. 8 of this book, and therefore, this chapter will concentrate on the effect of specific nanostructures on the performance and also discuss other members of the $LiMPO_4$ family.

7.2.1 *Lithium Diffusion*

Olivines crystallize in the Pmna space group. PO_4 tetrahedron is surrounded by FeO_6 octahedra, and lithium ions occupy channels that are parallel to the b axis [9] oriented in the [010] direction, as shown in Fig. 7.1. Thus, any defect present in those channels will impede the lithium diffusion. It is also reported based on first-principles calculation methods that lithium moves rapidly through 1D channels with little possibility for crossing between channels [10].

7.2.2 Synthetic Methods

7.2.2.1 Synthetic Methods for $LiFePO_4$

Given the importance of small particle size and carbon coating for a good performance of $LiFePO_4$, a lot of emphasis has been put on developing synthetic methods which can control the morphology and obtain a good carbon coating. Solid-state methods were amongst the first used to synthesize $LiFePO_4$. Precursors of Li_2CO_3, $Fe(CH_3COOH)_2$ and $(NH_4)_2HPO_4$ are thoroughly mixed and sintered in an argon atmosphere at 800°C for 24 h [4, 11]. The addition of a carbon precursor, namely, sucrose, and ball milling were however necessary to observe significant electrochemical activity [7]. Capacities of 160 mAh g^{-1} at room temperature with a carbonate-based electrolyte (1M $LiPF_6$ in PC:DMC) can be obtained [11]. Another method is based on the synthesis of $LiFePO_4$ ingot by melt casting from $FePO_4$ and Li_2CO_3 precursors [12]. The material was crushed in a disk mill, producing 20–100-μm particles, followed by planetary milling to obtain 10 μm and finally ball milling to produce 200–400-nm particles. While planetary-milled particles with PMAAO carbon coating showed a rapid capacity fade, from 112 mAh g^{-1} at C/4 rate and 60°C, to 80 mAh g^{-1} after 40 cycles, the nanomilled ones cycled much better, with 153 mAh g^{-1} initially to 151 mAh g^{-1} after 40 cycles under the same conditions. On the one hand, these 'top-down' methods, especially melt casting, are scalable and can be used to readily produce large quantities of material, but they offer less control on the nanostructure of the material. On the other hand, 'bottom-up' methods such as hydrothermal, solvothermal and ionothermal in which precursors and solvents are used to control the growth of the particles and morphology proved to work better in giving definite morphology of nanomaterial.

The first hydrothermal synthesis of $LiFePO_4$ was reported by Yang et al. [13] by the reaction of $FeSO_4$, H_3PO_4 and LiOH at 120°C in a hydrothermal reactor for several hours. The morphology of the material is microplatelets of ca. 3 μm diameter. The carbon-coated $LiFePO_4$ could reversibly intercalate 0.6 lithium ion per redox metal. Jin and Hu [14] used the same method and varied the reaction temperature and post-reaction heat treatment and obtained a capacity of 167 mAh g^{-1} at C/10. The investigation of the effect of the pH by Dokko et al. [15] led to the conclusion that needle-shaped particles with large facet in the bc plane are favoured in acidic solution (pH of 3.5). Weakly acidic solution (pH 4–6.5) favoured plate-like crystals with ac-plane facet, and random orientation is favoured at pH higher than 7.2. Figure 7.2 shows TEM images of a nanomaterial prepared at pH of 3.5, 5.1 and 7.2. The best electrochemical results are for the platelets with facets along the ac plane, as it decreases the diffusion length along the b axis. A capacity of 160 mAh g^{-1} was achieved at C/10 for the material synthesized at pH 5.1, while 120 mAh g^{-1} was achieved with the one synthesized at pH 7.2 and 110 mAh g^{-1} when the pH was 3.5.

An alternative to the hydrothermal method are solvothermal methods which involve the replacement of water by an organic solvent with a high boiling point

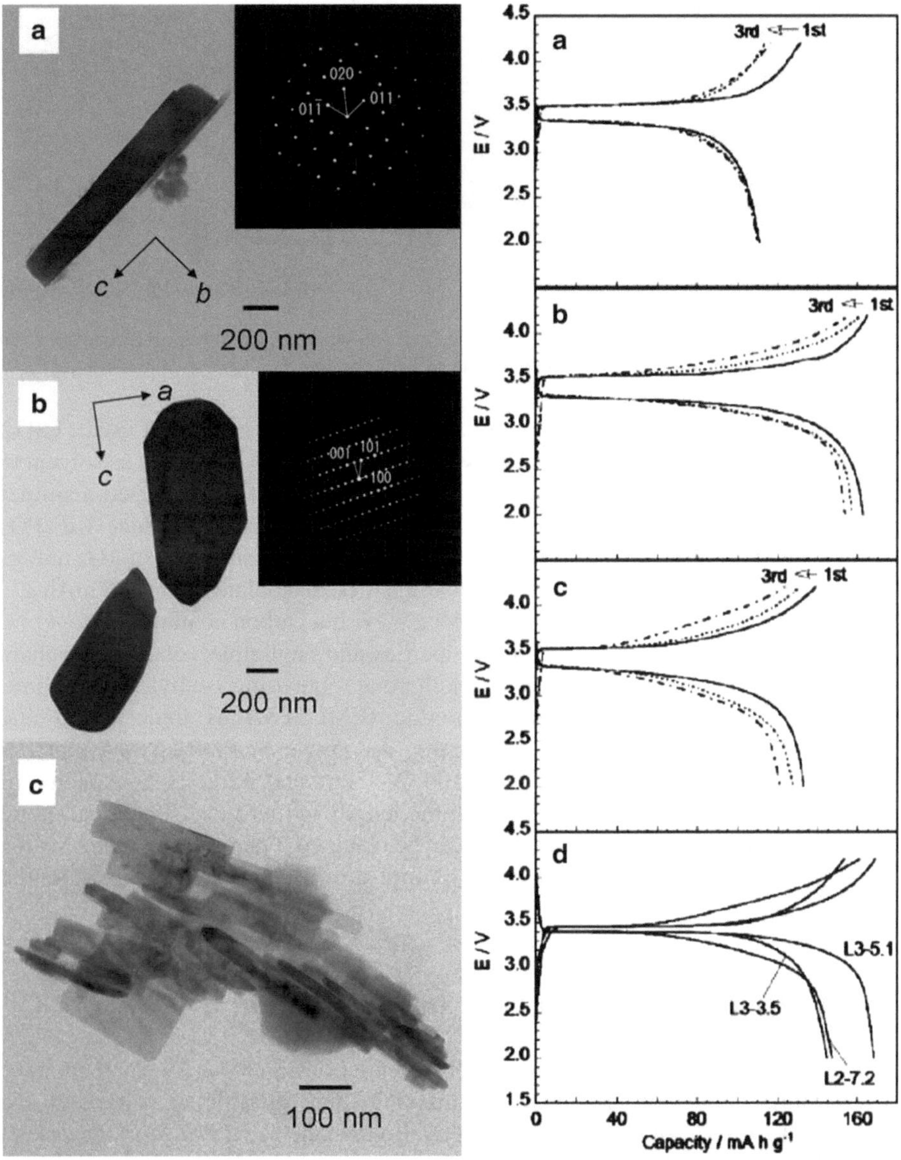

Fig. 7.2 TEM of $LiFePO_4$ obtained hydrothermally in pH of (**a**) 3.5, (**b**) 5.1 and (**c**) 7.2 (*left*). Charge and discharge curves of $LiFePO_4$ samples of L3-3.5 (**a**), L3-5.1 (**b**) and L2-7.2 (**c**) measured at a current density of 17 mA g^{-1} (*right*). (**d**) Charge and discharge curves of the three samples measured at a very low current density of 1.7 mA g^{-1} (Adapted from Ref. [15] with permission of The Royal Society of Chemistry)

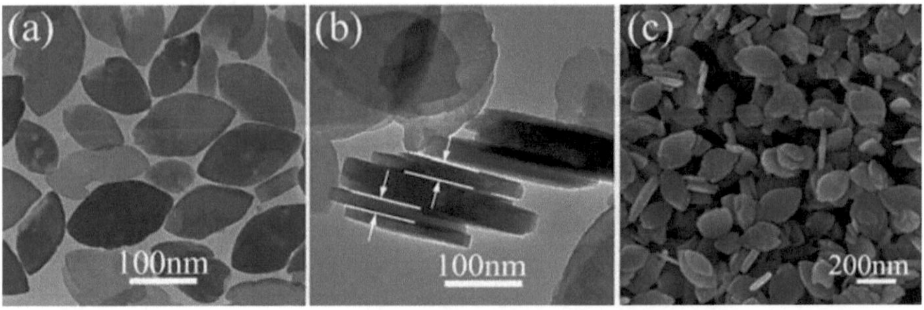

Fig. 7.3 SEM of $LiFePO_4$ nanoplates prepared by a solvothermal technique (Reproduced from Ref. [17] with permission of The Royal Society of Chemistry)

such as tetraethylene glycol or benzyl alcohol [16]. Yang et al. [16] used $FeSO_4$, H_3PO_4 and LiOH as precursors and polyethylene glycol 400 (PEG400) as solvent to synthesize nanoplates and microplates of $LiFePO_4$. Nan et al. [17] used a similar solvothermal technique with ethylene glycol as a solvent to produce $LiFePO_4$ nanoplates with a thickness of 30 nm, length of 200 nm and width of 100 nm, as shown in Fig. 7.3. Even without carbon coating, these nanoplates yield 128 mAh g^{-1} at 0.1C, which was improved to 165 mAh g^{-1} with a carbon coating (Fig. 7.4).

Ionic liquids have good solvating properties and negligible volatility. Recham et al. [18] synthesized $LiFePO_4$ by an ionothermal route using 1-ethyl-3-methylimidazolium *bis*(trifluoromethanesulfonyl)imide (EMI-TFSI) as ionic liquid and FeC_2O_4 $2H_2O$ and LiH_2PO_4 as precursors. As shown in Fig. 7.5, the material obtained has a particle size around 100–200 nm and exhibits a capacity of 150 mAh g^{-1} at C/10 rate. By changing the length and terminal functional group on the ionic liquid, different orientation can be obtained. For example, replacing the ethyl group on EMI-TFSI by a cyano-propyl group led to a (101) preferential orientation.

7.2.2.2 Synthetic Methods for Other Olivines: $LiMPO_4$, M = Mn, Co, Ni

Given the relatively low lithium insertion potential of $LiFePO_4$ (3.4 V), there have been efforts to develop $LiMnPO_4$ and $LiCoPO_4$ that offer higher voltage (4 and 4.7 V for $LiMnPO_4$ and $LiCoPO_4$, respectively) than the $LiFePO_4$. As in the case of $LiFePO_4$, the decrease in the crystallite size and carbon coating has been proven to increase the reversible capacity. Drezen et al. [19] synthesized $LiMnPO_4$ with different crystallite sizes by varying the calcination temperature. Subsequent dry ball milling and carbon coating gave capacities of 115 mAh g^{-1} at C/5 and 80 mAh g^{-1} at C for particle size of 140 nm, while it decreased to 80 mAh g^{-1} at C/5 and 50 mAh g^{-1} at C for a particle size of 200 nm. Using the polyol method, Wang et al. [20] and Martha et al. [21] obtained carbon-coated particles of 30 nm with a capacity of 140 mAh g^{-1} at C/5 and of 110 mAh g^{-1} at C rate. The material was cycled for 200 cycles and showed limited capacity fade. Choi et al. [22]

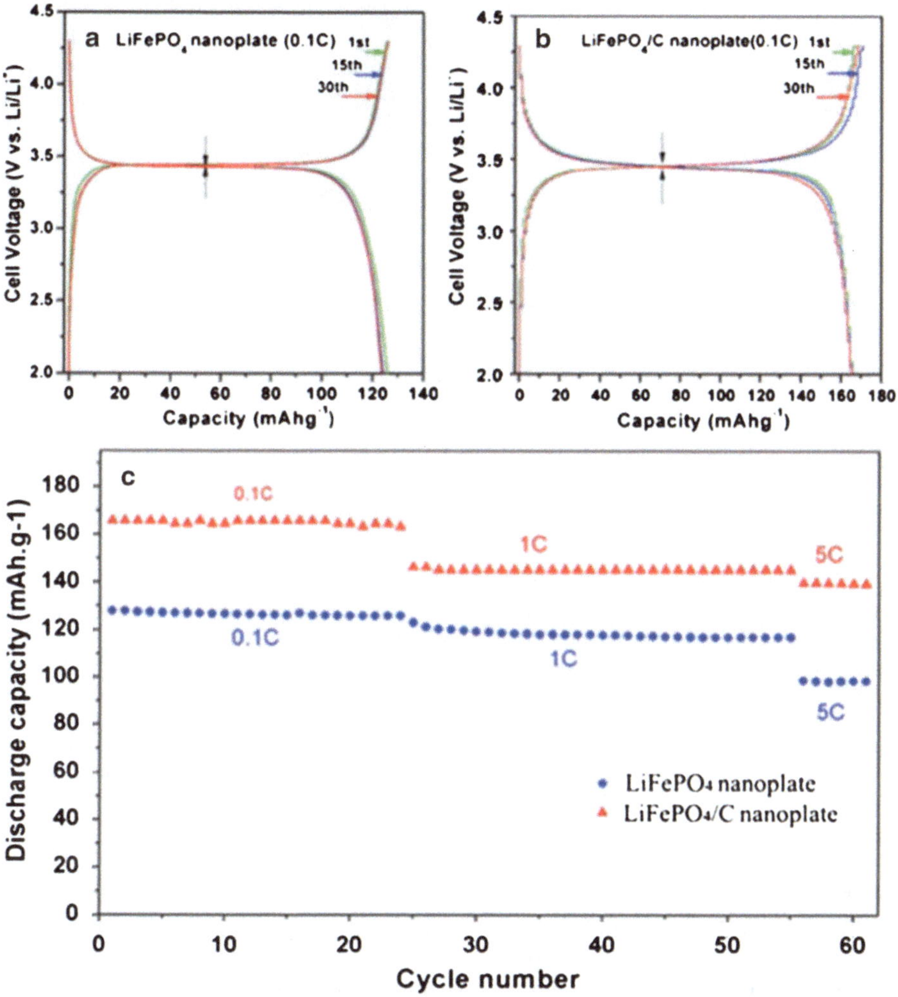

Fig. 7.4 Capacity voltage curves of (**a**) $LiFePO_4$ nanoplates and (**b**) carbon-coated $LiFePO_4$ nanoplates, (**c**) rate capability of $LiFePO_4$ nanoplates (Reproduced from Ref. [17] with permission of The Royal Society of Chemistry)

synthesized $LiMnPO_4$ from a solid-state reaction in molten hydrocarbons (paraffin) and obtained nanoplates which delivered 130 mAh g^{-1} at C/10 rate. However, the thermal stability of delithiated $LiMnPO_4$ is less than its Fe counterpart. Chen et al. [23] conducted thermal stability studies and showed that the release of oxygen occurs at 150°C, followed by the combustion of the electrolyte at 215°C for a total heat of 884 J g^{-1} which is comparable to the heat generated by $LiCoO_2$, $LiNi_{0.8}Co_{0.2}O_2$ and $Li(Ni_{0.8}Co_{0.15}Al_{0.05})O_2$. $LiFePO_4$, by comparison, does not decompose to evolve oxygen, and the onset of the exothermic reaction in the presence of electrolyte is at 250°C, and the heat generated is much lower at

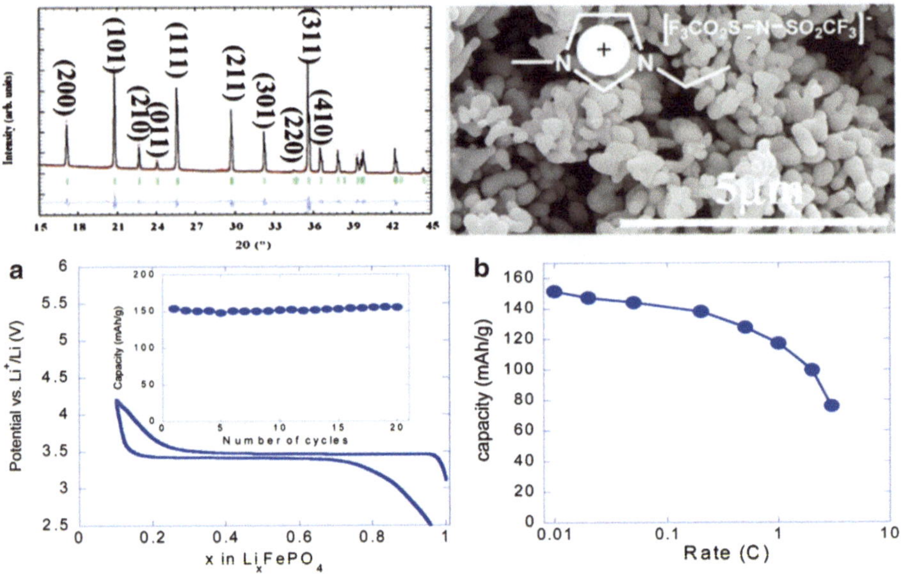

Fig. 7.5 Room-temperature voltage-composition curves together with the capacity retention (*inset*) are reported in (**a**) for samples made of an EMI-TFSI ionic liquid. The positive electrodes, containing 7–10 mg of active material per cm^2, were made by ball milling (Spex 800 for 15 min) $LiFePO_4$ powders and carbon SP (carbon black from MM, Belgium) mixtures in 85–15 % weight ratio. The power rate capabilities of such electrodes, determined using a 'signature curve', are shown in (**b**). *Top*, XRD pattern on SEM for powder presented in (**a**) and (**b**) prepared in EMI-TFSI ionic liquid (Adapted with permission from Ref. [18] Copyright 2009 American Chemical Society)

147 J g^{-1} [11]. Some of the Mn can be substituted for Fe to give $LiFe_{1-y}Mn_yPO_4$ in all the $0 < y < 1$ range. The voltage profile gives two plateaus, the first at 3.4 V due to the Fe^{2+}/Fe^{3+} couple and the other at 4 V due to the Mn^{2+}/Mn^{3+} couple. The capacity at the 4 V plateau is proportional to the amount of Mn in the material [24]. $LiCoPO_4$ is readily obtained and, like $LiMnPO_4$ and $LiFePO_4$, requires a carbon coating in order to yield interesting capacities. Initially, the capacities were low, 70 mAh g^{-1}, corresponding to 0.42 Li^+ [25]. However, carbon-coated $LiCoPO_4$ synthesized by a microwave method [26] yields a material with a capacity of 144 mAh g^{-1} at C/10 and a voltage plateau around 4.8 V and a capacity of 137 mAh g^{-1} at 1C rate. However, the capacity fades to 72.6 mAh g^{-1} after 30 cycles at a C/10 rate. The rapid capacity fade is attributed to the instability of the conventional electrolyte used in the cells and needs to be addressed if $LiCoPO_4$ is to be used at larger scale. There are also issues with the thermal stability of $LiCoPO_4$; in the delithiated state, Li_zCoPO_4 ($z = 0.6$) and $CoPO_4$ will convert to $Co_2P_2O_7$ combined with oxygen release when heated at 200°C [27]. This seems to be due to the instability of Co^{3+} in the olivine phase. Although these two cathode materials are promising in terms of capacity and rate capability, the thermal stability remains an issue, and work still needs to be done on that front. The substitution of part of Co for

Fe or Ni in an effort to improve the performance of $LiCoPO_4$ was attempted. In the case of $LiCo_{1-x}Fe_xPO_4$, x = 0.05, 0.1, the x = 0.1 doping improves the discharge capacity and the Fe content does not affect the 4.8 V plateau. The solid solutions of $LiNiPO_4$ and $LiCoPO_4$, $LiNi_{1-x}Co_xPO_4$, shortens the Co^{2+}/Co^{3+} plateau at 4.7 V as Ni content was increased and no plateau due to the Ni^{2+}/Ni^{3+} is observed [28], offering no advantage over pure $LiCoPO_4$. $LiNiPO_4$ has an extraction potential of 5.3 V and insertion potential of 5.1 V that are beyond the electrochemical stability window of conventional electrolytes [29].

7.3 Spinel Cathode Materials

7.3.1 $LiMn_2O_4$

The spinel structure has an advantage over the layered structure in that it permits extraction of all the lithium ions while having higher thermal stability (oxygen evolution occurs at higher temperature than $LiCoO_2$) [30], at the cost however of lower overall capacity. Compared to $LiCoO_2$, $LiMn_2O_4$ has the advantages of being more environmentally benign and less costly in a way similar to $LiFePO_4$. Additionally, $LiMn_2O_4$ exhibits a very low reactivity in the charged state with an onset temperature over 200°C and low heat release [31, 32]. Typically, the capacity of spinels is around 148 mAh g^{-1}, compared to capacities of 180–200 mAh g^{-1} for nickel-rich layered oxides. $LiMn_2O_4$ has a voltage plateau at 4 V and a practical capacity around 120 mAh g^{-1} [33–35]. The lithium diffusion coefficient is lower than that of $LiCoO_2$ by two orders of magnitude: 10^{-10} to 10^{-12} cm^2 s^{-1} [36]. While the capacity fade of $LiMn_2O_4$ is quite high, especially when cycled at 60°C due to the disproportionation of Mn^{+3} to Mn^{+2} and Mn^{+4}, species soluble in the electrolyte that can cross to reach the anode and deposit in the metallic state, this is not the case for lithium-rich and nickel doped $LiMn_2O_4$, $LiMn_{18.5}Li_{0.075}Ni_{0.075}O_4$ and $LiMn_{1.8}Li_{0.1}Ni_{0.1}O_4$, which exhibit excellent capacity retention at 60°C as well as excellent rate capability. However, the capacity is relatively low at 90–95 mAh g^{-1} [37]. It seems however that substituting graphite for $Li_4Ti_5O_{12}$ as anode material decreases the capacity fade by bypassing Mn^{2+} [0.2 V] plating at the anode because of the higher operating voltage of $Li_4Ti_5O_{12}$ compared to graphite [38]. Lithium-Ion cells with $LiMn_2O_4$ as cathode material and $Li_4Ti_5O_{12}$ were assembled [39]; the operating potential is between 2.75 and 1.5 V, and a capacity of 90 mAh g^{-1} after 160 cycles is achieved at a rate of C at 50°C [39]. Another cause of the capacity fade is the formation of the lithium-rich Li_2MnO_4 phase which leads to Jahn-Teller distortion and cracking at the surface of the electrode [40].

The spinel $LiMn_2O_4$ crystallizes in the Fd-3m space group. The three-dimensional crystal structure means that the Li^+ diffusion is also 3D. The synthesis of nanostructured materials with well-controlled shape, size and structure can improve the battery performance; however, in order to achieve good stability for

Fig. 7.6 (**a**) and (**b**) SEM of $LiMn_2O_4$ nanorods (*top*); (**c**) rate capability (*black circles*) compared to standard $LiMn_2O_4$ (*open circles*); (**d**) charge and discharge capacity of $LiMn_2O_4$ nanorods at 1C rate (Adapted with permission from Ref. [44]. Copyright 2009 American Chemical Society)

$LiMn_2O_4$ material, a good crystallinity is necessary [41–43]. While it is easy to synthesize nanosized $LiMn_2O_4$ by sol-gel and combustion methods, often the crystallinity will be poor and a heat treatment will induce coarsening of the powders. However, $LiMn_2O_4$ nanorods prepared by lithiation of β-MnO_2 nanorods can sustain 700°C of heat treatment which improves crystallinity without inducing particle coalescence. Figure 7.6 shows an SEM of $LiMn_2O_4$ nanorods and its electrochemical behaviour [44]. A capacity of 110 mAh g^{-1} at C/10 is achieved, similar to the commercial powder, but the rate capability is dramatically improved with nanorods, achieving 95 mAh g^{-1} at C rate compared to 50 mAh g^{-1} with commercial powder.

Shaju et al. also prepared a stoichiometric nano-$LiMn_2O_4$ using a one-pot resorcinol-formaldehyde route. As the material is composed of nanoparticles fused together and forming a porous morphology, the room and high temperature battery performance was improved compared to the material prepared by the sol-gel method [45].

Fig. 7.7 $LiMn_{1.5}Ni_{0.5}O_4$ prepared by a co-precipitation/sol-gel technique and its electrochemical performance

7.3.2 $LiMn_{1.5}Ni_{0.5}O_4$

The spinel $LiMn_{1.5}Ni_{0.5}O_4$ adopts either the Fd-3m space group of $LiMn_2O_4$ (so-called disordered structures, since Mn and Ni atoms are randomly mixed) or the $P4_332$ space group in which Ni and Mn occupy specific sites (the so-called ordered structure). The lithium extraction will occur at 4.7 V, corresponding to the oxidation of Ni^{2+} to Ni^{4+}, and the material has a theoretical capacity of 147 mAh g^{-1}. The synthesis of the ordered structure requires a careful control of the annealing conditions, and its electronic conductivity is several ordered of magnitude lower than its disordered counterpart. The cycling stability is also lower for the ordered structure [46]. For these reasons, the disordered structure is usually preferred. However, there has been a recent report of micron-sized ordered $LiMn_{1.5}Ni_{0.5}O_4$ that delivered very high capacity [47].

Like $LiMn_2O_4$, sol-gel, combustion, co-precipitation and solid-state techniques can be used to prepare this material. Also similar to $LiMn_2O_4$, the crystallinity of the material is critical to its stability as poorly crystallized material will lead to rapid deterioration in capacity [48] while a material with well-crystallized facet will be stable for hundreds of cycles [49], as shown in Fig. 7.7. However, an increase in the calcination time or the temperature can lead to oxygen loss, and the presence of Mn^{3+} with a discharge capacity around 4 V will decrease the energy density of the material.

The synthesis of well-crystallized $LiMn_{1.5}Ni_{0.5}O_4$ nanorods using β-MnO_2 nanorod precursor, nickel acetate and lithium hydroxide followed by heat treatment have been demonstrated [50, 51]. The obtained $LiMn_{1.5}Ni_{0.5}O_4$ nanorods, shown in Fig. 7.8, were assigned to the $P4_332$ (ordered) structure of an average diameter of 130 nm and a length of 1.2 μm. This material yielded 116 mAh g^{-1} capacity at C/2 rate and 111 mAh g^{-1} at C rate as shown in Fig. 7.9.

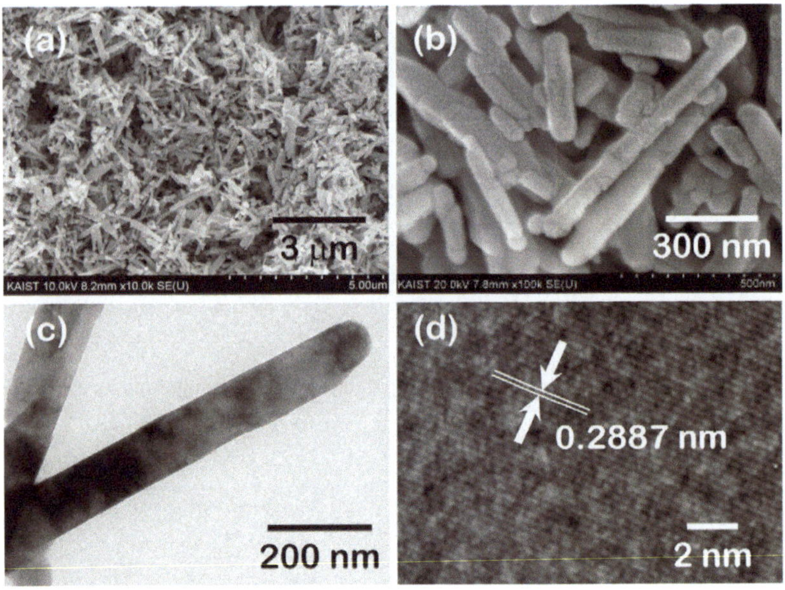

Fig. 7.8 FE-SEM (**a**, **b**), TEM (**c**, **d**) of $LiMn_{1.5}Ni_{0.5}O_4$ nanorods (Reprinted from Journal of Power Sources, Ref. [51], with permission from Elsevier)

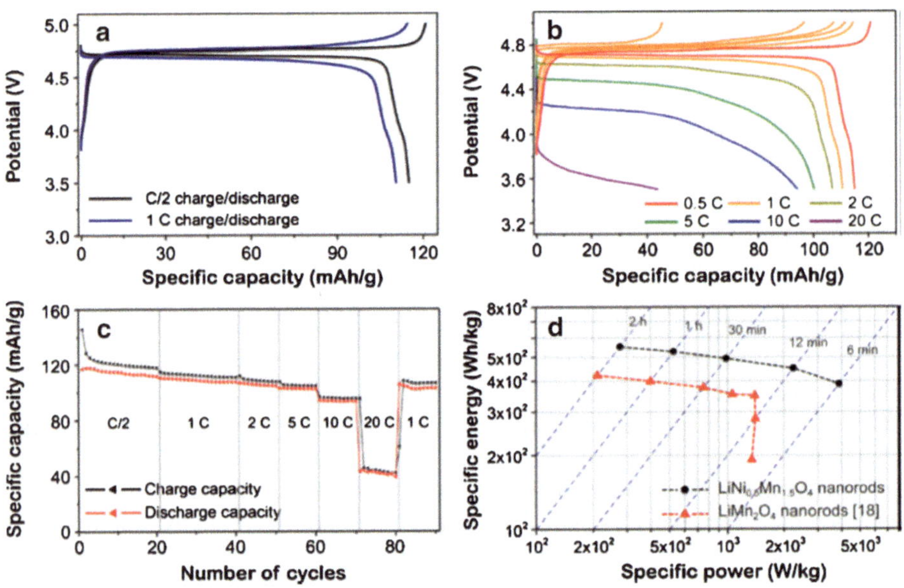

Fig. 7.9 (**a**, **b**) Voltage-capacity curve of $LiMn_{1.5}Ni_{0.5}O_4$ nanorod; (**c**) rate capability and (**d**) comparison of $LiMn_{1.5}Ni_{0.5}O_4$ and $LiMn_2O_4$ nanorods (Reprinted from Journal of Power Sources, Ref. [51], with permission from Elsevier)

7.4 Summary and Outlook

In this chapter, the benefits of fabricating cathode nanomaterials on their electrochemical and battery performance were discussed in depth. In olivines, it was found that the reduction in particle size to the nanoscale plays a critical role in obtaining an electrochemical activity in a lithium-ion battery. Also, owing to the 1D ion diffusion in this type of crystalline structure, the shape of the particles and direction of growth were found to play a major role in their ability to intercalate Li ions and perform well in a lithium-ion battery. The use of different synthetic routes in controlling the particle shape and size was discussed in details. It was shown that a carbon coating is still necessary to optimize performance in $LiFePO_4$ or to obtain any activity in $LiMnPO_4$ and $LiCoPO_4$. In spinels, nanosized particles need to be well crystallized in order to prevent rapid capacity fade. It was shown that well-crystallized $LiMn_2O_4$ and $LiMn_{1.5}Ni_{0.5}O_4$ nanorods yield higher performance than their micron-sized counterparts. The benefits of nanostructuring and nanosizing in overcoming the intrinsic limitations of diffusion in crystalline solid cathode materials and obtaining better battery performance are clearly demonstrated. These new design and synthetic methods can be easily adopted to make better cathode materials with different chemical and crystalline structures

References

1. Pitchai R, Thavasi V, Mhaisalkar S, Ramakrishna S (2011) Nanostructured cathode materials: a key for better performance in Li-ion batteries. J Mater Chem 21(30):11040–11051
2. Malik R, Burch D, Bazant M, Ceder G (2010) Particle size dependence of the ionic diffusivity. Nano Lett 10(10):4123–4127
3. Okubo M, Hosono E, Kim J, Enomoto M, Kojima N, Kudo T, Zhou H, Honma I (2007) Nanosize effect on high-rate Li-ion intercalation in LiCoO2 electrode. J Am Chem Soc 129(23):7444–7452
4. Padhi AK, Nanjundaswamy KS, Goodenough JB (1997) Phospho-olivines as positive-electrode materials for rechargeable lithium batteries. J Electrochem Soc 144(4):1188–1194
5. Ravet N, Chouinard Y, Magnan JF, Besner S, Gauthier M, Armand M (2001) Electroactivity of natural and synthetic triphylite. J Power Sources 97–98:503–507
6. Chung S-Y, Bloking JT, Chiang Y-M (2002) Electronically conductive phospho-olivines as lithium storage electrodes. Nat Mater 1(2):123–128
7. Ravet N, Goodenough JB, Besner S, Simoneau M, Hovington P, Armand M (1999) Improved iron based cathode material. ECS meeting abstracts 99(2): Abstract #127
8. Delacourt C, Poizot P, Levasseur S, Masquelier C (2006) Size effects on carbon-free LiFePO4 powders. Electrochem Solid State Lett 9(7):A352–A355
9. Fisher CAJ, Islam MS (2008) Surface structures and crystal morphologies of LiFePO4: relevance to electrochemical behaviour. J Mater Chem 18(11):1209–1215
10. Morgan D, Van der Ven A, Ceder G (2004) Li conductivity in LixMPO4 (M = Mn, Fe, Co, Ni) olivine materials. Electrochem Solid State Lett 7(2):A30–A32
11. Yamada A, Chung SC, Hinokuma K (2001) Optimized LiFePO4 for lithium battery cathodes. J Electrochem Soc 148(3):A224–A229

12. Gauthier M, Michot C, Ravet N, Duchesneau M, Dufour J, Liang G, Wontcheu J, Gauthier L, MacNeil D (2010) Melt casting LiFePO4. J Electrochem Soc 157(4):A453–A462
13. Yang S, Zavalij PY, Whittingham MS (2001) Hydrothermal synthesis of lithium iron phosphate cathodes. Electrochem Commun 3(9):505–508
14. Jin B, Gu H-B (2008) Preparation and characterization of LiFePO4 cathode materials by hydrothermal method. Solid State Ionics 178(37–38):1907–1914
15. Dokko K, Koizumi S, Nakano H, Kanamura K (2007) Particle morphology, crystal orientation, and electrochemical reactivity of LiFePO4 synthesized by the hydrothermal method at 443 K. J Mater Chem 17(45):4803–4810
16. Yang S, Zhou X, Zhang J, Liu Z (2010) Morphology-controlled solvothermal synthesis of LiFePO4 as a cathode material for lithium-ion batteries. J Mater Chem 20(37):8086–8091
17. Nan C, Lu J, Chen C, Peng Q, Li Y (2011) Solvothermal synthesis of lithium iron phosphate nanoplates. J Mater Chem 21(27):9994–9996
18. Recham N, Dupont L, Courty M, Djellab K, Larcher D, Armand M, Tarascon J-M (2009) Ionothermal synthesis of tailor-made LiFePO4 powders for Li-ion battery applications. Chem Mater 21(6):1096–1107
19. Drezen T, Kwon N-H, Bowen P, Teerlinck I, Isono M, Exnar I (2007) Effect of particle size on LiMnPO4 cathodes. J Power Sources 174(2):949–953
20. Wang D, Buqa H, Crouzet M, Deghenghi G, Drezen T, Exnar I, Kwon N-H, Miners J, Poletto L, Grätzel M (2009) High-performance, nano-structured LiMnPO4 synthesized via a polyol method. J Power Sources 189(1):624–628
21. Martha SK, Markovsky B, Grinblat J, Gofer Y, Haik O, Zinigrad E, Aurbach D, Drezen T, Wang D, Deghenghi G, Exnar I (2009) LiMnPO4 as an advanced cathode material for rechargeable lithium batteries. J Electrochem Soc 156(7):A541–A552
22. Choi D, Wang D, Bae I-T, Xiao† J, Nie Z, Wang W, Viswanathan V, Lee Y, Zhang J-G, Graff G, Yang Z, Liu J (2010) LiMnPO4 nanoplate grown via solid-state reaction in molten hydrocarbon for Li-ion battery cathode. Nano Lett 10(8):2799–2805
23. Chen G, Richardson TJ (2010) Thermal instability of Olivine-type LiMnPO4 cathodes. J Power Sources 195:1221–1224
24. Molenda J, Ojczyk W, Marzec J (2007) Electrical conductivity and reaction with lithium of LiFe1-yMnyPO4 olivine-type cathode materials. J Power Sources 174(2):689–694
25. Amine K, Yasuda H, Yamachi M (2000) Olivine LiCoPO4 as 4.8 V electrode material for lithium batteries. Electrochem Solid State Lett 3(4):178–179
26. Li H, Jin J, Wei J, Zhou Z, Yan J (2009) Fast synthesis of core-shell LiCoPO4/C nanocomposite via microwave heating and its electrochemical Li intercalation performances. Electrochem Commun 11(1):95–98
27. Bramnik N, Nikolowski K, M. Trotsc D, Ehrenberg H (2008) Thermal stability of LiCoPO4 cathodes. Electrochem Solid State Lett 11(6):A89–A93
28. Wolfenstine J, Allen J (2004) LiNiPO4-LiCoPO4 solid solutions as cathodes. J Power Sources 136(1):150–153
29. Wolfenstine J, Allen J (2005) Ni3+/Ni2+ redox potential in LiNiPO4. J Power Sources 142(1–2):389–390
30. Biensan Ph, Simon B, Pérès JP, de Guibert A, Broussely M, Bodet JM, Perton F (1999) On safety of lithium-ion cells. J Power Sources 81–82:906–912
31. MacNeil DD, Lu Z, Chen Z, Dahn J (2002) A comparison of the electrode/electrolyte reaction at elevated temperatures for various Li-ion battery cathodes. J Power Sources 108(1–2):8–14
32. MacNeil DD, Dahn JR (2001) The reaction of charged cathodes with nonaqueous solvents and electrolytes: II. LiMn2O4 charged to 4.2 V. J Electrochem Soc 148(11):A1211–A1215
33. Ohzuku T, Kitagawa M, Hirai T (1990) Electrochemistry of manganese dioxide in lithium nonaqueous cell. J Electrochem Soc 137(3):769–775
34. Tarascon JM, Guyomard D (1991) Li metal-free rechargeable batteries based on Li1+xMn2O4 cathodes ($0 <= x <= 1$) and carbon anodes. J Electrochem Soc 138(10):2864–2868

35. Tarascon J-M, Wang E, Shokoohi F, McKinnon W, Colson S (1991) The spinel phase of LiMn2O4 as a cathode in secondary lithium cells. J Electrochem Soc 138(10):2859–2864
36. Goonetilleke PC, Zheng JP, Roy D (2009) Effects of surface-film formation on the electrochemical characteristics of LiMn2O4 cathodes of lithium ion batteries. J Electrochem Soc 156 (9):A709–A719
37. Shin Y, Manthiram A (2003) High rate, superior capacity retention LiMn2-2yLiyNiyO4 spinel cathodes for lithium-ion batteries. Electrochem Solid State Lett 6(2):A34–A36
38. Amatucci G, Du Pasquier A, Blyr A, Zheng T, Tarascon J-M (1999) The elevated temperature performance of the LiMn2O4/C system: failure and solutions. Electrochim Acta 45 (1–2):255–271
39. Belharouak I, Sun Y-K, Lu W, Amine K (2007) On the safety of the Li4Ti5O12/LiMn2O4 lithium-ion battery system. J Electrochem Soc 154(12):A1083–A1087
40. Thackeray M, Shao-Horn Y, Kahaian A, Kepler K, Skinner E, Vaughey J, Hackney S (1998) Structural fatigue in spinel electrodes in high voltage (4 V) Li/LiMn2O4 Cells. Electrochem Solid State Lett 1(1):7–9
41. Gao Y, Dahn JR (1996) Synthesis and characterization of Li1+xMn2-xO4 for Li-ion battery applications. J Electrochem Soc 143(1):100–114
42. Blyr A, Sigala C, Amatucci G, Guyomard D, Chabre Y, Tarascon J-M (1998) Self-discharge of LiMn2O4/C Li-ion cells in their discharged state. J Electrochem Soc 145(1):194–209
43. Kanamura K, Dokko K, Kaizawa T (2005) Synthesis of spinel LiMn2O4 by a hydrothermal process in supercritical water with heat-treatment. J Electrochem Soc 152(2):A391–A395
44. Kim DK, Muralidharan P, Lee H-W, Ruffo R, Yang Y, Chan C, Peng H, Huggins R, Cui Y (2008) Spinel LiMn2O4 nanorods as lithium ion battery cathodes. Nano Lett 8(11):3948–3952
45. Shaju KM, Bruce PG (2008) A stoichiometric nano-LiMn2O4 spinel electrode exhibiting high power and stable cycling. Chem Mater 20(17):5557–5562
46. Shaju KM, Bruce PG (2008) Nano-LiNi0.5Mn1.5O4 spinel: a high power electrode for Li-ion batteries. Dalton Trans 40:5471–5475
47. Ma X, Kang B, Ceder G (2010) High rate micron-sized ordered LiNi0.5Mn1.5O4. J Electrochem Soc 157(8):A925–A931
48. Duncan H, Abu-Lebdeh Y, Davidson IJ (2010) Study of the cathode–electrolyte interface of LiNi0.5Mn1.5O4 synthesized by a sol–gel method for Li-ion batteries. J Electrochem Soc 157 (4):A528–A535
49. Niketic S, Whitfield P, Davidson I (2008) Synthesis and characterization of a high performance LiMn1.5Ni0.5O4 spinel cathode material. ECS meeting abstracts 801(5):155–155
50. Lee H-W et al (2011) Synthesis and electrochemical performance of spinel LiNi0.5Mn1.5O4 nanorods as high voltage cathode materials for Li-ion batteries. ECS meeting abstracts 1101 (11):536
51. Lee H-W, Muralidharan P, Mari C, Ruffo R, Kim D (2011) Facile synthesis and electrochemical performance of ordered LiNi0.5Mn1.5O4 nanorods as a high power positive electrode for rechargeable Li-ion batteries. J Power Sources 196(24):10712–10716

Chapter 8
Design and Properties of $LiFePO_4$ Nano-materials for High-Power Applications

K. Zaghib, A. Mauger, J.B. Goodenough, and C.M. Julien

Abstract This chapter presents a review of the structural and physicochemical properties of $LiFePO_4$ which is considered as the most advanced positive electrode for lithium-ion batteries. Depending on the synthesis, the fundamental properties can be modified because impurities poison this material. These impurities are identified, and a quantitative estimate of their concentrations is deduced from the combination of analytical methods. An optimized preparation provides materials with carbon-coated particles free of any impurity phase, insuring structural stability and electrochemical performance that justify the use of this material as a cathode element in new generation of lithium secondary batteries operating for powering hybrid electric vehicles and full electric vehicles.

8.1 Introduction

Since the discovery of Goodenough's group, the phospho-olivine $LiFePO_4$ (LFP) is considered as potential positive electrode material for use in lithium rechargeable batteries [1]. Yet, in this family, $LiFePO_4$ is currently the subject of many

K. Zaghib (✉)
Institut de Recherche d'Hydro-Québec (IREQ), 1800 boulevard Lionel-Boulet, Varennes, QC J3X 1S1, Canada
e-mail: zaghib.karim@ireq.ca

A. Mauger
Institut de Minéralogie et Physique de la Matière Condensée (IMPMC), Université Pierre et Marie Curie-Paris-6, 140 rue de Lourmel, Paris 75015, France

J.B. Goodenough
The University of Texas at Austin, Austin, TX 78712, USA

C.M. Julien
Physicochimie des Electrolytes, Colloïdes et Sciences Analytiques (PECSA), Université Pierre et Marie Curie, 4 place Jussieu, Paris 75005, France

Y. Abu-Lebdeh and I. Davidson (eds.), *Nanotechnology for Lithium-Ion Batteries*, Nanostructure Science and Technology, DOI 10.1007/978-1-4614-4605-7_8,

investigations because this cathode material realizes the highest capacity (≈170 mAh g^{-1}) at moderate current densities [2]. In addition, it is inexpensive and not toxic, two determinant advantages with respect to cobalt oxide–based materials for large-scaled applications such as hybrid electric vehicles (HEV).

Nevertheless, the bulk electronic conductivity of LFP is quite low, which may result in losses in capacity during high-rate discharge. To increase the electronic conductivity, it is a common practice in the production of lithium-ion battery cathodes to add carbon either by use of carbon additives to a LFP matrix [1, 3] or by surface coating of LFP particles with thin layers of carbon [4–6]. A seven-order-of-magnitude increase in the electronic conductivity can be reach by the addition of carbon such as shown using sucrose to produce carbon in $LiFePO_4$ raw materials by a spray pyrolysis technique [6]. The addition of carbon has then the advantage of combining much better electronic conductivity and high capacity. In particular, a capacity of about 160 mA h g^{-1} has been found in [4] for LFP coated with 1 wt.% carbon.

Ravet et al. reported two ways to carbon coat: mixing $LiFePO_4$ powder with sugar solution and heating the mixture at 700°C or synthesizing $LiFePO_4$ with some organic materials added before heating. Although the way to add carbon is not fully optimized yet [5, 7], the approach will consist in investigating deeply the synthesis and the effects of carbon deposition process on $LiFePO_4$ [8]. The electrochemical properties of LFP are known to be sensitive to the mode of preparation and the structural properties. This can be an advantage for potential applications since it allows for an optimization of the material if we can correlate the mode of preparation with the structural and the physical properties. Aiming to this problem, we have first investigated this correlation in $LiFePO_4$ that had been grown by three different techniques [9].

Evidence of different clustering effects has been established. A firing temperature larger than 800°C increases the fraction of Fe_2P [10], but Fe_2P nanoparticles in such large concentration that they drive superferromagnetism have been detected in samples that have not been heated to such high temperatures [9]. On the one hand, the presence of Fe_2P can increase the electronic conductivity, but on the other hand, it also decreases the ionic conductivity so that both the capacity and cycling rates are degraded with respect to the carbon-coated LFP. It is thus desirable to optimize the preparation of the samples so that such clustering effects do not occur. This can be done easily for Fe_2P clusters, but it is more difficult to avoid the presence of a small concentration (1.0×10^{-6} per chemical formula) of γ–Fe_2O_3 nanoparticles [9, 11] in carbon-free LFP. However, we know from the iron industry that hydrogen, carbon monoxide, or carbon can reduce Fe_2O_3 through different reduction steps that depend on temperature and other physical parameters such as particle sizes. Nearly all iron produced commercially is made using a blast furnace process covered by most chemistry textbooks. In essence, at high temperature, Fe_2O_3 is reduced with carbon (as coke) according to the reaction $2Fe_2O_3 + 3C \rightarrow 4Fe + 3CO_2$. This is one of the most significant industrial processes in history, and the origins of the modern process are traceable back to a small town called Coalbrookdale in Shropshire (England) around the year 1773. Although we can then expect that carbon might reduce Fe^{3+} ions directly over 1,000°C or through the formation of CO gas, thus preventing the

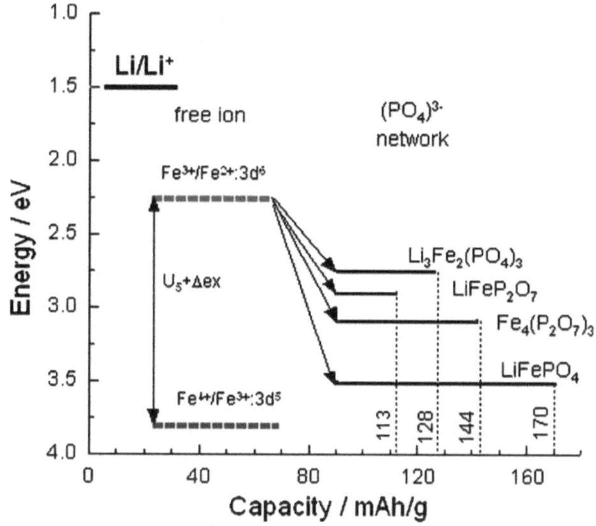

Fig. 8.1 Energy of the redox couples relative to lithium and iron in phosphate frameworks. The graph presents the theoretical capacity for each compound (Reprinted with permission from Elsevier)

formation of γ–Fe_2O_3, we believe that the carbon deposition process, using organic precursors used to make C-coated samples, generates reductive gas such as hydrogen that is more active kinetically to reduce Fe^{3+} impurities in the 500–700°C temperature range used. This is also favored by the fact that the organic precursor is usually mixed with the LFP material or with the LFP chemical precursors by solution processes at a molecular size level.

The aim of this chapter is to investigate the physicochemical properties of optimized LFP electrode materials. One approach to provide insight into the structural and electronic properties of electrode materials involves a systematic study by a combination of techniques including structural, magnetic, and spectroscopic measurements. Furthermore, advantage can be taken of the high sensitivity of some analytical tools for the detection of parasitic impurities that can be grown during synthesis of solid phases. These principles were fully exploited to optimize lithium iron phosphate compounds [1–3]. Both carbon-free and carbon-coated LFP samples are examined in order to investigate the effect of carbon on their structural properties.

Electrochemical extraction of Li from $LiFePO_4$ gives (Fe^{2+}/Fe^{3+}) redox potential at ca. 3.5 V versus Li. A small but first-order displacive structural change of the framework gives a two-phase separation over most of the solid-solution range $0 < x < 1$ for Li_xFePO_4 and therefore a flat V-x curve. A reversible capacity of 160 mAh g^{-1} is delivered by the nanostructured cathode particles coated with carbon. Electrochemical characteristics of $LiFePO_4$ are compared with those of other Fe-containing phosphates, as shown in Fig. 8.1. This graph presents the energy of the redox couples relative to lithium and iron in various phosphate frameworks. Electrochemical tests of optimized $LiFePO_4$ have been conducted under various conditions to assess the influence of the electrolyte on stability and the influence of electrode processing. Postmortem analysis, i.e., ICP, XRD, and SEM, showed that no iron species were detected at the separator–negative electrode interface in cells with anode of lithium metal, graphite, or C-$Li_4Ti_5O_{12}$ [12, 13]. This result is attributed to the high quality of the "optimized" $LiFePO_4$, impurity-free materials used as positive electrodes.

This chapter is organized as follows. First, we expose briefly in Sect. 8.2 the synthesis route and crystal chemistry of LFP. Section 8.3 presents the structure and morphology of optimized $LiFePO_4$ particles within data including those obtained by X-ray powder diffractometry (XRD), scanning electron microscopy (SEM), high-resolution transmission electron microscopy (HRTEM), Fourier transform infrared (FTIR), and Raman scattering (RS) spectroscopy. However, these techniques do not allow the detection of impurities or nanometer-sized clusters in concentration lower than 1%. We then complete the analysis (Sect. 8.4) with magnetic measurements: magnetization curves and electron spin resonance (ESR), since they are powerful tools to characterize the strongly ferrimagnetic nanoparticles of γ–Fe_2O_3 (both in size and concentration) even at concentrations as small as the ppm [9, 11]. Yet, this set of experiments is not sufficient to characterize the carbon compounds. First, they are nonmagnetic. Second, the sensitivity of FTIR spectroscopy, which is a probe of bulk properties, is not sensitive enough to detect the carbon. Therefore, we have added the investigation of Raman spectra. Since the penetration depth of the light inside the LFP particles in such experiments is very small, these experiments are a probe of the first layers at the surface of these particles and allow for the detection of carbon coating. They also give evidence that the carbon does not penetrate inside the particles, but that it is actually stuck at their surface. As a result, we find that the carbon-coated sample is free from γ–Fe_2O_3 and that the amount of Fe^{3+} ions, if any, does not exceed that of residual impurities. In Sect. 8.5, we explore the effects of the exposition of carbon-coated $LiFePO_4$ particles to H_2O. The deterioration of the carbon coat is found to be dependent on the synthesis process, either hydrothermal or solid-state reaction. In case the particles are simply exposed to humid air, the carbon coat protects more efficiently the particles. In this case, the exposure to H_2O mainly results in the delithiation of the surface layer, due to the hydrophilic nature of Li. Again, however, this process only affects the surface layer, at least for a reasonable time (weeks) of exposure to humid air. In addition, within this timescale, the surface layer can be chemically lithiated again, and the samples can be dried to remove the moisture, restoring the electrochemical properties that are then reversible. Finally, Sect. 8.6 shows the electrochemical performance of the optimized LFP particles. We demonstrate the electrochemical ability of the materials operating at high temperature, ca. 60°C, that justifies its use as a cathode element in new generation of lithium secondary batteries powering electric vehicles (EV). We report a lithium-ion battery that can be charged within few minutes, passes the safety tests, and has a very long shelf life. The active materials are nanoparticles of $LiFePO_4$ and $Li_4Ti_5O_{12}$ for the positive and negative electrodes, respectively. The "18650" battery prepared under such conditions delivers a capacity of 800 mAh. It retains full capacity after 20,000 cycles performed at charge rate 10C and discharge rate 5C and retains 95% capacity after 30,000 cycles at charge rate 15C and discharge rate 5C.

8.2 Synthesis, Route, and Crystal Chemistry of LFP

8.2.1 Preparation of LFP Particles

Several $LiFePO_4$ samples were synthesized by solid-state reaction. The A-type $LiFePO_4$ specimens were prepared by mixing iron(II) oxalate [$Fe(C_2O_4)\cdot 2H_2O$], ammonium dihydrogen phosphate [$NH_4H_2PO_4$], and lithium carbonate [Li_2CO_3] in the molar ratio (1:1:1). The B-type samples were prepared from $FePO_4(H_2O)_2$ and Li_2CO_3. A stoichiometric amount of precursors was thoroughly mixed together in isopanol. After drying, the blend was heated at 700°C for 8 h under reducing atmosphere. Carbon-coated C-$LiFePO_4$ was prepared with sucrose and cellulose acetate as the carbon precursors in acetone solution according to the following procedure: The carbon-free powder was mixed with the carbon precursors. The dry additive corresponded to 5 wt.% carbon in $LiFePO_4$. After drying, the blend was heated at 700°C for 4 h under argon atmosphere. The quantity of carbon coat represents about 1 wt.% of the material (C-detector, LECO Co., CS 444). It should be noted that the choice of this moderate sintering temperature minimizes the amount of Fe^{3+} ions present in the powder since the presence of Fe^{3+} has been detected by Mössbauer experiments at sintering temperatures below 500°C, and both trivalent Fe_2O_3 and $Li_3Fe_2(PO_4)_3$ are formed in such large quantities that they are detected by X-rays by sintering above 800°C [14]. Nevertheless, we know from our prior work [9, 10] that B-type $LiFePO_4$, even with an intermediate sintering temperature in the range 500–800°C, does contain Fe_2O_3 nanoparticles, although in such small quantities that they can be detected only by investigation of magnetic properties.

The chemical homogeneity and purity of the powders appear to be sensitive to the conditions of preparation. A comparison of the physical and electrochemical properties of a series of carbon-coated LFP samples allows optimization as the materials for battery application. It appears that a severe control of synthesis conditions is needed to obtain materials with good performance under high current density. The choice of the raw materials, carbon precursor, temperature, and atmosphere is required not only for the reduction of Fe(III) but also for the carbon-coating formation.

Nevertheless, a major effect of the carbon deposition process has been to reduce Fe^{3+}, most probably through a gas-phase reduction process involving hydrogen from the organic carbon precursor. The hydrogen prevents formation of γ–Fe_2O_3 nanoparticles in which iron is in the trivalent state. The cartoon shown in Fig. 8.2 summarizes the conditions of synthesis of LFP powders as positive materials for lithium-ion batteries. Although the heating of our sample in the preparation process did not exceed 700°C, the a-C carbon film is expected to have properties similar to carbon pyrolyzed photoresists prepared at a pyrolysis temperature $Tp = 830 \pm 30$°C. This feature is essential to explain the performance of the carbon coating since increasing the sintering temperature degrades LFP, as is evidenced by the increase of the amount of Fe_2O_3 clusters (or even Fe_2P clusters in some cases) in the material. On the other hand, the decrease of the pyrolysis

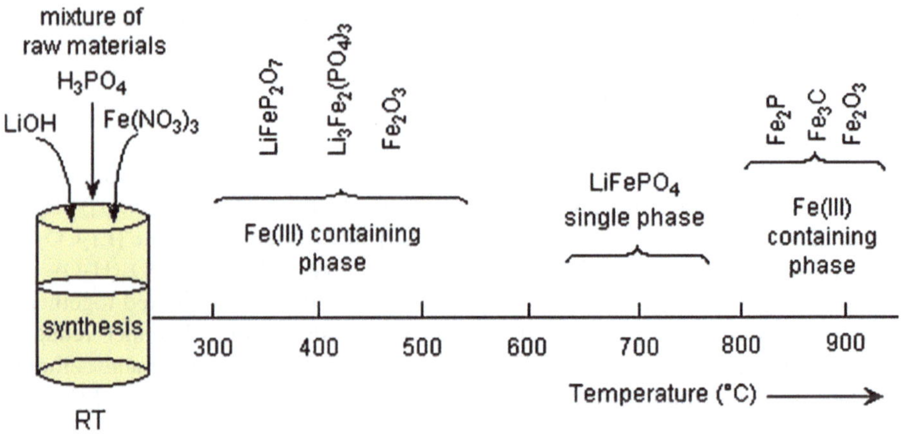

Fig. 8.2 Cartoon of the synthesis of $LiFePO_4$ powders as positive materials for lithium-ion batteries (Reprinted with permission from Elsevier)

temperature below 800°C degrades dramatically the electronic conductivity of the carbon. The carbon deposit can be viewed as a film of irregular thickness, 30 nm thick on average, with gaps.

The gaps might actually be fortunate as the lithium can pass through them without having to tunnel through the carbon film, which could be one reason why the ionic conductivity is not affected by the coating. The magnetic measurements indicated that nanoparticles of γ–Fe_2O_3 at a concentration of 0.7 ppm are present in the A-type sample; such clusters are simply not observed in the new generation of carbon-coated sample (B type). Thus, this study will demonstrate that magnetic measurements (combination of $M(H)$ and $\chi(T)$ data) are beneficial for detecting ferric and/or ferrous impurities, as well as for the quality control of LFP.

8.2.2 The Crystal Chemistry of $LiFePO_4$

Triphylite is a rather scarce orthophosphate primary mineral found in phosphatic pegmatites and pegmatitic dikes. Its formula is $Li(Mn,Fe)PO_4$ and differs from the other mineral lithiophilite by being rich in iron instead of manganese. The structures of the two minerals are the same and form a solid solution, referred as the triphylite series, isomorphous with olivine. Therefore, any differences in physical properties between the two would be related to the iron/manganese percentage. These differences are then best evidenced by comparing the physical properties of the two members at the opposite edges of the triphylite family, namely, $LiFePO_4$ and the often-associated material $LiMnPO_4$, which, at contrast with triphylite and lithiophilite, are artificial ceramics [15]. In addition, triphylite's name in Greek means "family of three" (referring to iron, manganese, and lithium). Any confusion between $LiFePO_4$ and triphylite met recently in the literature should then be avoided.

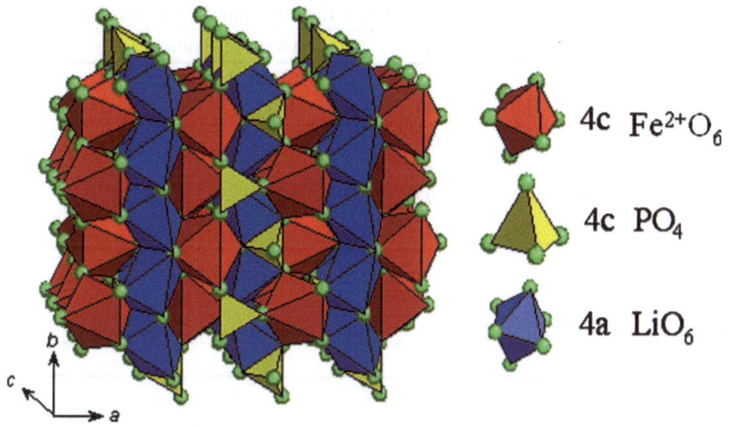

Fig. 8.3 Crystal structure of $LiFePO_4$ olivine. Corner-shared FeO_6 octahedra are linked together in the *bc*-plane, while LiO_6 octahedra form edge-sharing chains along the *b*-axis. The tetrahedral PO_4 groups bridge neighboring layers of FeO_6 octahedra by sharing a common edge with one FeO_6 octahedra and two edges with LiO_6 octahedra

Triphylite alters easily into other phosphate minerals, and geologists show a lot of respect to it, for making the other phosphate minerals possible. This, however, means that this material is not very stable and that it is difficult to make samples of a good quality and well crystallized. This feature, which makes geologists happy, is thus only bad news for physicists. Despite this drawback, physicists also devote a lot of attention to the triphylite series, although their effort has been focused on the definite compounds $LiFePO_4$ and $LiMnPO_4$ rather than their solid solutions, as already mentioned. They belong to the rich family of olivines of the Mg_2SiO_4-type with the general formula B_2AX_4 [16].

The crystal structure of lithium iron phosphate materials has been studied by several authors [16–24]. As a member of the olivine family, $LiFePO_4$ crystallizes in the orthorhombic system (No. 62) with *Pnma* space group. It consists of a distorted hexagonal close-packed oxygen framework containing Li and Fe located in half the octahedral sites and P ions in one-eighth of the tetrahedral sites [16]. The FeO_6 octahedra, however, are distorted, lowering the regular octahedral O_h to the C_s symmetry. This structure is illustrated in Fig. 8.3 showing the channels via which the lithium ions can be removed. Corner-shared FeO_6 octahedra are linked together in the *bc*-plane, while LiO_6 octahedra form edge-sharing chains along the *b*-axis. The tetrahedral PO_4 groups bridge neighboring layers of FeO_6 octahedra by sharing a common edge with one FeO_6 octahedra and two edges with LiO_6 octahedra.

The $LiFePO_4$ structure consists in three nonequivalent O sites. Most of the atoms of the olivine structure occupy the 4*c* Wyckoff position except O(3) which lies in the general 8*d* position and Li^+ ions occupying only the 4*a* Wyckoff position (*M1* site on an inversion center). The Fe magnetic ions are in the divalent Fe^{2+} state and occupy only the 4*c* Wyckoff position (*M2* site in a mirror plane), i.e., the center of the FeO_6 units. As a consequence, Fe is distributed so as to form FeO_6 octahedra

Table. 8.1 Lattice parameters for stoichiometric $LiFePO_4$ materials in the *Pnma* (62) structure

a (Å)	*b* (Å)	*c* (Å)	Unit cell volume (Å^3)	Reference
10.332(4)	6.010(5)	4.692(2)	291.4(3)	[9]
10.334	6.008	4.693	291.39	[14]
10.329(0)	6.006(5)	4.690(8)	291.02	[16]
10.31	5.997	4.686	289.73	[17]
10.3298	6.0079	4.6921	291.19	[19]
10.322(3)	6.008(1)	4.690(2)	290.8(4)	[22]

Table 8.2 Fractional coordinates and site symmetry of atoms in $LiFePO_4$ (*Pnma*)

Atom	*x*	*y*	*z*	Site symmetry
Li	0	0	0	$\bar{1}$(4*a*)
Fe	0.28222	¼	0.97472	m (4*c*)
P	0.09486	¼	0.41820	m (4*c*)
O(1)	0.09678	¼	0.74279	m (4*c*)
O(2)	0.45710	¼	0.20602	m (4*c*)
O(3)	0.16558	0.04646	0.28478	1 (8*d*)

isolated from each other in $TeOc_2$ layers perpendicular to the [001] hexagonal direction [20]. In addition, the lattice has a strong two-dimensional character, since above a $TeOc_2$ layer comes another one at the vertical of the previous one, to build (100) layers of FeO_6 octahedra sharing corners and mixed layers of LiO_6 octahedra and PO_4 octahedra [20]. The lithium iron phosphate material differs from the primary mineral triphylite Li(Mn,Fe)PO_4 by the fact that triphylite is only rich in iron, with some manganese ions also in the *M*2 site. However, while the triphylite is a naturally occurring mineral, $LiFePO_4$ is an artificial product.

On a fundamental point of view, the main interest lies in the fact that the olivine structure generates frustration effects [19]. However, three olivine structure classes can be distinguished as a function of the site occupation by magnetic ions. In Mn_2SiS_4 and Fe_2SiS_4, magnetic ion (Mn, Fe) lies in the *M1* and the *M2* site [22], while in $NaCoPO_4$ and $NaFePO_4$, magnetic ion lies on the *M1* site only [23]. The third class concerns with the phospho-olivine Li*M*PO_4 (M = Ni, Co, Mn, Fe) where the magnetic ion lies in the *M2* site with the *M1* site occupied by a nonmagnetic ion (Li^+).

We used the recent structure determination by Streltsov et al. [18] as a standard reference (Table 8.1). The orthorhombic unit cell of the olivine structure contains 28 atoms ($Z = 4$). Structural parameters and interatomic distances are listed in Tables 8.2 and 8.3. Fe–O distances range from 2.064 to 2.251 Å. The Fe–Fe separation in $LiFePO_4$ is large (3.87 Å), to our knowledge larger than any other iron oxides. The magnetic interactions between Fe ions are believed to be a superexchange type of the form –Fe–O–Fe– or –Fe–O–P–O–Fe–, consistent with the antiferromagnetic ordering observed below $T_N = 52$ K [15].

Table 8.3 Interatomic distance (in Å) in $LiFePO_4$ (*Pnma*)

Fe octahedron		Li octahedron		P tetrahedron	
Fe–O(1)	1 × 2.204(2)	Li–O(1)	2 × 2.171(1)	P–O(1)	1.524(2)
Fe–O(2)	1 × 2.108(2)	Li–O(2)	2 × 2.087(1)	P–O(2)	1.538(2)
Fe–O(3)	2 × 2.251(1)	Li–O(3)	2 × 2.189(1)	P–O(3)	2 × 1.556(1)
Fe–O(3)	2 × 2.064(2)				

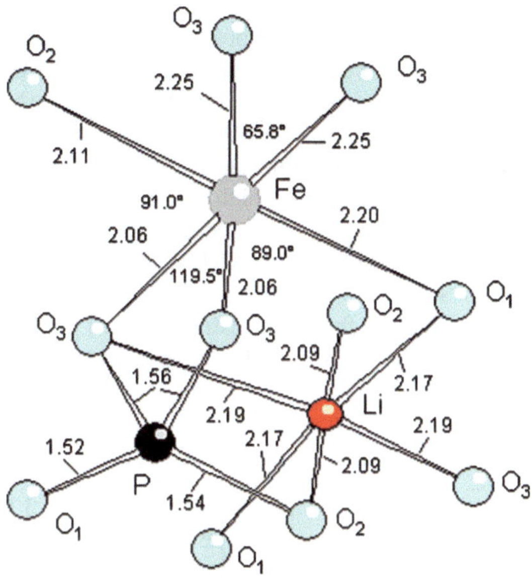

Fig. 8.4 Schematic view of the cation coordination in the $LiFePO_4$ olivine lattice. There are three nonequivalent oxygen atoms noted O_1 to O_3. The distorted FeO_6 octahedron lowers its symmetry from O_h to C_s in the strong crystal field of oxygen atoms (Data on the atomic positions are from Streltsov et al. [18]. Reprinted with permission from Elsevier)

The $LiFePO_4$ lattice consists of six oxygen atoms that surround the Fe 3*d* transition metal atom in an octahedral environment. The primarily O_h local symmetry decreases to C_s symmetry due to the split of the 3*d* levels into e_g and t_{2g} states under the crystal field of oxygen. As shown in Fig. 8.4, the oxygen atoms can roughly be grouped into axial (O_{ax}) and equatorial (O_{eq}) types. The angle of O_{ax}–Fe–O_{ax} is roughly 180°; on the plane perpendicular to the O_{ax}–Fe–O_{ax}, the O_2FeO_2 forms roughly a scissor structure. In the equatorial plane, the Fe–O bond length differs by as much as 0.2 Å, and the O–Fe–O angle is far from 90°. Remarkable structural features include the short O–O distances in the PO_4 tetrahedron, and that three of the six edges are shared with the metal octahedral.

8.3 Structure and Morphology of Optimized $LiFePO_4$ Particles

8.3.1 Crystal Structure

The XRD of the carbon-free and carbon-coated samples is reported in Fig. 8.5. The pattern of the carbon-free sample is characteristic of $LiFePO_4$. The introduction of carbon by spray pyrolysis generates a broad amorphous hump in the count baseline centered at about $2\theta = 22°$ [6]. This broad peak, which is also observed in the XRD pattern of our carbon-coated $LiFePO_4$, gives evidence of the amorphous nature of the carbon deposit. On the other hand, the introduction of carbon by our process does not alter the crystallinity of the $LiFePO_4$ particles, which is evidenced by the crystalline peaks superposed on the amorphous background. The position of the peaks is the same in both samples, which means that the lattice parameters are unaffected by the carbon, a first evidence that the carbon does not penetrate into $LiFePO_4$. In addition, the width of the XRD peaks is about the same in the two samples. According to Scherrer's law, this width is inversely proportional to the size *d* of the crystallites. We can then infer that the size of the $LiFePO_4$ crystallites, hereafter called primary particles, is roughly the same in both samples. To be more

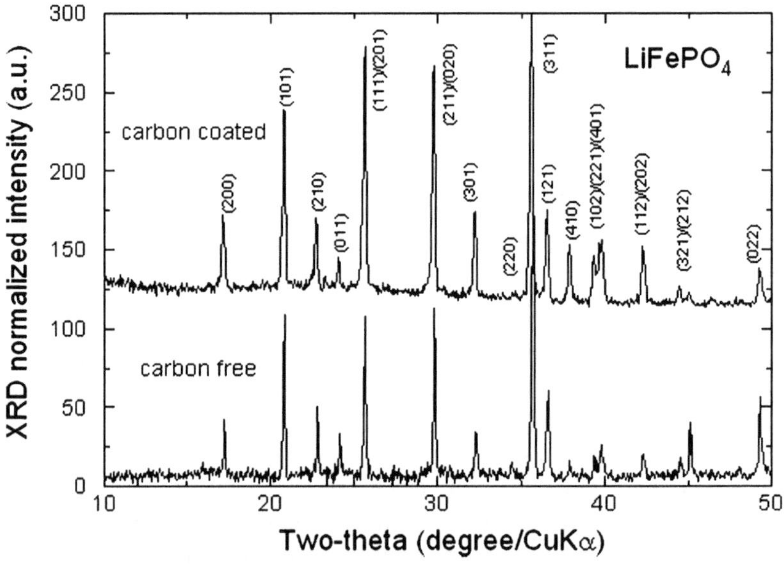

Fig. 8.5 X-ray diffraction diagrams for the carbon-free and carbon-coated $LiFePO_4$ samples. XRD lines are indexed in the rhombohedral system, *Pnma* space group. XRD features are dominated by the four (101), (111) + (201), (211) + (020), and (311) lines. According to the Scherrer analysis, the particle size is estimated to be 36 and 32 nm for the CF and the CC samples, respectively (Reprinted with permission from Elsevier)

specific, the size estimated from Scherrer's law is 36 and 32 nm for the carbon-free and the carbon-coated samples, respectively.

There are two equivalent ways to index the lines of the XRD, depending on the choice of space group *Pnma* or *Pnmb* and therefore of what is called the *a*- or the *b*-axis. Both are listed in the X-ray powder diffraction date files (88-2092, 40-1499) by the American Society for Testing and Materials (ASTM). We have chosen the notation corresponding to *Pnma* ($a = 10.33$, $b = 6.010$, $c = 4.693$ Å). The only difference between the XRD pattern of the carbon-free and carbon-coated samples is the relative intensity of the Bragg peaks. The spectrum of $LiFePO_4$ is dominated by the four (101), (111) + (201), (211) + (020), and (311) lines. Which one of the four lines has the largest intensity is sample dependent. It may be the (211) + (020) line as in the case of the carbon-free sample (also the case reported in the ASTM file 40-1499) or it may be the (311) line, as in the XRD pattern of the carbon-coated sample (also the case reported in the ASTM file 83-2092, and in [13]). These differences, however, relate the disorder of the Li; the common feature of all the $LiFePO_4$ materials investigated in the past is that these four lines have comparable intensities [25].

8.3.2 *Morphology of Optimized $LiFePO_4$*

The surface morphology of the $LiFePO_4$ powders and the shape of the carbon coat have been investigated by scanning electron microscopy (SEM) and high-resolution transmission electron microscopy (HRTEM). Typical SEM images for the carbon-coated sample are reported in Fig. 8.6 a and b. The powders are composed of well-dispersed secondary particles that are slightly agglomerated and show a small quantity of fragments as displayed in the SEM images (Fig. 8.6a, b). The SEM observation shows similar images at any part of the sample, which is homogeneous at a scale large with respect to the area investigated. Figure 8.6a and b is then representative of the particle size distribution, and the average size is 200 nm. Each of the secondary particles is made of a large number of small primary particles that are observed by HRTEM. The HRTEM images for the carbon-coated sample are illustrated in Fig. 8.6c and d. They show polydispersed primary particles with a mean size $\approx$90 nm, which is larger by a factor 3 than the average size of the monocrystallite grains deduced from the application of Scherrer's law on the XRD pattern. Therefore, the primary particles are polycrystallites of $LiFePO_4$ made of a few (three on average) monocrystallites of $LiFePO_4$. The amorphous carbon layer is well illustrated in the TEM pictures (Fig. 8.6c, d). A network formation of carbon appears in the interstitial grain-boundary region, which could explain the electrical continuity between $LiFePO_4$ crystallites. In the micrographs, the $LiFePO_4$ crystallites appear as the darker regions while the carbon coating surrounding the primary particle as the grayish region. An average thickness is estimated to be 30 nm. The carbon film is highly porous, which results in an irregular coating of the crystallites well observed on the SEM and HRTEM images, but the important

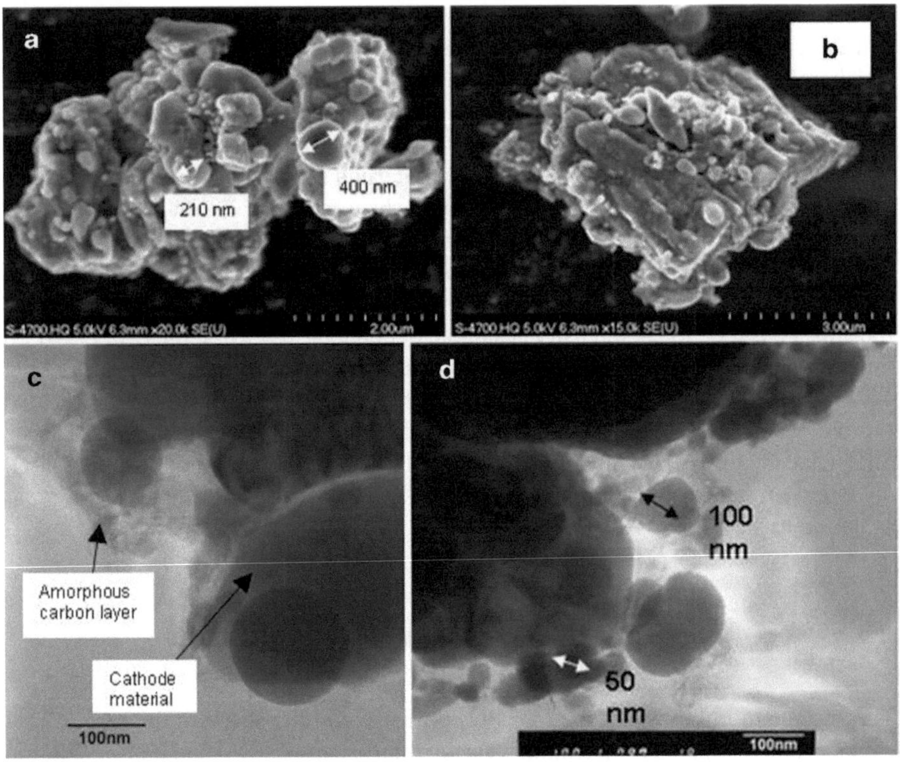

Fig. 8.6 SEM image (**a**) and (**b**) showing the shape of the secondary particles. There are slight agglomeration and small quantity of fragments. Values of the grain size are given in nm. HRTEM images (**c**) and (**d**) showing the amorphous carbon layer deposited onto the $LiFePO_4$ crystallite

point for the electronic conductivity is that it connects the particles. To summarize these results, the SEM and HRTEM images clearly depict a carbon layer coating the $LiFePO_4$ crystallites. XRD and HRTEM data are consistent [26].

8.3.3 Local Structure

Fourier transform infrared (FTIR) spectroscopy probes bulk properties [27, 28], while Raman scattering (RS) spectroscopy is the tool to perform surface analysis [29, 30]; for instance, the amount of carbon on $LiFePO_4$ is too small to be detected by FTIR, but it is well characterized by RS experiments [31]. The vibrational modes of $LiFePO_4$ are primarily due to motion associated with phosphate and iron; the other modes show some lithium contribution [32].

The FTIR spectra of the samples are reported in Fig. 8.7. We have also reported the position of the peaks intrinsic to this material, already identified

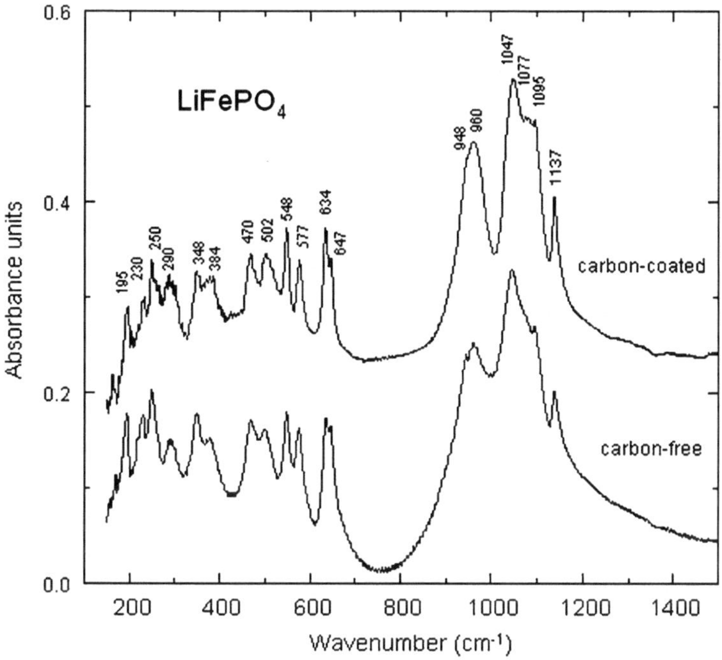

Fig. 8.7 FTIR absorption spectra of the carbon-free and carbon-coated $LiFePO_4$ samples. Peak positions are marked (in cm^{-1}). Infrared spectra were recorded on pellet of $LiFePO_4$ powders diluted into ICs matrix (1:300)

in earlier works [27, 28]. Let us recall that the spectra result from absorption measurements, so that they are a probe of the bulk properties, and the amount of carbon in the material is too small to be detected by such experiments. This is the basic reason why the FTIR spectra are characteristics of the $LiFePO_4$ part. The position of all the IR bands is in agreement with those listed in Table I in [27]. No extra line is observed with respect to pure $LiFePO_4$. The bands in the range 372–647 cm^{-1} are bending modes (ν_2 and ν_4) involving O–P–O symmetric and asymmetric modes, and Li vibrations [32]. In particular, the line at 230 cm^{-1} corresponds to the same cage mode of the lithium ions that undergo translation vibrations inside the cage formed by the six nearest-neighbor oxygen atoms [11]. The bands in this range 372–647 cm^{-1} are thus the part of the spectrum that is sensitive to the local lithium environment. This is also the part of the spectrum that is the same in both the carbon-free and carbon-coated samples. We can then infer from this result that the lithium ions do not "see" the carbon ions, one other evidence that the carbon did not penetrate inside the $LiFePO_4$ particles.

The part of the spectrum in the range 945–1,139 cm^{-1} corresponds to the stretching modes of the $(PO_4)^{3-}$ units. They involve symmetric and asymmetric modes of the P–O bonds, at frequencies closely related to those of the free molecule [32], which explains that the frequencies of these modes are the same in both

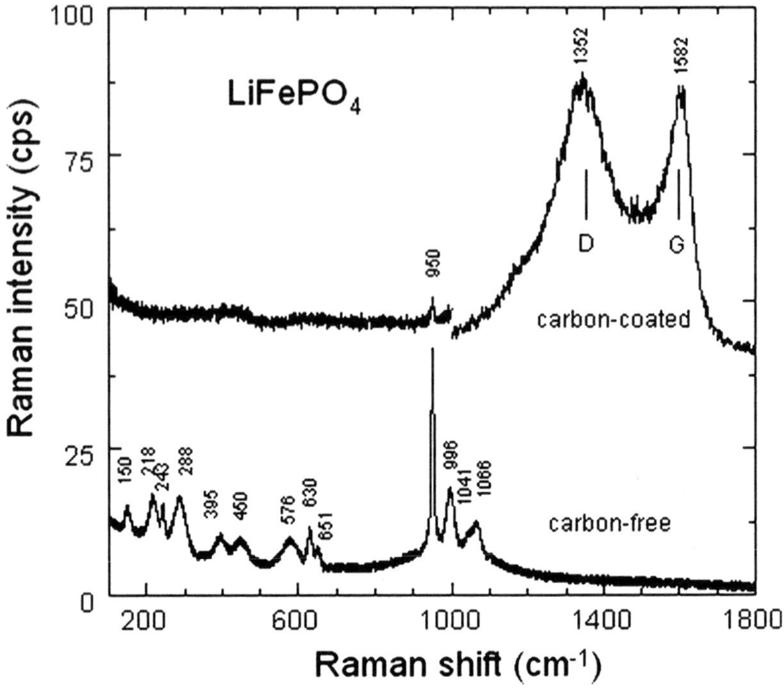

Fig. 8.8 Raman spectra of the carbon-free and carbon-coated $LiFePO_4$ samples. Spectra were recorded using the 514.5-nm laser line at the spectral resolution 2 cm^{-1}. RS features of the $LiFePO_4$ bulk material are screened by the carbon deposit for which the G and D bands are observed (Reprinted with permission from Elsevier)

samples. However, the modes in the carbon-free sample are significantly broader than in the carbon-coated sample. This broadening gives evidence of a decrease in the lifetime of the phonons, and thus the existence of defects breaking the periodicity of the lattice sites inside the $LiFePO_4$ crystallites of the carbon-free sample. The analysis of magnetic properties in the next section will allow us to identify these defects as γ–Fe_2O_3 nanoparticles.

To explore the surface properties of the $LiFePO_4$ particles, Raman spectra have been measured; the penetration depth for carbon with Raman spectroscopy is approximately 30 nm [31]. This is one order of magnitude larger than the thickness of the carbon coat deposited at the surface of the $LiFePO_4$ particles in case of a uniform carbon distribution. Therefore, any screening effect of carbon on the $LiFePO_4$ spectra is not expected. The penetration depth inside $LiFePO_4$ is unknown, but it should be small, so that the detector in the Raman experiments collects the signal within the light penetration depth, which basically represents the total amount of carbon and a few percent of the amount of $LiFePO_4$. Since the total amount of carbon is itself 5 wt.% of $LiFePO_4$, we can expect that comparable amounts of carbon and $LiFePO_4$ are probed by the sampling depth. This is confirmed by the Raman spectra reported in Fig. 8.8. The part of the spectrum in the wave number range 100–1,100 cm^{-1} is the same in the carbon-free and the carbon-coated sample, and only the lines characteristics of $LiFePO_4$ are detected in this range. The peak positions reported in Fig. 8.6 in this range are within a few reciprocal centimeters the same as those that have been reported in [27], and we refer to this prior work for their assignment. The largest difference is for the line at 395 cm^{-1}, which is reported at 410 cm^{-1} in [27]. This line is associated with the PO_4 bending modes ν_2 and ν_4 which are strongly coupled. However, we cannot consider this difference as significant since all the other lines associated with PO_4 have the same position. This is the case in particular for the lines at 620, 940, 986, and 1,058 cm^{-1} associated with ν_4, ν_1, ν_3, and ν_2 intramolecular stretching modes of PO_4, respectively. The only difference in this range of wave numbers is a shift of the Raman lines by about 10 cm^{-1} toward lower frequencies in the carbon-coated sample. This shift of the Raman lines is in contrast with the absence of any shift of the FTIR lines, which gives evidence that it is a surface effect. This shift of the Raman lines is attributable to the increase of the bonding length in the first layers of $LiFePO_4$ particles near the interface with the carbon, taking its origin in the strain induced by the adhesion of the carbon film. For samples with a different mode of preparation, carbon was reported to be responsible for a screening of the signal from $LiFePO_4$, so that only a weak band at 942 cm^{-1} associated with $LiFePO_4$ could still be detected [29]. Again, such a screening is not expected for the reasons above mentioned, and it is not observed in the present case.

8.3.4 Characterization of the Carbon Coating

The main difference between the Raman spectra is at larger wave numbers. Two broad lines at 1,345 and 1,583 cm^{-1} are evidenced in the carbon-coated sample only, as can be seen in Fig. 8.8. These broad lines are a fingerprint of amorphous carbon films. Since they constitute protective optical or tribological coatings [33], a tremendous amount of work has been devoted to amorphous carbon films deposited by a wide variety of methods; this work is reviewed and referenced, for instance, in [28]. These different methods affect both local bonding and intermediate-range orders, so that they lead to a wide variety of films, including amorphous diamond, hydrogenated "diamond-like" carbon, and plasma polymers [34]. All these films have in common the existence of these two broad lines in the Raman spectra, derived from the corresponding features in the spectrum of graphite. The structure at 1,583 cm^{-1} mainly corresponds to the G line associated with the optically allowed E_{2g} zone-center mode of crystalline graphite. The structure at 1,345 cm^{-1} mainly corresponds to the D line associated with disorder-allowed zone-edge modes of graphite. The exact position of the structures in amorphous films depends on the probe laser wavelength [35, 36], so that a quantitative comparison between spectra in the literature is possible only between experiments using the same wavelength. Tamor and Vassell have compared Raman spectra of nearly 100 amorphous carbon films obtained with the same probe laser wavelength (argon line) as the one we have chosen [37]. We have also paid attention to the fact that all the other Raman spectra to which we refer for direct comparison [28], including those obtained for carbon-coated $LiFePO_4$ [29], have also been measured with this probe laser wavelength.

First of all, we note that the Raman spectra of hydrogen-free carbon films can be distinguished from those of hydrogenated films by an additional broad feature centered at 600 cm^{-1} [38]. Since this structure never exists in hydrogenated carbon and always exists in hydrogen-free films, this criterion is considered to be robust [37]. In the present case, this structure is not observed. Therefore, the carbon is hydrogenated, which is actually not surprising, since the preparation process involved different organic additives. As we shall see, however, the amount of hydrogen is only small. Second, the spectrum is characteristic of amorphous graphitic carbon, meaning that the carbon atoms are essentially three-coordinated and bound by sp^2-type hybrid orbitals, in opposition to diamond-like carbon [39]. This result was actually expected since the graphitic carbon is the only carbon type that can be conductive, the reason why the carbon coating was found to be efficient to increase the electronic conductivity in our material. We report hereunder a more complete analysis of the D and G bands since they have been recognized as predictive of the structural as well as physical properties [37]. In particular, a comprehensive study to relate the D and G features in the Raman spectrum to the structure of the disordered graphitic films can be found in [40], while the relation to the physical properties can be found in [37].

The analysis of the D and G lines in such films is always done by fitting the Raman curves in the region from 1,000 to 2,000 cm^{-1} with Gaussians. The number of Gaussians varies from two [37] to four [28]. In our case, we found that the deconvolution of the Raman spectra with two Gaussians (one for the D line, one for the G line) did not give good results, and four Gaussians were necessary to account for the Raman spectra. The result of this fit is shown in Fig. 8.9 and Table 8.4. The Gaussians are centered at 1,344, 1,378, 1,569, and 1,612 cm^{-1}. These lines compare well with the corresponding lines at 1,344, 1,367, 1,591, and 1,622 cm^{-1}, respectively, found by using the same fitting procedure for pyrolyzed photoresists [29] and have then the same origin. The band at 1,569 cm^{-1} can be assigned to the E_g mode of graphite, while the very broad dominant band centered at 1,378 cm^{-1}, which extends over the entire spectral range of carbon vibrations, is the disorder-induced peak characteristic of highly defective graphite [41]. If two Gaussians only are used in the fitting procedure, those are the only structures identified. Among the two extra structures identified in the fit of the spectra by four Gaussians, the band at 1,612 cm^{-1} is typical for severely disordered carbonaceous materials [39, 42]. The origin of the other line at 1,344 cm^{-1} is more questionable. Such a line has been observed in the Raman spectra of polyparaphenylene (PPP)-based carbon prepared at low heat-treatment temperature below 750°C [43]. For this PPP-based carbon, this line was attributed to a quinoid-like inter-ring stretching mode due to a contraction in inter-ring bond length as the PPP chains

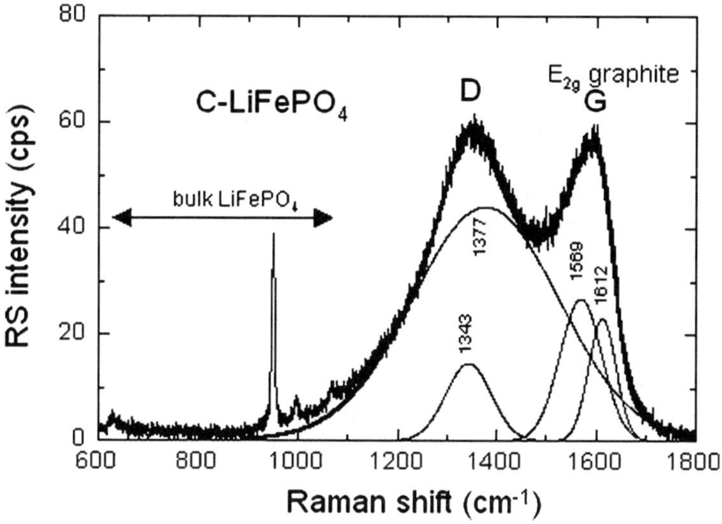

Fig. 8.9 Fit (*thick line*) showing the deconvolution of the Raman spectrum by Gaussians (*thin lines*, identified by their position) of the diamond-like (*D*) and graphite-like (*G*) carbon structures of the Raman spectrum of the carbon-coated $LiFePO_4$ sample (Reprinted with permission from Elsevier)

Table 8.4 Parameters of the Gaussians which fit the G and D lines of the Raman spectrum

Amplitude	Position (cm^{-1})	Width (cm^{-1})
14.6	1,343.7	107.4
44.1	1,377.7	347.6
26.7	1,569.4	99.4
22.9	1,612.7	64.6

are converted into graphitic ribbons, or to a "bridging" of the aromatic rings along the chain by more than one C–C bond [43]. The initial idea that this line was related somehow to the nature of the initial polymer comes from the fact that it is not observed in PPP-based carbon films heated at higher temperatures ($T > 750$ °C), which suggests the reminiscence of some PPP domains at lower temperature. However, the fact that the same peak is observed in carbon films prepared by pyrolyzed photoresists [29] and now in our C-$LiFePO_4$ shows that it is not related to the existence of PPP and should be related to some aromatic rings preferentially formed, irrespective of the original polymer, in the course of the conversion of the carbon into disordered graphite. The fact that the original polymer is unimportant is also evident by the fact that the peak has been observed in many pyrolyzed photoresists, irrespective of the pyrolysis temperature, which could be as high as 1,000°C [29].

8.3.5 Quality of the Carbon Layer

Let us now analyze the other parameter of interest to characterize the carbon film, namely, the intensity of the Raman lines. The ratios of the Raman intensities, defined as the integral of the Gaussians in Fig. 8.9 and Table 8.4, are $I_{1,343.7}/I_{1,377} = 0.102$ and $I_{1,569}/I_{1,612} = 1.789$. If we compare these intensities with the values determined for pyrolyzed photoresists [29], we find that the carbon coating of $LiFePO_4$ has the Raman spectrum of a carbon film deposited on silicon wafers by spin coating and then pyrolyzed at a temperature in the range 800–860°C. The remarkable result is that our carbon film in the present case has been obtained by heating at 700°C only. This temperature difference is critical for the electric conductivity of the carbon film since the sheet resistance of a pyrolyzed carbon sheet is highly resistive when the carbon film has been prepared at a pyrolysis temperature $T_p = 700$°C; the resistivity decreases dramatically for higher pyrolysis temperatures to reach a sheet resistivity of 10 ohms per square at $T_p = 1{,}000$°C. We can then expect, on the basis of the Raman spectra, that the conductivity of the carbon in the carbon-coated $LiFePO_4$ is comparable to that of the carbon deposited by pyrolysis at 850°C, which means reasonably good. It explains the successful increase in the electronic conductivity that has been reported in the literature for carbon-coated $LiFePO_4$. Incidentally, it shows that if the efficiency of the carbon-coating process in $LiFePO_4$ would have not been improved with respect to the pyrolysis technique, the result would have been a total failure because it is not possible to heat $LiFePO_4$ above 800°C without damaging the material, resulting in the growth of inclusions of different chemical compositions mentioned earlier in this work.

The width of the G line at 1,569 cm^{-1} is 99.3 cm^{-1}, characteristic of hydrogen-free (a-C) carbon layers and markedly larger than the width of this line in the hydrogenated amorphous carbon (a-C:H) [37]. This gives evidence that although there is some hydrogen in the carbon deposited on $LiFePO_4$ for the reason already mentioned, the H/C ratio is very small. This is actually consistent with the fact that the dramatic increase in the electronic conductivity after pyrolysis at temperatures above 700°C is due to a decrease in the H/C ratio [29]. Since the Raman spectrum is that of pyrolyzed carbon obtained at a significantly higher temperature of 800–860°C, the ratio H/C must be small indeed. For the same reason, we expect the hardness of the carbon deposit is comparable to that of a-C films. However, the carbon films investigated in the literature are thick, so that the hardness is an intrinsic property that does not depend on the substrate. We have already noticed that the average thickness of the deposit is 30 nm. This thickness is not large enough to guarantee that the adherence on the $LiFePO_4$ particles does not quantitatively affect the hardness of the a-C film since the strain interactions are long range, but it is large enough to insure that the order of magnitude of the hardness is unaffected. Although the hardness of the a-C:H films increases from 0 up to 20 GPa when the G-line width increases from 50 to 80 cm^{-1}, the hardness of a-C film with a G-line width 100 cm^{-1} is just in the middle, namely, 10 GPa [37]. This hardness can be qualified as small. For instance, a hardness up to 80 GPa has been reported for "diamond-like i-C" carbon films. We then qualify the hardness of the carbon deposited on the $LiFePO_4$ as small. This is actually expected, especially as the substrate on which the carbon is deposited is not flat as in the case of silicon wafers, but is the bent surface of nanoparticles. We can even consider that this a-C structure chosen by the carbon is an example of self-adaptation to allow for an adhesion on such a surface which would be impossible with a strong hardness.

The D/G intensity ratio is used in the literature to determine the size of the graphite particles in polycrystallite carbon [39]. Some extension has often been made to use the same relation to determine the correlation length of the graphitic order. This is, however, a confusion already outlined in [39]. The D/G intensity ratio gives the size of particles in the absence of any disorder and should not be confused with the loss of long-range order in amorphous materials. In disordered carbon, the information on the disorder is provided by the optical gap, according to the Robertson and O'Reilly law, which allows for an estimate of the number of carbon rings inside a local cluster [44]. In particular, the simultaneous study of both the optical gaps and the Raman D/G ratio has revealed contradictions that show the D/G intensity ratio is determined by factors other than the graphitic cluster size in amorphous carbon [45–47]. This point is sometimes missed, and we can find recent analyses on C-$LiFePO_4$ that postulate that decreasing D/G intensity ratio is related to the carbon disorder [29]. In this same analysis, it is postulated that decreasing D/G intensity also means decreasing sp^3/sp^2 ratio. This is not justified either, and it is not possible to evaluate the content of sp^2 and sp^3 coordinated carbon in a material that is dominantly graphitic. This is because the intrinsic Raman intensity of the graphite spectrum is 50 times that of the diamond spectrum. Therefore, Raman spectroscopy is a sensitive tool to detect residual sp^2 bonds in diamond, but it is not a reliable test of the presence of sp^3 bonds in a dominantly graphitic carbon [37, 39]. In the carbon-coated samples investigated in this work, we did not investigate the optical gaps in the present study, but we note that the width (not the intensity) of the Raman lines is related to the degree of carbon disorder, which shows that in the present case, the carbon is amorphous. We do not know the sp^3/sp^2 ratio, but we know that the amount of sp^3 is small. In addition, this is always the case for disordered carbon. Even in diamond-like carbon films, the percentage of tetrahedral carbon is small [22, 29]. In the present case, however, the percentage should be even smaller than in most cases because the position of the D and G lines is quite close to those of graphite.

Note a bending of a graphite sheet is expected to induce some sp^3 character into the sp^2 bonds, which are planar. Therefore, the small amount of sp^3 gives evidence that the bending is small, i.e., that the radius of curvature is large at the scale of the bond length. This is consistent with the HRTEM spectra in Fig. 8.4c and d, which show that the carbon coats the secondary particles with a typical radius of 30 nm and does not penetrate into the $LiFePO_4$ particles.

8.4 Magnetic and Electronic Features

8.4.1 Magnetic Properties

Well-crystallized $LiFePO_4$ is antiferromagnetic (AF), with a Néel temperature $T_N = 52$ K [17, 48, 49]. The topology of the AF order has been determined by neutron experiments [48, 49], from which the magnetic interactions have been determined [50]. The dominant interactions that fully account for this structure is the intralayer superexchange Fe–O–Fe interaction J_1 and two super-superexchange interactions Fe–O–O–Fe, namely, an interlayer interaction J_2 and an intralayer interaction J_b. Other interactions envisioned in earlier works [51] turn out to be negligible. All the interactions J_1, J_2, and J_b are antiferromagnetic, and their estimated values are the following [50]:

$$J_1 = -1.08\,\text{meV}, \quad J_2 = -0.92\,\text{meV}, \quad J_b = -0.4\,\text{meV}. \tag{8.1}$$

Interesting enough, this recent result shows that the FeO_4 layers are strongly coupled antiferromagnetically. This is in essence why the system undergoes a true transition to three-dimensional antiferromagnetic ordering, while a 2D magnetic system does not order because of enhanced quantum spin fluctuations. In addition, the J_2 and J_b cannot cause geometric frustration of the magnetic interactions, in contrast with prior claims, because J_b is significantly smaller than J_1.

Figure 8.10 shows the isothermal plots of the magnetic moment versus applied magnetic field for A-type and optimally synthesized $LiFePO_4$ (B type). From these results, we clearly observe a difference between the magnetic properties of the phospho-olivine materials. The B-$LiFePO_4$ sample displays a linear variation in its $M(H)$ curve, while a nonlinear behavior is observed at low magnetic field for the A-type $LiFePO_4$. This curvature is signature of ferromagnetic impurities [9, 11, 52, 53]. While nano-sized ferromagnetic particles were evidenced in previously prepared $LiFePO_4$, such clusters do not exist in the optimized LFP. These results illustrate that the magnetization $M(H)$ is the superposition of two contributions:

$$M(H) = \chi_m H + M_{\text{extrin}}; \tag{8.2}$$

the intrinsic part, $\chi_m H$, that is linear in the applied magnetic H and an extrinsic component,

$$M_{\text{extrin}} = Nn\mu\,£(\xi), \tag{8.3}$$

that is easily saturated by the application of H are due to ferromagnetic impurities. Here, $£(\xi) = £(n\mu H/k_B T)$ is the Langevin function, N is the number of magnetic clusters, and each cluster is made of n magnetic moments μ. T is the absolute temperature and k_B the Boltzmann constant.

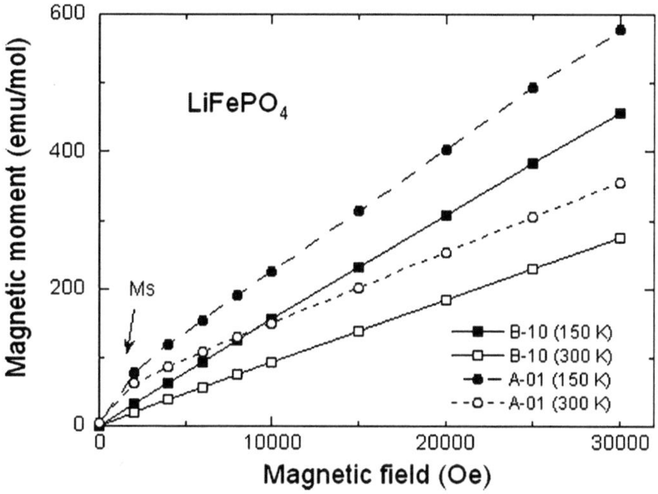

Fig. 8.10 Isothermal curves of the magnetic moment versus applied magnetic field as a function of temperature for A-type and optimized (B-type) $LiFePO_4$ sample. *Symbols* are experimental data; the *straight continuous lines* are guides for the eyes (Reprinted with permission from Elsevier)

Analysis of the A-type samples of Fig. 8.10, however, gives an average separation of the magnetic clusters that is too large for interaction between particles (superparamagnetic model). This hypothesis must be abandoned where the number *n* of magnetic clusters is so large that magnetic interactions between the ferrimagnetic particles become important [9]. At high fields, M_{extrin} saturates to $Nn\mu$ so that this quantity is readily determined as the ordinate at $H = 0$ of the intersection of the tangent to the magnetization curves at large fields. As a result, we find that $Nn\mu$ does not depend significantly on temperature below 300 K. We are in the situation where the cluster magnetization is temperature-independent, which amounts to saying that the Curie temperature Tc inside the clusters is much larger than 300 K. This is important information on the nature of the ferromagnetic clusters. In particular, this feature precludes the existence of Fe_2P clusters in some $LiFePO_4$ samples prepared according to a different procedure [52] since the Curie temperature of these clusters is only 220 K. The nature of the strongly ferromagnetic clusters in the present case is most likely maghemite (γ–Fe_2O_3).

The combination of ESR spectroscopy and SQUID magnetometry has shown that the fraction of iron in the Fe^{3+} configuration falls to a residual impurity concentration lower than a few tenths of a ppm. This is the net result of the thermal effect, the carbon coating, and also the careful selection of the Fe^{3+} precursor [53]. The other structural properties show that carbon does not penetrate significantly inside the $LiFePO_4$ nano-sized particles.

Figure 8.11 shows the temperature dependence of the reciprocal magnetic susceptibility of $LiFePO_4$ samples. The $\chi_m^{-1}(T)$ curves are in quantitative agreement with a prior work [17]. The insert shows the cusp of the transition from

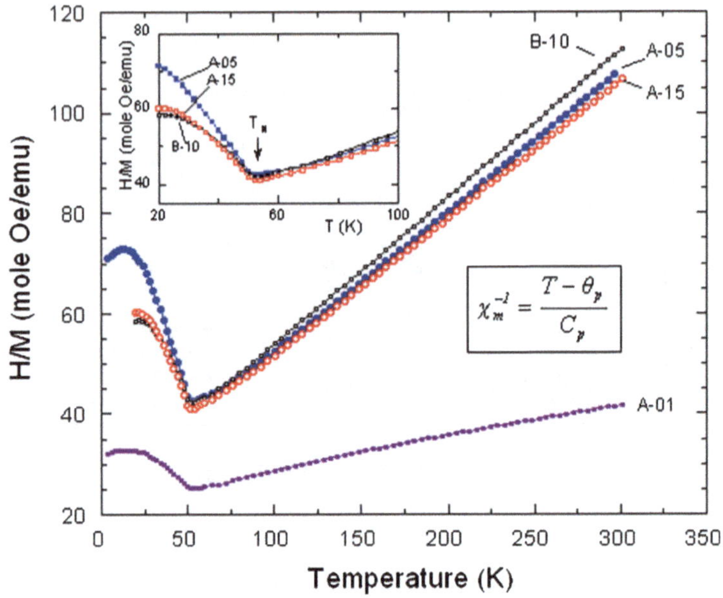

Fig. 8.11 Temperature dependence of the reciprocal magnetic susceptibility of $LiFePO_4$ samples investigated in Fig. 8.8. The best material (B-10) displays the lowest Curie constant 3.41 emu K mol^{-1}. Insert shows the cusp of the transition from antiferromagnetic ordering to the paramagnetic range at $T_N = 52$ K

antiferromagnetic ordering to the paramagnetic range at $T_N = 52$ K. It is remarkable that the A-type sample displays different magnetic features due to the existence of ferrimagnetic impurities. The first consequence is an ambiguity in what is called the magnetic susceptibility χ_m since M/H is distinct from dM/dH. The magnetic susceptibility measured with a SQUID at $H = 10$ kOe shows the nonlinearity of the magnetic moments, which is due to the presence of γ–Fe_2O_3. The best material (B-20 sample) shows the lowest Curie constant 3.41 emu K mol^{-1}. The effective magnetic moment $\mu_{eff} = 5.22\ \mu_B$ is close to the theoretical value 4.90 μ_B calculated from the spin-only value of Fe^{2+} in its high-spin configuration. Departure from the spin-only value may reflect the presence of Fe^{3+} ions and/or an orbital-momentum contribution from the Fe^{2+} ions [53].

8.4.2 Electron Paramagnetic Resonance (EPR)

The derivative signals of the absorption spectra of the carbon-free sample are reported in Fig. 8.12. Let us recall that no EPR signal is detected in $LiFePO_4$ in the absence of magnetic clusters [15]. The EPR signal is then an evidence of magnetic clustering [9, 11]. For uncorrelated magnetic clusters, one expects a signal characteristics of a gyromagnetic factor $g = 2$. At the frequency used in the

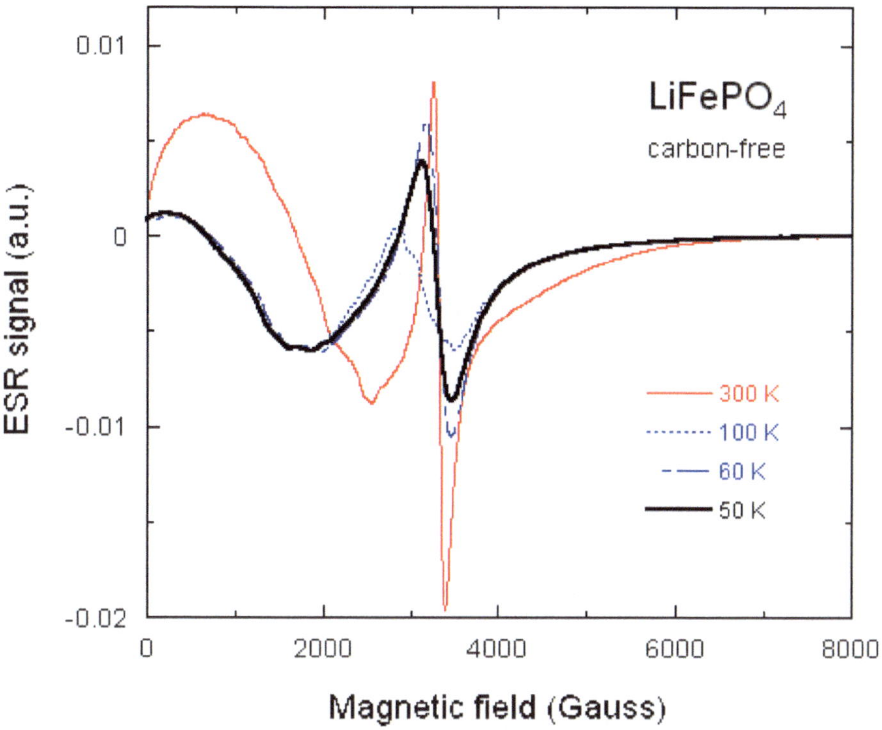

Fig. 8.12 Electron spin resonance spectrum for the carbon-free $LiFePO_4$ sample at several temperatures indicated in the figure. Note the unit is arbitrary, but it is the same one as in the previous figure, so that the relative intensity between the spectra of the two samples is given by the ratio of the ordinates between the spectra in the two figures (Reprinted with permission from Elsevier)

experiments, such a signal is centered at $H = 3{,}300$ G. Indeed, this signal, already detected in other $LiFePO_4$ samples that contained ferrimagnetic particles [9, 11], is also detected in the present work and has a comparable shape. The structure at 3,300 G has the same width. In addition, a large signal is also detected at smaller magnetic fields; it is associated with a strong rise in the magnetization curves. The spectrum of the carbon-coated sample is reported in Fig. 8.13; it shows a dramatic decrease in the ESR signal. The integrated strength of the structure at 3,300 G is 100 times smaller than in the carbon-free sample. This feature is clearly related to the fact that we did not observe any ferrimagnetic nanoparticles in the carbon-coated sample. This very small signal is the signature of uncoupled spins in such a small concentration that it may be due to any residual impurities or defects. The structure at 3,300 G is much broader than in the carbon-free sample, which suggests that the origin of the signal is different. Instead of being due to ferrimagnetic clusters, the signal might be due to some defects, such as an unpaired electron spin associated with an oxygen vacancy or to a defect such as an Fe^{3+} ion associated with a lithium vacancy. This kind of defect is very sensitive to its local neighborhood, which might explain the large broadening of the ESR line.

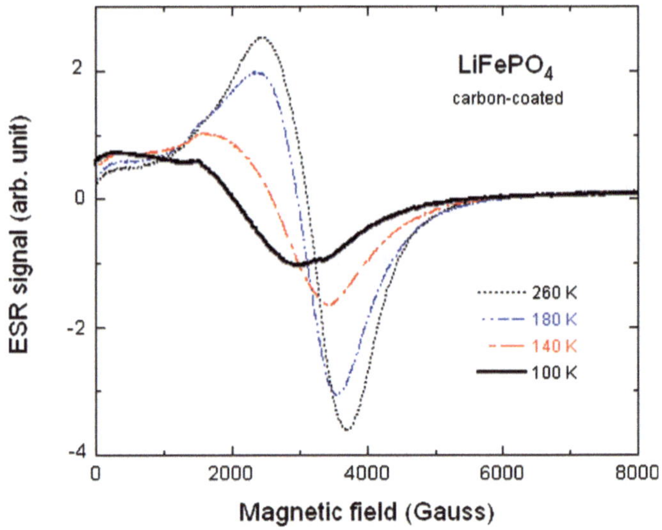

Fig. 8.13 Electron spin resonance spectrum for the carbon-coated $LiFePO_4$ sample at several temperatures indicated in the figure (Reprinted with permission from Elsevier)

8.5 Aging of $LiFePO_4$ upon Exposure to H_2O

8.5.1 Introduction

It is well known since decades that all the lithium-ion batteries need to be protected against humidity [53]. The main reason is that lithium is very reactive with water, producing lithium hydroxide (LiOH), which is a corrosive alkali hydroxide. When crystallized, it is a white hygroscopic material. It is also soluble in water, a property that has been used to investigate aqueous lithium hydroxide as a potential electrolyte in lithium-ion batteries with $LiFePO_4$ cathode [54]. Since the carbon coat is not a barrier for Li^+ ion transport (the reason for the success of C-$LiFePO_4$ as a cathode element of lithium-ion batteries), we expect the reaction with the water to be effective, implying extraction of Li ions from the $LiFePO_4$ to interact with water. As we shall see in the present work, this delithiation is the only effect that is observed in case of exposure to H_2O in air, and yet, if affects only the disordered surface layer of the particles. Thus, the molecules formed by reactivity of water molecules with Li^+ ions at the surface of powders will be detected by Raman spectroscopy.

Despite this reaction of lithium to humidity, the drop of the $LiFePO_4$ particles in water is used in the laboratory to check the carbon-coating process by separating coated and uncoated particles [55]. As a result, when the C-$LiFePO_4$ powder is dropped into water, part of the carbon that links the particles unties and floats at the surface, retaining with it some of the particles, and the major part sinks. More recently, Porcher et al. have determined that the exposure of C-$LiFePO_4$ particles to water results in the formation of Li_3PO_4 thin layer (few nm thick) at the surface of the particles [56], as a result of the migration of Fe in the water. In the present work, we investigate the effect of water on carbon-coated $LiFePO_4$ particles and analyze both the particles that have sunk and the floating part. We report that the water attacks the particles and that the carbon coat is not a protection because it detaches and is not watertight. We find that Fe is not the only element that reacts with water that contains also P and Li species after immersion of the $LiFePO_4$. However, quantitative analysis shows that only a thin layer near the surface reacts with water, for reasons that are discussed in this part.

A strong interaction between $LiFePO_4$ with H_2O molecules was not necessarily expected. After all, parkerization of iron is an industrial process that amounts to drop iron into a hot bath with manganese phosphide, which results in the formation of a thin layer of $FePO_4$ at the surface. Since iron phosphate is hydrophobic, this layer protects the iron against oxidation and corrosion. An intuition would then have suggested that, upon immersion of $LiFePO_4$ to water, a delithiation in a thin layer at the surface would lead to the formation of a $FePO_4$ layer that would protect the particles against any other damage. Our investigations, however, show that the situation is slightly more complicated.

In this section, we present the characterization of the sample before and after immersion in water. Both the floating part and the sinking part have been investigated as a function of time spent in water. We also report the chemical analysis of the aqueous solution that gives us some insight on the reaction of $LiFePO_4$ on water. Since the immersion in water is a rather dramatic event that does not give any opening to the carbon coat to isolate the $LiFePO_4$ surface from H_2O molecules, we have also explored the case where the material only suffers the exposure to humidity of ambient air. Finally, we also report on hygrometry of $LiFePO_4$ and its consequence on the aging of the electrochemical performance of this material upon exposition to ambient atmosphere at the scale of few months.

8.5.2 *Reaction of Water on $LiFePO_4$*

$LiFePO_4$ reacts with H_2O, particularly with the SSR sample, which is evident by the yellow color in the water. The HTR sample appears to be less reactive, since the water remains uncolored. Note, however, that this attack is not related to impurity phases because we know how to detect impurity phases when they exist, by magnetic plus FTIR experiments (see our prior work [9] and references herein). From these analyses that we performed before exposure to moisture, we know that there is no impurity phase in these samples (except for a residual concentration of Fe_2O_3 for one sample only, as will be shown below).

Table 8.5 Inductively coupled mass spectroscopy analysis of the liquid in which the SSR- and HTR-$LiFePO_4$ samples have been immersed

	NMP after 24 h		H_2O after 24 h	
Sample	Fe (ppm)	P (ppm)	Fe (ppm)	P (ppm)
SSR	6.2	6.3	150	400
HTR	0.3	2.3	130	300

A more quantitative characterization of the reaction of the samples with water is provided by inductively coupled mass (ICM) spectroscopy analysis of the liquid in which the SSR- and HTR-$LiFePO_4$ samples are immersed. For comparison, the same experiments were performed with the samples immersed in *N*-methylpyrrolidone (NMP), since this solvent is generally used to dissolve the polyvinylidene fluoride (PVDF) most widely adopted as a binder for electrodes in the lithium-ion cell manufacturing processes. The results are reported in Table 8.5. The same amount of product and same volume of liquid (either NMP solvent or water) have been used in all the experiments, so the results can be compared quantitatively. The amount of P and Fe ions in the liquid is negligible (few ppm only) in the case of immersion in NMP, so that we can conclude the $LiFePO_4$ does not react with this product. On another hand, the amount of P and Fe dissolved in the liquid is increased by almost two orders of magnitude in case NMP is replaced by water.

The water in which the HTR sample has been immersed remains transparent despite the presence of Fe and P elements. This feature suggests that Fe and P mainly belong to solvable species. If the PO_4^{3-} phosphate ion is conserved in the dissolution process, we can envision the formation of lithiophosphate Li_3PO_4. However, it is nonsoluble in water. There are different iron phosphate ions, either ferric or ferrous. As mentioned in the introduction, ferric phosphate $FePO_4$ is waterproof, and should stay at the surface of the particles to protect it against further dissolution. This will be confirmed hereunder from the analysis of the physical properties. Ferrous phosphate $Fe_3(PO_4)_2$ which gives rise to numerous minerals $Fe_3(PO_4)_2/nH_2O$, like ludlamite ($n = 2.4$) and vivianite (up to $n = 8$), is also unexpected because the ratio of the P over Fe concentrations dissolved in water is $[P]/[Fe] = 2.5 \pm 0.2$ after Table 8.5, while this ferrous phosphate would lower this ratio to a value smaller than one. The large value of [P]/[Fe] does not preclude the formation of such materials, but it actually requires that P enters dominantly in one of the three less-charged ions that can be formed in aqueous solution: hydrogen phosphate ion, HPO_4^{2-}, and dihydrogen phosphate ion $H_2PO_4^-$. Aqueous orthophosphoric acid H_3PO_4 is ruled out by the fact that the pH of the aqueous solution in which both the SSR and HTR samples have been immersed is basic, namely, 9.5 in both cases. Indeed, $Fe(H_2PO_4)_2$ is soluble in water. In addition, this material is formed during the phosphatation of iron, an industrial process used to passivate the surface of iron compounds. The lithium reacts with the water to form LiOH that is soluble. In addition, Li can also react with the carbon to form lithium carbonate $LiCO_3$ that is also soluble. Note that the formation of $Fe(H_2PO_4)_2$ leads to $[P]/[Fe] = 2$. This value is only slightly smaller than the experimental value, but

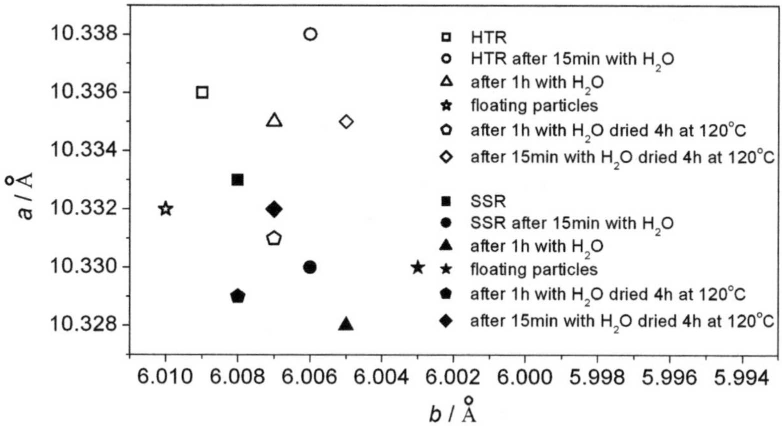

Fig. 8.14 Lattice parameters *a* and *b* for both SSR- and HTR-$LiFePO_4$ particles. The data labeled "SSR" and "HTR" are obtained before immersion in water. The other results have been obtained after different treatments indicated in the figure. When the label "floating" is indicated, it means that the result has been obtained for the floating part of the sample, otherwise the result is for the sinking part

still the difference exceeds the experimental uncertainty. There should then be a residual concentration of another phosphate that does not involve the iron element. We then preclude a small amount of Li_3PO_4 particles in suspension in the aqueous solution. For instance, in the aqueous solution for HTR, where [P]/[Fe] = 2.3, we can envision a corrosion that would satisfy this ratio

$$3Fe(H_2PO_4)_2(aq) + [Li_3PO_4]. \tag{8.4}$$

Note, however, that this equation is certainly not sufficient to summarize the whole corrosion, as the reaction in Eq. 8.4 associated to the formation of LiOH will also hold, among others. Note also the amount of lithium phosphate particles, if any, must be small enough so that it does not affect the transparency of the solution.

The XRD patterns of SSR-$LiFePO_4$ particles indicate samples from the floating and sinking part after 1 h in water do not differ significantly from the XRD spectrum before immersion in water. The lattice parameters are not significantly modified either, as illustrated in Fig. 8.14, where the crystallographic parameters *a* and *b* are reported after different times of exposure to water. For all samples, irrespective of the time spent in water, $a = 10.333(5)$ Å, $b = 6.006(4)$ Å, and $c = 4.703(5)$ Å, while the unit cell volume is 291.8(7) $Å^3$. These parameters are evidence that the bulk of the LFP particles is unaffected by water, only the surface layer is affected. However, XRD is not sufficiently sensitive to probe this surface layer for two reasons: the surface/volume ratio is too small for these big particles and the surface layer is disordered. Therefore, detecting changes in the surface by water will require other types of experiments.

8.5.3 Quantitative Characterization

Magnetic measurements have been used to detect surface effects [57]. The magnetization $M(H)$ of the SSR sample before immersion is linear in field H up to 30 kOe, and the magnetic susceptibility is trivially defined as $\chi = M/H$. The $\chi^{-1}(T)$ curve (not shown here) is similar to the curves that we have already published for this material, and the Curie–Weiss law is satisfied in the paramagnetic regime. For the HTR sample, the magnetization curves are linear in H for $H > 1$ kOe, but at contrast with the SSR case, they extrapolate to a finite magnetization M_0 in the limit $H \rightarrow 0$. This feature is characteristic of a residual amount of γ–Fe_2O_3 impurity. Although the amount of γ–Fe_2O_3 is very small (less than 0.1% of the product), this extrinsic contribution to the magnetization must be subtracted from the total magnetization to determine the intrinsic magnetic susceptibility that is now defined as $\chi = (M - M_0)/H$ that satisfies the Curie–Weiss law. The effective magnetic moment μ_{eff} carried by the iron ions can then be deduced. The result is $\mu_{eff} = 5.38$ and 5.36 for SSR and HTR samples, respectively. The value of μ_{eff} for the samples that do not contain any Li vacancy is in agreement with the theoretical value 4.9 of Fe^{2+} in the high-spin state, but values such as those of SSR and HTR are often met due to a small concentration (lower than 1%) of Li vacancy in the bulk of the LFP particles. After immersion in water, the magnetic moment of the sinking part increases by 0.04 μ_B in both SSR and HTR samples, to reach $\mu_{eff} = 5.42$ and 5.40, respectively. This value of the magnetic moment is obtained very fast. For technical reasons, the shortest time in which the samples have been investigated is 15 min, where this limit for μ_{eff} was already achieved. The magnetic moment stays as this value even when the samples stay in water for longer times (up to 1 h). This increase of μ_{eff} in the short time limit is the signature of an oxidation of iron from Fe^{2+} to Fe^{3+} at the surface of the samples, the evidence of a delithiation of the surface layer.

To quantify this effect, we first need to model the size distribution of the particles [54]. If d is the effective diameter for an equivalent spherical particle, the surface/volume ratio is approximated in the large d limit by $\pi d^2/(\pi d^3/6) = 6/d$ (with d measured in a-unit). Then it is straightforward to compute the surface to volume ratio for our sample r_s:

$$r_s = \int \mathrm{d}d q_e(d) r(d). \tag{8.5}$$

The result for HTR sample is $r_s = 5 \times 10^{-3}$. From the fraction x of iron ions at the surface that switched from Fe^{2+} to Fe^{3+}, we estimate the thickness of the surface layer that has been delithiated $xa/r_s = 3.3$ nm. The thickness of the layer damaged by the exposition to water, however, is slightly underestimated here because this calculation deduced from magnetic properties does not take into account the dissolution of Fe and P ions in the water. Converting the amount of Fe and P ions reported in Table 8.5 in mass product, we find that 1.3% of iron and 3% of phosphorous (actually PO_4 ions since this unit is very stable) in the SSR sample have been dissolved in the water. This is comparable to the value $x = 4\%$ we have

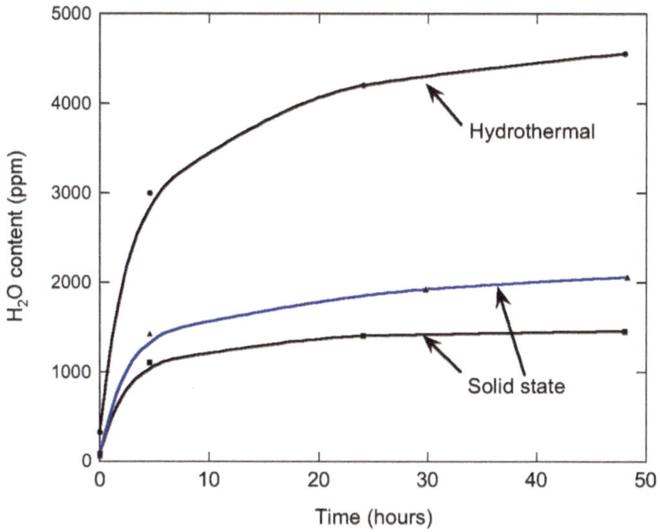

Fig. 8.15 Moisture content of several C-$LiFePO_4$ samples, expressed in ppm of H_2O with respect to $LiFePO_4$ as a function of time exposure in ambient air at 21°C (55 % relative humidity). The *symbols* are experimental results. The *lines* are only guide for the eyes obtained by smooth fit of the data. None of these samples, however, correspond to the HTR and SSR samples investigated (Reprinted with permission from The Royal Society of Chemistry)

found for the fraction of iron ions converted from Fe^{2+} to Fe^{3+} valence states, but still, it is smaller. This result corroborates that the main effect of exposure to H_2O has been the delithiation of the surface layer to form a $FePO_4$ layer that protected the particles against further reaction.

The evolution of the $LiFePO_4$ surface with time during exposure to H_2O is better investigated in experiments where the material is exposed to ambient air rather than immersion in water. The reason is obviously that the concentration of H_2O that interacts with the surface of the particles is much lower in air. In addition, we expect that degradation of the particle surface layer is lower, at least in the timescale of the experiments, because iron ions dissolve in the aqueous solution, but they cannot evaporate. This is another reason why the main effect will be delithiation in the surface layer, as a consequence of the hydrophilic nature of Li. The measurement of the moisture content of the samples exposed to ambient air is an indirect means of investigating this reaction. The results are illustrated in Fig. 8.15 for several samples.

8.5.4 Voltammetry of Water-Exposed LFP Samples

The voltammetry measurements of the SSR and HTR samples (sinking part) after immersion for 1 h in water are reported in Fig. 8.16. In these measurements, an initial 3.2 V working potential is applied. Then, the voltage was varied at the rate

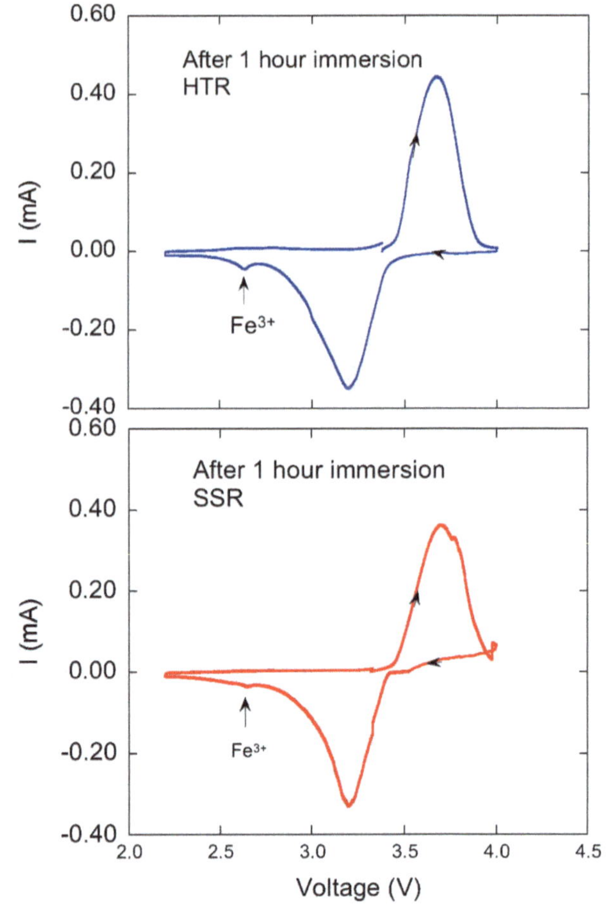

Fig. 8.16 Electrochemical performance of the C-$LiFePO_4$ (HTR sample)/$LiPF_6$-EC-DEC/Li cells at room temperature. The results are shown before immersion of this sample in water, and after immersion during 63 h, then dried during 48 h at 85°C

1.25 mV min^{-1}, as shown in the figure: increase of the voltage up to 4 V, followed by a decrease to 2.2 V, and an increase again up to 3.2 V. Besides the peak associated with Fe^{2+}, the part of the curve obtained by decreasing the voltage shows a secondary peak at 2.63 V that is characteristic of the Fe^{3+} in iron oxide (versus more than 3.5 V in phosphate) [58]. The presence of Fe^{3+} ions in both SSR and HTR samples confirms the delithiation of the surface layer evident in the previous sections. On the other hand, upon increasing the voltage again, this signal disappeared, which shows that the voltammetry curve of the samples before exposure to H_2O was recovered. Therefore, the surface layer was lithiated again during Li insertion, and the effect of immersion in water was reversed. The same holds true for longer immersion times of a few days. In the following experiments, the samples were immersed for 63 h. Then, the samples were dried for 48 h at 85°C. Moreover, it should be noted that the open-circuit voltage (OCV) decreased by 2.3% by immersion in water. Since the OCV is directly related to the state of charge of the battery, it can be viewed as an indirect measurement of the delithiation rate of

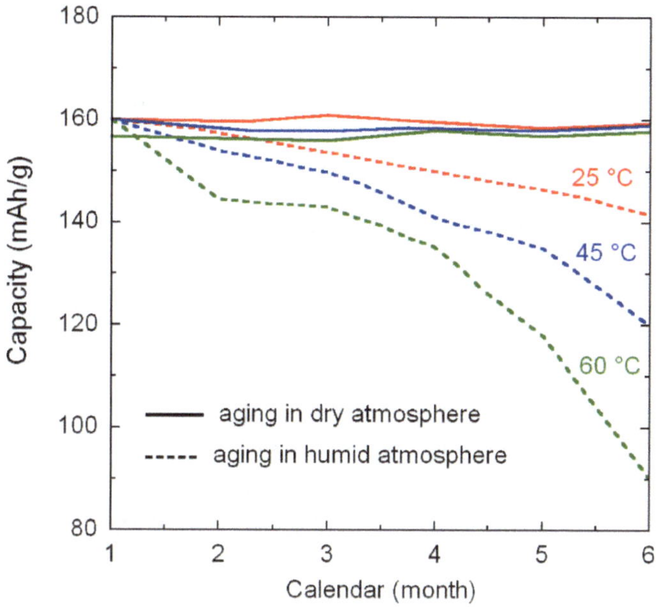

Fig. 8.17 Capacity of the C-$LiFePO_4$ (HTR sample)/$LiPF_6$-EC-DEC/Li cells as a function of time spent in dry atmosphere and in ambient atmosphere (55% relative humidity), at three different temperatures. The temperatures at which the full curves (in dry atmosphere) have been obtained can be distinguished by the fact that they do not overlap and the property that the lower the temperature, the higher the capacity is (Reprinted with permission from Elsevier and The Royal Society of Chemistry)

the battery. Indeed, this result is fully consistent with the 4% delithiation rate deduced from the magnetization measurements and the 1% and 3% loss of Fe and P in the immersion process estimated from the physical and chemical analyses. It thus fully confirms that the delithiation process is located in the surface layer.

The effect of H_2O on the electrochemical properties was also evaluated by exposure of the sample to ambient air. This is illustrated for the HTR sample in Fig. 8.17 that shows the change of the capacity as a function of time at different temperatures in dry atmosphere and in ambient air (55% relative humidity).

8.6 Electrochemical Performance of Optimized $LiFePO_4$

8.6.1 Postmortem Analysis

Here, we present an overview of the high-temperature performance for an optimized B-type sample, i.e., carbon-coated (C-LFP). The coffee-bag cell was charged and discharged at C/8 for the first cycle followed by 12 cycles at C/4 with

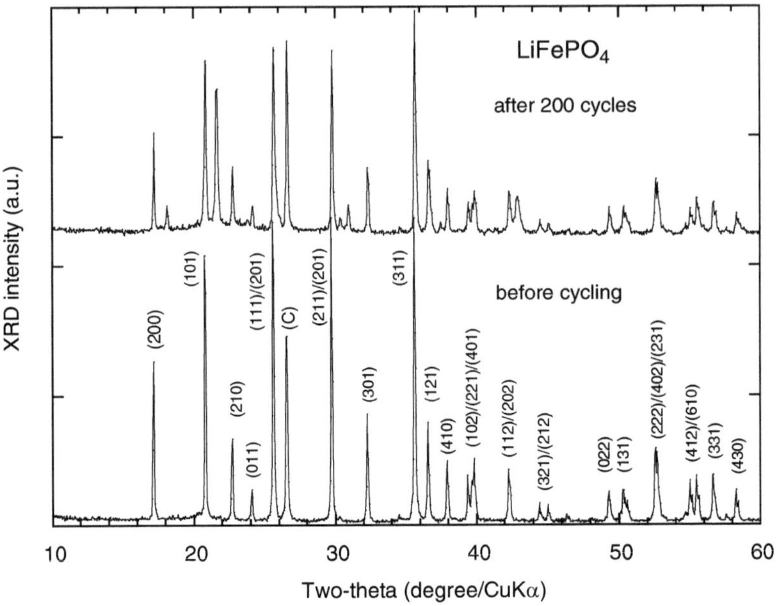

Fig. 8.18 XRD pattern of C-$LiFePO_4$ positive electrode before and after 200 cycles. Notice that the olivine framework (*Pnma* space group) remains intact after cycling at 60°C (Reprinted with permission from Elsevier)

1 h rest before each charge and discharge. This high-temperature test was made at 60°C, which is the appropriate condition to investigate possible iron dissolution in nonaqueous electrolytes. Figure 8.18 shows the XRD patterns of the new generation of $LiFePO_4$ after 200 cycles (47 days) at 60°C. There is no change in the olivine structure after cycling at 60°C. We observed Bragg lines with the same intensity as that for the pristine material. The capacity loss was below 3% in 100 cycles for this optimized electrode material.

A close examination was made for the detection of any iron dissolution that could occur after long-term cycling. The analysis of iron species was investigated at the separator/lithium (SL) interface by SEM cross section (slice view) as shown in Fig. 8.19a and c. The micrograph (Fig. 8.19a) obtained from evaluation of the earlier generation material shows the presence of iron islands at the SL interface. Obviously, some iron particles (or ions) migrate through the electrolyte from the $LiFePO_4$ positive electrode to the lithium negative. The net effect of this migration is a large decrease in capacity retention of the Li/$LiFePO_4$ cell. Figure 8.19b shows a micrograph obtained from tests with an optimized electrode in a Li cell with a lithium-foil anode. In this case, there is no iron detected at the SL interface, which remained intact after 100 cycles. In fact, this high performance was possible not only because the optimized synthesis of the $LiFePO_4$ powders but also because of strict control of the structural quality of the materials. Several physical methods were utilized to analyze the local structure and the electronic properties of the

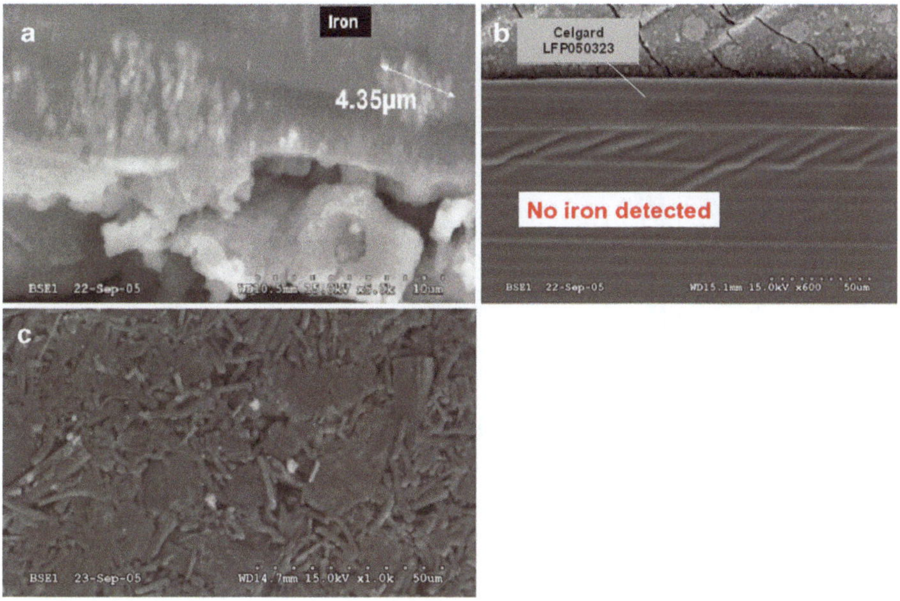

Fig. 8.19 Postmortem SEM images of the detection of iron species at the separator/lithium interface. (**a**) Image showing the formation of iron islands at the interface with an earlier generation of $LiFePO_4$. (**b**) No iron was detected at the surface of lithium foil with the optimized $LiFePO_4$. (**c**) lithium-ion graphite electrode after 200 cycles does not show the presence of iron particle (Reprinted with permission from Elsevier)

phospho-olivine framework. Figure 8.19c shows the SEM picture of graphite taken from a lithium-ion cell after 200 cycles. No iron was observed at the surface of the electrode material. EDX analysis of the graphite electrode confirms this last observation. Iron dissolution, if it occurred, should be evident in the EDX spectrum on the graphite side. In addition, elemental analysis of the electrolyte in the lithium-ion cell was carried out by the ICP technique to detect any iron after 200 cycles. No iron, even at the ppm level, was found in the electrolyte solution. Thus, all these data converge to the conclusion that $LiFePO_4$ is not soluble at 60°C with the optimized material.

The electrochemical properties of $LiFePO_4$ are known to be sensitive to the mode of preparation and the structural properties. This can be an advantage for potential applications since it allows for an optimization of the material if we can correlate the mode of preparation with the structural and the physical properties.

To address this issue, we investigated this relationship in LFP sample that was grown under different conditions. Undesirable impurities in the lattice can be introduced during the growth process. For instance, the presence of Fe_2P can increase the electronic conductivity, but on the other hand, it also decreases the ionic conductivity so that both the capacity and cycling rates are degraded with respect to C-$LiFePO_4$. Furthermore, we know that hydrogen, carbon monoxide, or carbon can reduce Fe_2O_3 through different reduction steps that depend on

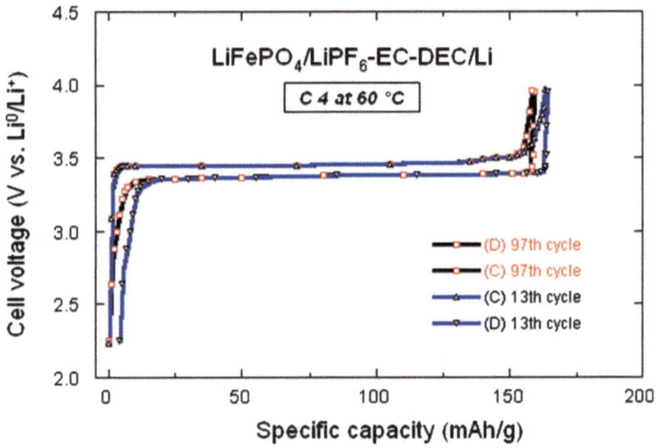

Fig. 8.20 The charge–discharge profiles of the Li/$LiFePO_4$ cells with optimized carbon-coated electrode (13th and 97th cycle) at 60°C. Experiments were carried out at C/4 rate in the voltage range 4.0–2.2 V. The 13th and 97th cycles show similar specific capacity of 160 mAh g^{-1} (Reprinted with permission from Elsevier)

temperature and other physical parameter such as particle size. Although we anticipate that over 1,000°C carbon might reduce Fe^{3+} ions through the formation of CO gas to prevent the formation of γ–Fe_2O_3, other factors may be involved. We believe that the carbon deposition process, which was by organic precursors to make C-coated samples, generates a reductive gas such as hydrogen that is more kinetically active and reduces Fe^{3+} impurities in the 400–700°C temperature range used in our studies. This model is also favored by the fact that the organic precursor is usually mixed with the $LiFePO_4$ material or with the $LiFePO_4$ chemical precursors by solution processes at a molecular-size level [53].

8.6.2 Electrochemical Performance at 60°C

Figure 8.20 presents the typical electrochemical performance of C-$LiFePO_4$/$LiPF_6$-EC-DEC/Li cells at 60°C. The charge–discharge curves were obtained by cycling at C/4 rate (about 35 mA g^{-1}) in the voltage range 2.2–4.0 V versus Li^0/Li^+. The optimized $LiFePO_4$ exhibits a reversible capacity that is maintained over many charge–discharge cycles. The 13th and 97th cycles show a similar specific capacity 160 mAh g^{-1}. These results illustrate the excellent electrochemical performance of the carbon-coated olivine material. The electrode can be fully charged up to 4 V, which is its most reactive state. This remarkable performance is attributed to the optimized carbon-coated particles and their structural integrity under a large current in the electrode. Even at such a high cycling rate, C-$LiFePO_4$ exhibits rapid kinetics of lithium extraction and realizes most of its theoretical

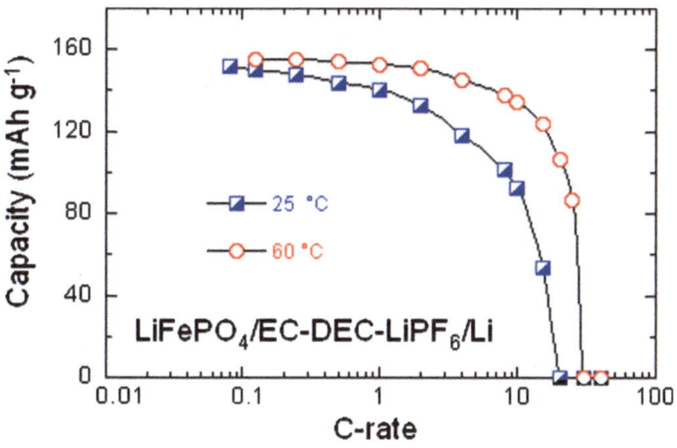

Fig. 8.21 Ragone plots of the Li/$LiFePO_4$ cells cycled at 25°C and 60°C (Reprinted with permission from Elsevier)

capacity (170 mAh g^{-1}). The discharge profile appears with the typical voltage plateau (at ca. 3.45 V vs. Li^0/Li^+) attributed to the two-phase reaction of the (1-x) $FePO_4$ + $x LiFePO_4$ system. Figure 8.21 shows the Ragone plots of the Li/$LiFePO_4$ cells cycled at 25°C and 60°C. From C/12 to 6C, the capacity is almost maintained constant at 150 mAh g^{-1}; from 6C to 25C, the capacity decreases with increasing current density. The cell still has a good performance at 10C with 134 mAh g^{-1}, which represents 86% of the capacity at C/12.

A comparison of the specific capacity between A-type $LiFePO_4$ and B-type $LiFePO_4$ electrodes during long-term cycling has shown that the cycling performance for the new generation of CC-$LiFePO_4$ material is excellent at 60°C [59]. After 100 cycles at C/4 rate and with a typical cutoff voltage 4.0–2.2 V, a constant capacity was observed. This best performance is due to the improved technology used in electrode fabrication, i.e., improvements in the nature and the morphology of the carbon coating and the optimization of the particle size of the olivine phase. These are the two main factors controlling the electrode performance. In our previous exploration of the surface properties of the $LiFePO_4$ particle, we have shown by Raman spectroscopy that the deposit is a disordered graphite-type carbon [26]. The small amount of carbon (<2 wt.%) can be viewed as a film of irregular thickness, 30 nm thick on average, with gaps. The above experimental condition (ca. 60°C) has a severe impact on the kinetics of the Fe^{2+}/Fe^{3+} redox reaction, showing that this type of C-$LiFePO_4$ electrode can be cycled at 60°C without significant capacity loss for over 200 cycles [60]. Optimized particle size in the range 200–300 nm agrees well with the average grain diameter L that validates the characteristic diffusion time $\tau = L^2/4\pi^2 D^*$ [55], where D^* is the chemical diffusion coefficient of Li^+ ions in the $LiFePO_4$ matrix (typically 10^{-12} $cm^2\ s^{-1}$) when compared with the experimental discharge rate up to 5C.

Figures 8.22 and 8.23 display the cycling behavior of cells with three different materials as negative electrodes, lithium foil, graphite, and carbon-coated lithium

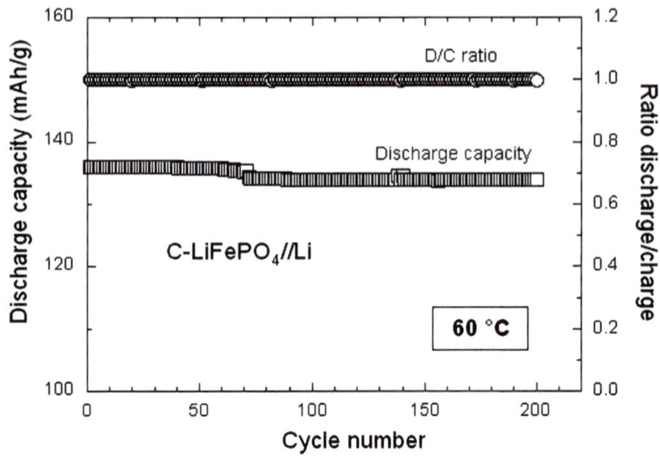

Fig. 8.22 Discharge capacity and discharge/charge ratio (electrochemical utilization) versus cycle number for C-$LiFePO_4$/$LiPF_6$-EC-DEC/Li cells at 60°C. Cells were discharged at 1C and charged at C/6 in the potential range 2.5–4.0 V (Reprinted with permission from Elsevier)

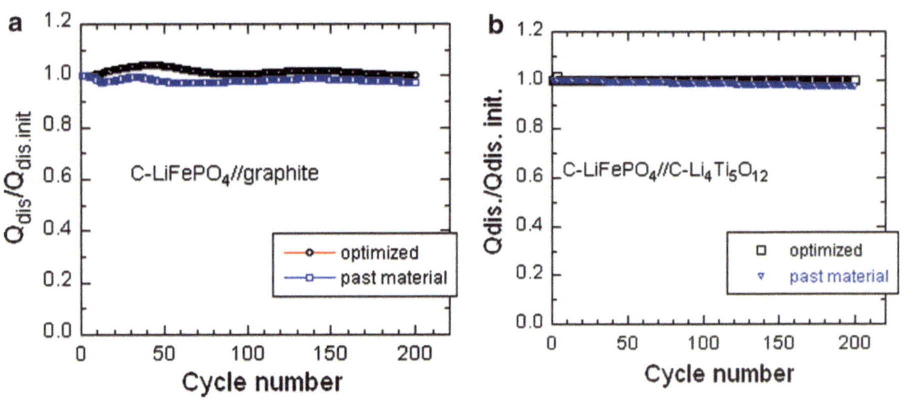

Fig. 8.23 Plots of the ratio between the current discharge capacity and the initial capacity versus cycle number for lithium-ion cells cycled at 60°C with graphite (**a**) and C-$Li_4Ti_5O_{12}$ (**b**) electrodes. Cells were discharged at 1C in the potential range 2.5–4.0 V (Reprinted with permission from Elsevier)

titanate C-$Li_4Ti_5O_{12}$ (LTO). The cells were cycled at 60°C in the potential range 2.5–4.0 V. The discharge capacity and electrochemical utilization, i.e., the ratio discharge/charge, versus cycle number are excellent for the C-$LiFePO_4$/$LiPF_6$-EC-DEC/Li cells discharged at 1C and charged at C/6 (Fig. 8.22). Similar results are shown for cells with graphite (Fig. 8.23a) and C-$Li_4Ti_5O_{12}$ negative electrodes (Fig. 8.23b). These lithium-ion cells provide coulombic efficiencies 99.9% and 100%, respectively. Plots of the ratio between the current discharge capacity and the initial capacity versus cycle number indicate constant capacity retention at

60°C. No capacity fade was observed after 200 cycles with lithium, LTO, and graphite. Another study showed the impact of different strategies to alleviate the problem of poor conductivity and slow diffusion in the $LiFePO_4$ solid phase [25]. We made a similar investigation to evaluate the performance of C-$LiFePO_4$ at different rates of charge and discharge. Measurements carried out in the range from C/4 to 10C (based on a capacity of 170 mAh g^{-1}) show that at the high rate of 4C, the Li/CC-$LiFePO_4$ cell delivers a discharge capacity of 130 mAh g^{-1}. Clearly, some treatment of the C-$LiFePO_4$, i.e., deliberate in situ carbon coating, is necessary for adequate performance of $LiFePO_4$.

8.6.3 Safe and Fast-Charging Lithium-Ion Battery with Long Shelf Life for Power Applications

In recent work, we have shown that lithium-ion battery can be charged within few minutes, passes the safety tests, and has a very long shelf life. The active materials are nanoparticles of $LiFePO_4$ and $Li_4Ti_5O_{12}$ for the positive and negative electrodes, respectively. The $LiFePO_4$ particles are covered with 2 wt.% carbon to optimize the electrical conductivity, but not the $Li_4Ti_5O_{12}$ particles. The electrolyte is the usual carbonate solvent. The binder is a water-soluble elastomer. The 18650-type battery prepared under such conditions delivers a capacity of 800 mAh. It retains full capacity after 20,000 cycles performed at charge rate 10C (6 mn) and discharge rate 5C (12 mn) and retains 95% capacity after 30,000 cycles at charge rate 15C (4 mn) and discharge rate 5C.

The electrochemical performance of the $LiFePO_4$ and $Li_4Ti_5O_{12}$ materials has been tested separately in half cell with respect to Li metal anode, using the same electrolyte mentioned above. The voltage window is 2–4 V for $LiFePO_4$ and 1.2–2.5 V for $Li_4Ti_5O_{12}$. For $LiFePO_4$, the first coulombic efficiency is 100% and the reversible capacity is 148 mAh g^{-1}. For $Li_4Ti_5O_{12}$, the first coulombic efficiency is 98% and the reversible capacity is 157 mAh g^{-1}. The well-known plateaus at 3.4 and 1.55 V are characteristics of the topotactic insertion/deinsertion of lithium in the two-phase systems $LiFePO_4$-$FePO_4$ and $Li_4Ti_5O_{12}$-$Li_7Ti_5O_{12}$, respectively. The Ragone plots in Fig. 8.24 show that the high-rate capacity of both materials is stable up to 5C. At 10C, $LiFePO_4$ still delivers 73% of the rated capacity against 53% for $Li_4Ti_5O_{12}$.

The $LiFePO_4$/1 mol.L^{-1} $LiPF_6$ in EC-DEC/$Li_4Ti_5O_{12}$ batteries (18650 type) has been made with these products. The cycling life is illustrated in Fig. 8.25 for charge rate at 10C (6 mn). Even after 20,000 cycles, the cell did not lose its initial capacity. Another cell cycled over 30,000 cycles for faster charge rate at 15C (4 mn), and the same discharge rate at 5C shows the capacity (910 mAh in the first cycles) decreases with the number of cycles, but the loss at the end of the process is only 9%, which can be considered as negligible based on one charge–discharge cycle each day (a rather intense use for portable applications would take 50 years of life).

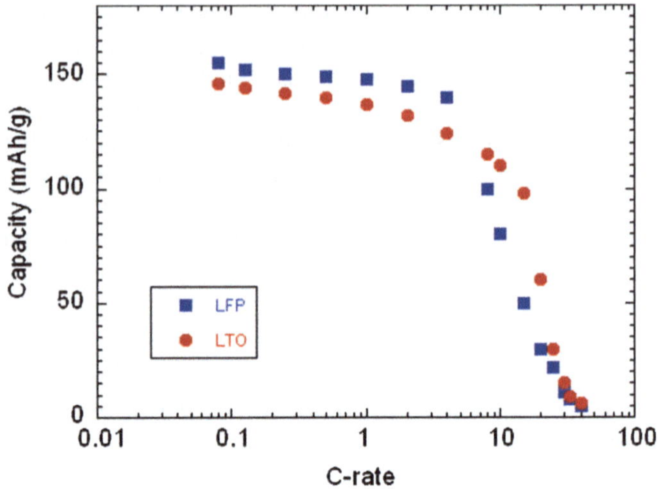

Fig. 8.24 Ragone plots of the cells $LiFePO_4$/EC-DEC-1 M $LiPF_6$/Li (LFP) and $Li_4Ti_5O_{12}$/EC-DEC-1 M $LiPF_6$/Li (LTO)

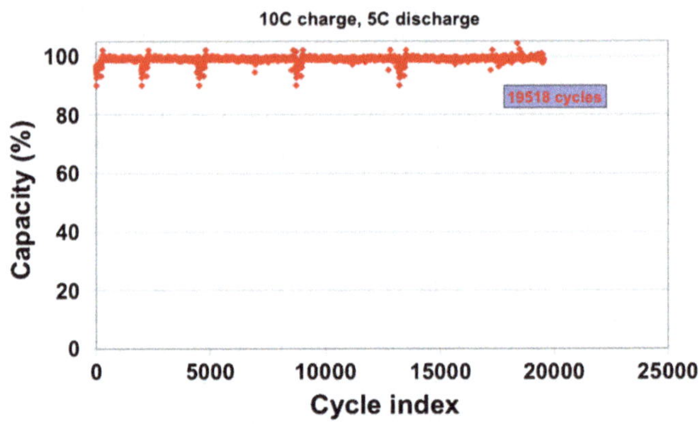

Fig. 8.25 Cycle life of a $LiFePO_4$/EC-DEC-1 M $LiPF_6$/$Li_4Ti_5O_{12}$ "18650" cell. The cycle charge rate is 10C (6 mn); the discharge rate is 5C (12 mn)

Note that full depth of discharge (DOD) has been achieved for each cycle in these experiments. This is to our knowledge a breakthrough in the performance of lithium-ion batteries since their poor cycling life was considered as a major defect.

During the transportation tests performed on these batteries, no flame and no smoke have been detected, and the maximum temperature that has been reached is 72°C. For comparison, in the same tests, the $LiCO_2$/graphite 18650-type cell goes on fire with heavy smoke, the temperature reaching 350°C. $LiMn_2O_4$/graphite cells also go on fire in the short circuit test.

8.7 Summary and Outlook

This chapter is a thorough report on $LiFePO_4$ synthesized by two methods, solid state and hydrothermal. We have followed a logical structure starting with synthesis, crystallographic properties, the characterization of the crucial carbon-coating layer at the surface of LiFePO4, magnetic properties, reaction to water and humidity, and finally electrochemical performance on optimized material at room temperature, 60°C, and in full lithium-ion cells.

These studies have shown that, depending of the synthesis, the fundamental properties can be modified because of impurities poisoning this material. These impurities are identified, and a quantitative estimate of their concentrations is deduced from the combination of analytical methods. The most powerful technique used so far is the SQUID magnetometry, which is very sensitive to detect Fe^{3+} ions. Note that such iron-based impurities may be at the origin of iron dissolution in the electrolyte producing short circuit of the battery. Thus, an optimized preparation provides materials with carbon-coated particles free of any impurity phase, insuring structural stability and electrochemical performance. The carbon-coated particles were studied by Raman scattering spectroscopy, which informs on the quality of the deposit, especially its degree of graphitization.

The structural properties have been correlated with the electrochemical performance of the positive electrode materials. It appears that a severe control of synthesis conditions is needed to obtain materials with good performance under high current density.

The product reacts with water, but not with dry atmosphere, so that storage of the $LiFePO_4$ powder in a dry chamber at 5% relative humidity is needed, but it is also sufficient to guarantee that the product does not age before manufacturing the battery. The carbon coat is not protective. Exposure to humidity induces delithiation of the surface layer. However, the resulting $FePO_4$ layer that is formed within minutes of exposure to humidity is very protective, so that the damage remains limited to the surface layer (about 3 nm thick) unless the time of exposure to humid atmosphere is very large (months).

All these researches justify the use of the $LiFePO_4$ material as a cathode element in new generation of lithium secondary batteries operating for powering hybrid electric vehicles and full electric vehicles.

References

1. Padhi K, Nanjundaswamy KS, Goodenough JB (1997) Phospho-olivines as Positive-Electrode Materials for Rechargeable Lithium Batteries. J Electrochem Soc 144:1188
2. Huang H, Yin SC, Nazar LF (2001) Approaching Theoretical Capacity of $LiFePO_4$ at Room Temperature at High Rates. Electrochem Solid State Lett 4:A170
3. Dominko D, Gaberscek M, Drofenik J, Bele M, Jamnik J (2003) Influence of carbon black distribution on performance of oxide cathodes for Li ion batteries. Electrochim Acta 48:3709

4. Ravet N, Goodenough JB, Besner S, Simoneau M, Hovington P, Armand M (1999) Improved Iron based Cathode Material. Proceedings of the 196th ECS meeting, Honolulu, extended abstract n° 127
5. Ravet N, Chouinard Y, Magnan JF, Besner S, Gauthier M, Armand M (2001) Electroactivity of natural and synthetic triphylite. J Power Sources 97:503
6. Bewlay SL, Konstantinov K, Wang GX, Dou SX, Liu HK (2004) Conductivity improvements to spray-produced LiFePO4 by addition of a carbon source. Mater Lett 58:1788
7. Chen Z, Dahn JR (2002) Reducing Carbon in LiFePO4/C Composite Electrodes to Maximize Specific Energy, Volumetric Energy, and Tap Density. J Electrochem Soc 149:A1184
8. Ravet N, Besner S, Simoneau M, Vallée A, Armand M, Magnan JF (2005) Electrode materials with high surface conductivity. US Patent 6,962,666; (2005) 6,855,273
9. Ait-Salah A, Mauger A, Julien CM, Gendron F (2006) Nano-sized impurity phases in relation to the mode of preparation of LiFePO4. Mater Sci Eng B 129:232
10. Subramanya-Herle P, Ellis B, Coombs N, Nazar LF (2004) Nano-network electronic conduction in iron and nickel olivine phosphates. Nat Mater 3:147
11. Ait-Salah A, Dodd J, Mauger A, Yazami R, Gendron F, Julien CM (2006) Structural and magnetic properties of LiFePO4 and lithium extraction effects. Z Allg Inorg Chem 632:1692
12. Zaghib K, Shim J, Guerfi A, Charest P, Striebel KA (2005) Effect of Carbon Source as Additives in LiFePO4 as Positive Electrode for Lithium-Ion Batteries. Electrochem Solid State Lett 8:A207
13. Zaghib K, Armand M (2002) Electrode covered with a film obtained from an aqueous solution containing a water soluble binder, manufacturing process and uses thereof. Canadian Patent CA 2,411,695
14. Yamada A, Chung SC, Hinokuma K (2001) Optimized LiFePO4 for Lithium Battery Cathodes. J Electrochem Soc 148:A224
15. Arcon D, Zorko A, Dominko R, Jaglicic Z (2004) A comparative study of magnetic properties of LiFePO4 and LiMnPO4. J Phys C 16:5531
16. Geller S, Durand JL (1960) Refinement of the structure of LiMnPO4. Acta Crystallogr 13:325
17. Santorro RP, Newnham RE (1967) Antiferromagnetism of LiFePO4. Acta Crystallogr 22:344
18. Streltsov VA, Belokoneva EL, Tsirelson VG, Hansen NK (1993) Multipole analysis of the electron density in triphylite LiFePO4, using X-ray diffraction data. Acta Crystallogr B 49:147
19. Andersson AS, Thomas JO (2001) The source of first-cycle capacity loss in LiFePO4. J Power Sources 97–98:498
20. Rousse G, Rodriguez-Carvajal J, Patoux S, Masquelier C (2003) Magnetic structures of the triphylite LiFePO4 and its delithiated form FePO4. Chem Mater 15:4082
21. Losey A, Rakovan J, Huges J, Francis CA, Dyar MD (2004) Structural variation in the lithiophilite-triphylite series and other olivine-group structures. Can Miner 42:1105
22. Junod A, Wang KQ, Triscone G, Lamarche G (1995) Specific heat, magnetic properties and critical behaviour of Mn2SiS4 and Fe2GeS4. J Magn Magn Mater 146:21
23. Moring J, Kostiner E (1986) The crystal structure of NaMnPO4. J Solid State Chem 61:379
24. Nakamura T, Miwa Y, Tabuchi M, Yamada Y (2006) Structural and Surface Modifications of LiFePO4 Olivine Particles and Their Electrochemical Properties. J Electrochem Soc 153: A1108
25. Striebel K, Shim J, Srinivasan V, Newman J (2005) Comparison of LiFePO4 from different sources. J Electrochem Soc 152:A664
26. Ait Salah A, Mauger A, Zaghib K, Goodenough JB, Ravet N, Gauthier M, Gendron F, Julien CM (2006) Reduction Fe3+ of Impurities in LiFePO4 from Pyrolysis of Organic Precursor Used for Carbon Deposition. J Electrochem Soc 153:A1692
27. Paques-Ledent MT, Tarte P (1974) Vibrational studies of olivine-type compounds—II Orthophosphates, -arsenates and -vanadates A*I*B*II*X*V*O4. Spectrochim Acta A: Molecular Spectroscopy 30:673
28. Burma CM, Frech R (2004) Raman and FTIR Spectroscopic Study of Li x FePO4 ($0 \leq x \leq 1$). J Electrochem Soc 151:A1032

29. Kostecki R, Schnyder B, Alliata D, Song X, Kinoshita K, Kotz R (2001) Surface studies of carbon films from pyrolyzed photoresist. Thin Solid Films 396:36
30. Julien CM, Massot M (2004) Vibrational spectroscopy of electrode materials for rechargeable lithium-ion batteries. III. Oxide frameworks. In: Stoynov Z, Vladikova D (eds) Proceedings of the international workshop on advanced techniques for energy sources investigation and testing, Bulgarian Academy of Sciences, Sofia
31. Julien CM, Zaghib K, Mauger A, Massot M, Ait-Salah A, Selmane M, Gendron F (2006) Characterization of the carbon coating onto LiFePO4 particles used in lithium batteries. J Appl Phys 100:63511
32. Ait-Salah A, Jozwiak P, Zaghib K, Garbarczyk J, Gendron F, Mauger A, Julien CM (2006) FTIR features of lithium iron-phosphates as electrode materials for rechargeable lithium batteries. Spectrochim Acta A: Molecular and Biomolecular Spectroscopy 65:1007
33. Hu Y, Doeff MM, Kostecki R, Finones R (2004) Electrochemical Performance of Sol-Gel Synthesized LiFePO4 in Lithium Batteries. J Electrochem Soc 151:A1279; Doeff MM, Hu Y, McLarnon F, Kostecki R (2003) Effect of Surface Carbon Structure on the Electrochemical Performance of LiFePO4. Electrochem Solid-State Lett 6:A207
34. Tamor MA, Vassell WC, Carduner KR (1991) Atomic constraint in hydrogenated diamond-like carbon, Appl Phys Lett 58: 592
35. Robertson J (1992) Properties of diamond-like carbon. Surf Coat Technol 50:185
36. Ramsteiner M, Wagner J (1987) Resonant Raman scattering of hydrogenated amorphous carbon: Evidence for π-bonded carbon clusters. Appl Phys Lett 51:1355
37. Yoshikawa M, Katagani G, Ishida H, Ishitami A, Akamatsu T (1988) Resonant Raman scattering of diamond-like amorphous carbon films. Appl Phys Lett 52:1639
38. Tamor MA, Vassell WC (1994) Raman "fingerprinting" of amorphous carbon films. J Appl Phys 76:3823
39. Wada N, Gaczi PJ, Solin SA (1980) "Diamond-like" 3-fold coordinated amorphous carbon. J Non-Cryst Solids 35–36:543
40. Knight DS, White WB (1989) Characterization of diamond films by Raman spectroscopy. J Mater Res 4:385
41. Lespade P, Marchand A, Couzi M, Cruege F (1984) Caractérisation de matériaux carbonés par microspectrométrie Raman. Carbon 22:375
42. Dresselhaus MS, Dresselhaus G (1982) Light scattering in graphite intercalation compounds. In: Cardona M, Gunterhodt G (eds) Light scattering in solids III, vol 51, p. 3, Topics in applied physics. Springer, Berlin
43. Nakamizo M, Tamai K (1984) Carbon Raman spectra of the oxidized and polished surfaces of carbon 22:197
44. Matthews MJ, Bi XX, Dresselhaus MS, Endo M, Takahashi T (1996) Raman spectra of polyparaphenylene-based carbon prepared at low heat-treatment temperatures. Appl Phys Lett 68:1078
45. Robertson J, O'Reilly EP (1987) Electronic and atomic structure of amorphous carbon. Phys Rev B 35:2946
46. Ramsteiner M, Wagner J, Wild Ch, Koidl P (1987) Raman scattering from extremely thin hard amorphous carbon films. J Appl Phys 62:729
47. Tamor MA, Haire JA, Wu CH, Hass KC (1989) Correlation of the optical gaps and Raman spectra of hydrogenated amorphous carbon films. Appl Phys Lett 54:123
48. Santoro RP, Newnham RE, Nomura S (1966) Magnetic properties of Mn2SiO4 and Fe2SiO4. J Phys Chem Solids 27:655
49. Santoro RP, Segal DJ, Newnham RE (1966) Magnetic properties of LiCoPO4 and LiNiPO4. J Phys Chem Solids 27:1192
50. Dal D, Koo HJ, Rocquefelte X, Jobic S (2005) Analysis of the spin exchange interactions and the ordered magnetic structures of lithium transition metal phosphates LiMPO4 (M = Mn, Fe, Co, Ni) with the olivine structure. Inorg Chem 44:2407

51. Mays JM (1963) Nuclear magnetic resonances and Mn-O-P-O-Mn superexchange linkages in paramagnetic and antiferromagnetic LiMnPO4. Phys Rev 131:38
52. Ait-Salah A, Zaghib K, Mauger A, Gendron F, Julien CM (2006) Magnetic studies of the carbothermal effect on LiFePO4. Phys Status Solidi A 203:R1
53. Zaghib K, Dontigny M, Charest P, Labrecque JF, Guerfi A, Kopec M, Mauger A, Gendron F, Julien CM (2008) Aging of LiFePO4 upon exposure to H2O. J Power Sources 185:698
54. Manickam M, Singh P, Thurgate S, Prince KJ (2006) Redox behavior and surface characterization of LiFePO4 in lithium hydroxide electrolyte. J Power Sources 158:646
55. Zaghib K (2003) BATT review meeting, Berkeley
56. Porcher W, Moreau P, Lestriez B, Jouanneau S, Guyomard D (2008) Is LiFePO4 Stable in Water?: Toward Greener Li-Ion Batteries. Electrochem Solid State Lett 11:A4
57. Zaghib K, Mauger A, Gendron F, Julien CM (2008) Surface effects on the physical and electrochemical properties of thin LiFePO4 particles. Chem Mater 20:462
58. Liu HS, Zhang ZR, Gong ZL, Yang Y (2004) Origin of Deterioration for LiNiO2 Cathode Material during Storage in Air. Electrochem Solid State Lett 7:A190
59. Zaghib K, Ravet N, Gauthier M, Gendron F, Mauger A, Goodenough JB, Julien CM (2006) Optimized electrochemical performance of LiFePO4 at 60°C with purity controlled by SQUID magnetometry. J Power Sources 163:560
60. Zaghib K, Battaglia V, Charest P, Srinivasan V, Guerfi A, Kostecki R (2006) Extended abstract 35 of the IBA-HBC meeting, Hawaii

Chapter 9
Effect of Nanoparticles on Electrolytes and Electrode/Electrolyte Interface

Nuha Salem and Yaser Abu-Lebdeh

Abstract The addition of nano-sized inorganic fillers such as SiO_2 to solid and liquid electrolytes to enhance their electrochemical and physical properties has been recently the focus of great deal of research. In this chapter, we review the work done in this area where various types of nanoparticles including ceramics and clay were used as additives to electrolytes commonly used in lithium-ion batteries research such as polymer electrolytes (gel and solid form), ionic and organic liquid electrolytes and plastic crystals.

9.1 Introduction

Electrolytes are one of the key components in any electrochemical cell including batteries [1]. In addition to facilitating ion transfer between the two electrodes to complete the electrochemical cell, they also provide the physical separation necessary to prevent any short circuit. A good electrolyte for the application in batteries and other electrochemical devices must have [1–3] (a) high overall ionic conductivity over a temperature range $-30°C$ to $100°C$ and high transport number for both cations and anions to sustain high cycling rate; (b) high chemical, thermal and electrochemical stability to produce systems with high voltage and more safety; (c) excellent compatibility with other components in the cell; and (d) low cost and less environmental impact. Most commercial lithium batteries utilise liquid electrolytes that consist of lithium salt (e.g. $LiPF_6$) dissolved in a mixture of cyclic and linear carbonate-based solvent(s) (e.g. ethylene carbonate, EC and dimethyl carbonate, DMC) [3]. These electrolytes are highly conductive towards ions, with conductivity values above 10^{-2} S cm^{-1} by virtue of the high dielectric constant of EC and low viscosity of DMC which leads to well-solvated Li^+ and PF_6^- ions. They are also

N. Salem • Y. Abu-Lebdeh (✉)
National Research Council of Canada, Ottawa, ON K1A 0R6, Canada
e-mail: Yaser.Abu-Lebdeh@nrc.gc.ca

Y. Abu-Lebdeh and I. Davidson (eds.), *Nanotechnology for Lithium-Ion Batteries*, Nanostructure Science and Technology, DOI 10.1007/978-1-4614-4605-7_9,

Table 9.1 Effect of nanoparticles on the properties of different types of electrolytes

	Conductivity		Mechanical strength	
	Without	With	Without	With
Liquids [6–13]	↑	↑↑	↓	↑
Ionic liquids [30–32]	↔	↑↑	↓	↑
Ionic plastic crystals [5, 33–42]	↔	↑↑	↑	↑↑
Gel electrolytes [19–21, 27, 28]	↑	↑↑	↑	↑↑
Polymers [16–18, 22–26]	↓	↑	↑	↑↑

↓ low, ↔ acceptable, ↑ high, ↑↑ higher

stable within a voltage window of ~5 V which make them compatible with wide range of electrode materials. Liquid electrolytes, however, have no mechanical strength which impose a limitation on battery design and are not safe because of possible leakage and flammability. This led to an increased interest in solid electrolytes such as polymer electrolytes which attracted a lot of attention in the last three decades. Solid electrolytes are safe and have excellent mechanical properties that make them ideal for wide range of battery design and fabrication. One significant limitation of solid electrolytes, however, is their poor ionic conductivities that lie in the range of 10^{-6}–10^{-4} S cm^{-1} at ambient conditions and low lithium-ion transport number (<0.3).

It can be concluded that liquid electrolytes have the right ion transport properties but lack the mechanical strength, while solid electrolytes have the right mechanical properties but lack the acceptable ion transport. Recently, in an attempt to combine the advantages of both liquid and solid electrolytes, an intense research has been devoted towards materials that have high ionic conductivity and good mechanical properties simultaneously. This can be achieved by using additives to make solid electrolytes have liquid-like ionic conductivities while retaining their desired mechanical properties and to make liquid electrolytes have better mechanical properties while retaining their desired ionic conductivities. In an effort to enhance the ionic conductivity of polymer electrolytes, for example, the addition of ceramic nanoparticles such as SiO_2, Al_2O_3 or TiO_2 in small quantities led to enhancement of the transport phenomena in poorly conducting polymers such as salt/poly(ethylene oxide) (PEO) complexes [4]. Same effect was observed when nano-sized SiO_2 particles were added to some organic ionic plastic crystals such as *N,N′*-ethylmethylpyrrolidinium bis(trifluoromethanesulfonyl)-amide ([C_2mpyr][NTf_2]) resulting in two orders of magnitude increase in ionic conductivities [5]. It was also reported that the same ceramics can be added to conventional organic liquid electrolytes (e.g. EC/DMC) and to ionic liquids, producing what is called ‘soggy sand’, to enhance their mechanical properties while retaining the favourable ionic transport properties [6–15]. Table 9.1 summarises the effect of adding nanoparticles on the conductivity and mechanical properties of different types of electrolytes.

In this chapter, we will discuss the mechanism and the outcome of adding various nanoparticles to different types of electrolyte materials including solid and gel polymer electrolytes, ionic/organic liquid electrolytes and plastic crystals.

9.2 Nanocomposites of Polymer Electrolytes

In this section, we illustrate some of the attempts made to improve electrochemical properties of polymer electrolytes and even, in some cases, their mechanical properties by adding different types of inorganic fillers. The most common polymer electrolyte used in lithium-ion batteries is polyethylene oxide (PEO) mixed with a lithium salt (LiX) [4, 16, 17]. As previously mentioned, the main problem associated with solid electrolytes is their poor conductivities at temperatures <70°C, low T_{Li+} (~0.3) and reactivity towards lithium metal. For successful application of this type of electrolyte materials in electrochemical cells, these problems must be solved. It was shown that polymer nanocomposites obtained by dispersing ceramics, layered clays and mesoporous materials in the polymer matrix exhibit better ion transfer number, ionic conductivities (up to ten folds), mechanical properties and electrolyte–electrode interface compared to the host polymer [4, 16–23]. Inorganic fillers in this case act as plasticisers that decrease the crystalline structure of the polymer, which enhances chain mobility in the amorphous region. It also provides conductive pathways for ions, mainly in the interface region, and act as Lewis acid–base centres that promote salt dissociation which free more cations (e.g. Li^+). These effects combined improve the ionic conductivity and the cation transference number of the new hybrid electrolytes which could prove satisfactory for the increased demand on new battery materials for commercial applications. The degree of enhancement depends on the type, size, concentration and morphology of these fillers. For example, it was found that mesoporous materials, which has larger surface area, showed larger improvement in ionic conductivities of polymer electrolytes compared to the nonporous ones and that larger pores facilitate the intercalation of the polymer chains more than the smaller ones [18].

9.2.1 Nanocomposites of Solid Polymer Electrolytes

Interest in solid polymer electrolyte nanocomposites goes back to as early as 1988 [24, 25] when Skaarup et al. used Li_3N with poly(ethylene oxide) (PEO) and poly (ethylene) (PE) polymers. A decade later, Scrosati et al. [16] used TiO_2, Al_2O_3 and SiO_2 to prepare PEO-based composite electrolytes (PEO_8-$LiClO_4$ (8:1)-ceramic). The composite was simply made by making a slurry of the ceramic, $LiClO_4$ and PEO in acetonitrile followed by thin film casting and solvent evaporation.

Temperature dependence of the conductivity of ceramic-free polymer (PEO_8-$LiClO_4$), which was used as a reference, showed that the conductivity increased significantly above 60°C which is caused by the transition of the polymer chains from crystalline to amorphous phase (Fig. 9.1). A cooling curve for the same polymer was reproduced confirming that the chains transfer back to crystalline state at around the same temperature. Same experiment for PEO_8-$LiClO_4$-Al_2O_3

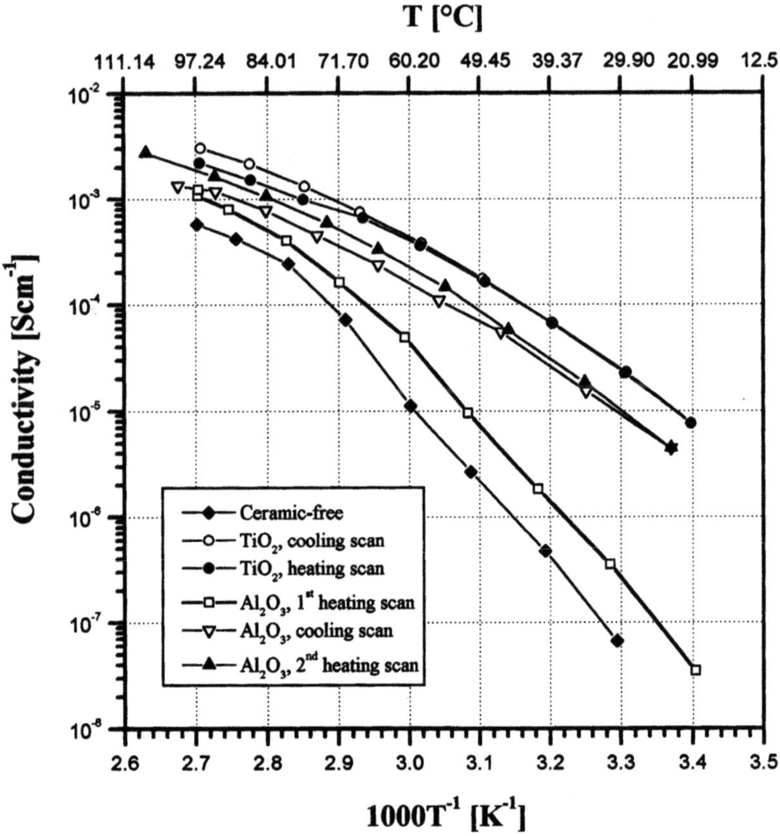

Fig. 9.1 Arrhenius plots of the conductivity of PEO_8-$LiClO_4$, PEO_8-$LiClO_4$-TiO_2 and PEO_8-$LiClO_4$-Al_2O_3 obtained by impedance spectroscopy measurements (Reprinted from [16], Copyright (2011), from Elsevier)

followed a similar pattern when the composite was heated for the first time with no significant enhancement in conductivity. A cooling curve, however, was different. Upon cooling, no break occurred at 60°C and conductivities remained significantly higher at lower temperatures compared to the ceramic-free polymer. This enhancement in conductivity remained the same for a second heating and cooling scan. One reason for conductivity increase is that after annealing the polymer at a temperature above the crystalline to amorphous transition (>70°C), the ceramic particles prevented polymer chains from reorganising to crystalline structure once they were cooled.

Lithium-ion transference number was also measured and found to be in the order of 0.5–0.6 in a temperature range 45–90°C which is higher than most common PEO electrolytes (~0.3). The increase in Li^+ transfer was attributed to the Lewis acid property of the added ceramic which competes with lithium ion in forming

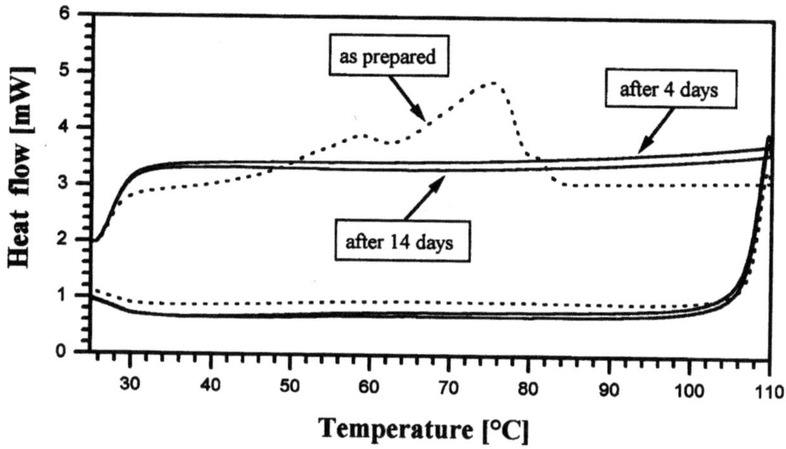

Fig. 9.2 DSC of PEO_8-$LiClO_4$-Al_2O_3 polymer nanocomposite as prepared and after 4 and 14 h of storage. Heating and cooling rate was 10 °C min^{-1} (Reprinted from [16], Copyright (2011), from Elsevier)

complexes with the polymer chains, causing them to cross-link. This phenomenon increases the stiffness of the polymer and reduces its ability to maintain the crystalline structure and thus provides pathways for ionic transfer. The effect of ceramics on the crystalline to amorphous transition of PEO_8-$LiClO_4$ was confirmed by DSC measurements which showed the absence of the transition for the composite after the first heating (annealing) cycle (Fig. 9.2).

In another study by Scrosati et al. [23], nano-sized sulphated zirconia (S-ZrO_2) was used to prepare PEO-based polymer electrolyte nanocomposite. Zirconia nanoparticles were treated with ammonium sulphate solution to produce modified nanoparticles with high degree of acidity, about 100% stronger than sulphuric acid. The coordination of the sulphate group with the electron-deficient zirconium cation increases its electron accepting ability and thus its acidity, leading to stronger Lewis acid–base interaction, a factor considered crucial for better ion transport properties in polymer electrolyte nanocomposites. As a result, lithium-ion transference number was enhanced from 0.42 for PEO_{20}-$LiBF_4$ to 0.81 for PEO_{20}-$LiBF_4$-10% S-ZrO_2, a value of almost 100%. Cells made of PEO_{20}-$LiClO_4$-5% ZrO_2 exhibited higher capacity with little loss and better efficiency on cycling compared to S-ZrO_2-free polymer electrolyte (Fig. 9.3) [26].

Chen-Yang et al. [18] prepared a composite of polyacrylonitrile, $LiClO_4$ and silica aerogel powder (PAN/$LiClO_4$/SAP) by suspending all components in DMF followed by thin film casting and solvent evaporation. Variable SAP concentrations (1, 2 and 3 wt.%) were used to study the concentration effect. SAP content higher than 3 wt.% exhibited reduced effect on the electrochemical properties of the host material. This happens due to the aggregation of SAP particles at higher loads which has less SAP/$LiClO_4$ Lewis acid–base interaction resulting in less salt dissociation. Ionic conductivity was the highest for the composite with 3 wt.%

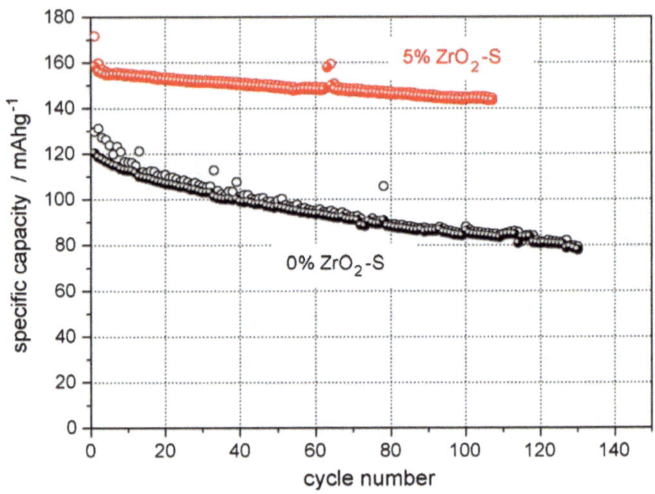

Fig. 9.3 Capacity versus charge–discharge cycles of the Li/P(EO)$_{20}$LiClO$_4$ + 5% S-ZrO$_2$/LiFePO$_4$ battery and of the Li/P(EO)$_{20}$LiClO$_4$/LiFePO$_4$ battery. Temperature, 90°C; rate, *C*/5, corresponding to a current density of 0.20 mA cm^{-2} for the S-ZrO$_2$ electrolyte battery and of *C*/7 (0.16 mA cm^{-2}) for the ceramic-free electrolyte battery. The capacity values are referred to the cathode (Reprinted from [26], Copyright (2011), from Elsevier)

SAP with a value 12.5 times higher than PAN/LiClO$_4$. A proposed conduction mechanism is illustrated in Fig. 9.4. This mechanism of ion transport suggests that the process happens mainly at the PAN/SAP interface where the Lewis basic sites interact with Li$^+$, weakening its binding to the C ≡ N group of the PAN chain which facilitates its transport. This mode of transport provides shorter pathways due to the high surface area and porosity of SAP.

The composite with 3 wt.% SAP had an electrochemical stability window of 4 V versus Li/Li$^+$ as determined by cyclic voltammetry. The charge/discharge of a coin cell made with composite of 3 wt.% SAP (Li/PAN-LiClO$_4$-3 wt.% SAP/LiFePO$_4$) produced discharge capacity of 120 mAh.g^{-1}. One hundred per cent of this value was retained after 20 cycles (Fig. 9.5).

9.2.2 Nanocomposites of Gel Polymer Electrolytes

Nano-sized inorganic fillers can be also used with gel polymer electrolytes made of solid polymers and liquid electrolytes [4, 19–21, 27, 28]. TiO$_2$ was used by Kim et al. [21] in poly(vinylidene fluoride-co-hexafluoropropylene) (P(VdF-HFP)). In this study, the effect of the concentration of added TiO$_2$ was investigated. Concentrations >50% were excluded since they have showed poor mechanical properties of the cast. As expected, the crystallinity of P(VdF-HFP) was reduced upon the addition of TiO$_2$ particles. Furthermore, the reduction in crystallinity

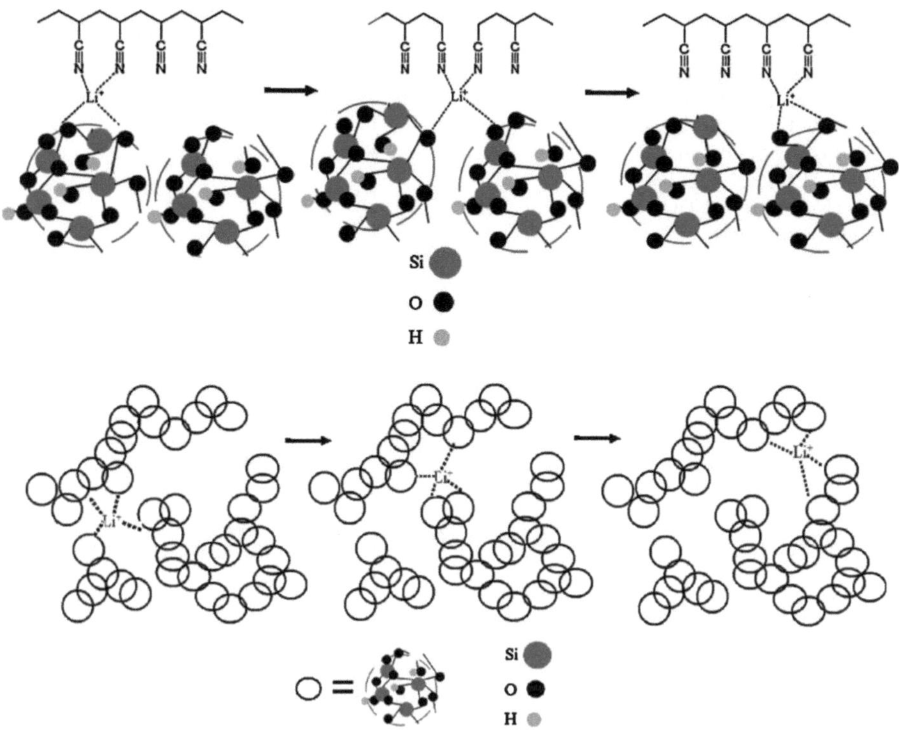

Fig. 9.4 Proposed conduction mechanism in PAN/$LiClO_4$/SAP polymer nanocomposite (Reprinted from [18], Copyright (2011), from Elsevier)

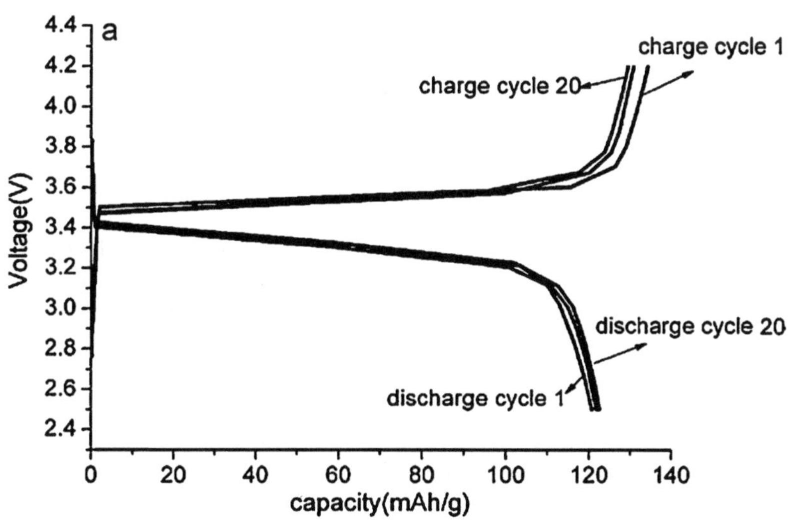

Fig. 9.5 Charge/discharge curves of Li/PAN-$LiClO_4$-3 wt.% SAP/$LiFePO_4$ battery. The discharge rate was 0.5 C at voltage range of 2.5–4.2 V (Reprinted from [18], Copyright (2011), from Elsevier)

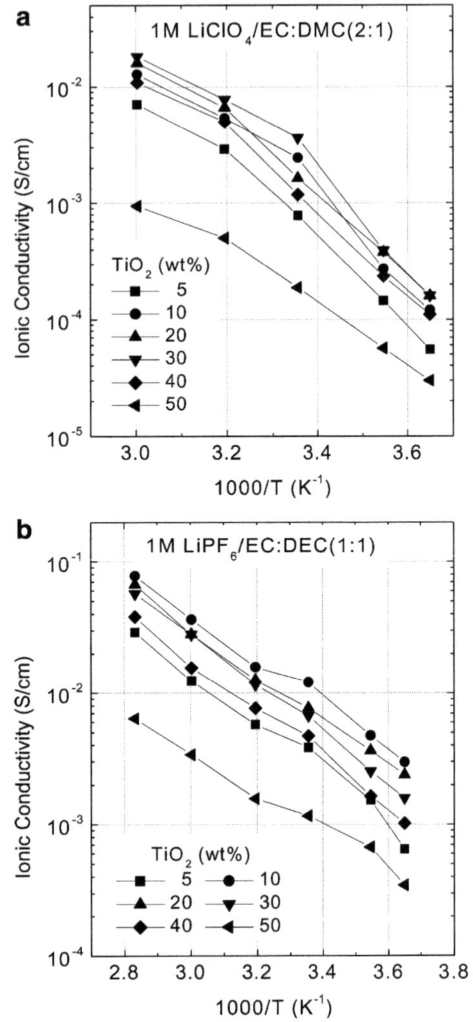

Fig. 9.6 Temperature dependence of ionic conductivity of polymer gel electrolytes, P(VdF-HFP)-TiO_2/1 M $LiClO_4$/EC-DMC (2:1) (**a**) and P(VdF-HFP)-TiO_2/1 M $LiPF_6$/EC-DEC (1:1) (**b**) at different TiO_2 weight percentages (Reprinted from [21], Copyright (2011), from Elsevier)

increased with increasing TiO_2 concentration up to 30 wt.%. Electrochemical properties such as conductivity (Fig. 9.6), electrochemical stability (Fig. 9.7) and compatibility with lithium electrode for systems made of P(VdF-HFP)-TiO_2 saturated with 1 M $LiClO_4$/EC-DMC(2:1) or 1 M $LiPF_6$/EC-DEC (1:1) were optimum for TiO_2 contents of ~20 wt.% and 30 wt.%, respectively.

In another similar study [27], ^1H and ^7Li-NMR were used to prove that adding TiO_2 to poly(methyl methacrylate) gel made of $LiClO_4$, EC, PC and PMMA in a ratio 4.5:46.5:19:30 has improved the ionic conductivity and lithium-ion transference number (T_{Li+}). Results showed that up to 50% increase in T_{Li+} was observed at 80°C after TiO_2 was added. The change in lithium-ion environment due to TiO_2–Li^+ interaction was shown in the line width and T_2 relaxation data obtained

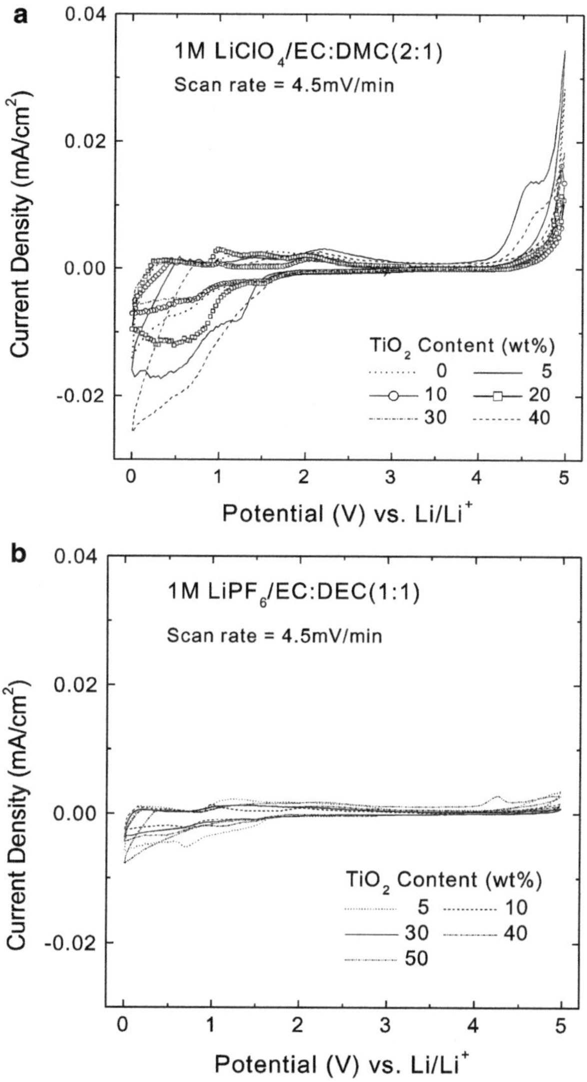

Fig. 9.7 Cyclic voltammograms of polymer gel electrolytes, P(VdF-HFP)-TiO_2/1 M $LiClO_4$/EC-DMC (2:1) (**a**) and P(VdF-HFP)-TiO_2/1 M $LiPF_6$/EC-DEC (1:1) (**b**) at different TiO_2 weight percentages using stainless steel working electrode (Reprinted from [21], Copyright (2011), from Elsevier)

by ^{7}Li-NMR. The results suggest that the interaction between the inorganic filler and lithium ions breaks up the cross-linking between polymer chains originally caused by Li^+-polymer interaction and by H bonding from the urethane functional group which allow for higher lithium-ion mobility.

In all previously mentioned systems, ionic transfer phenomenon happens within the salt/polymer component of the nanocomposite, while the added ceramic does not actually participate in the process but rather facilitate it. These fillers are called 'passive' fillers. Other types of fillers, however, are active and play a crucial role in the ionic transfer. These fillers are called 'active' fillers. Common examples of active fillers are clays [20, 22] such as hectorites, silicates, etc. In a study performed

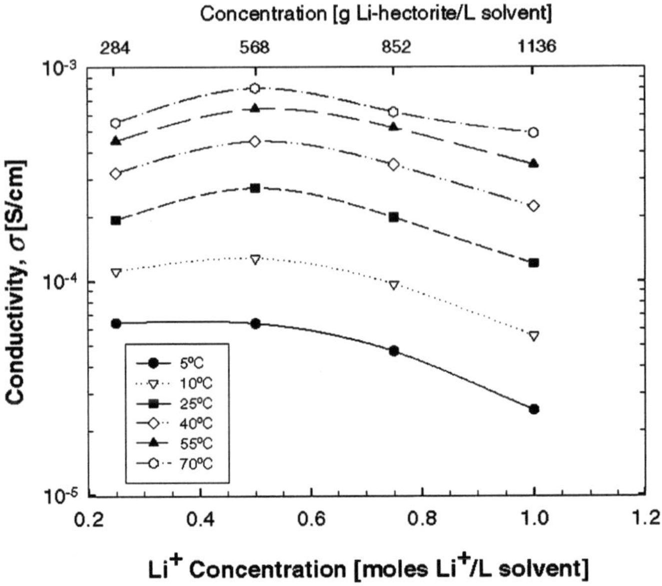

Fig. 9.8 Dependence of conductivity on Li-hectorite concentration at various temperatures. Solvent is 1:1 v/v EC/PEGdm (Reprinted from [4], Copyright (2011), from John Wiley and Sons)

by Khan et al. [4], lithium-exchanged hectorite filler was suspended in 1:1 v/v mixture of EC in poly(ethylene glycol dimethyl ether) (EC/PEGdm) to form a physical gel. The ionic conductivities of the gel (Fig. 9.8) were found to be in the 10^{-4} S cm^{-1} range at 25°C and in the 10^{-3} S cm^{-1} range at 70°C. In this case, the hectorite platelets play an active role by acting as the anion causing the lithium-ion transport number to reach as high as unity. A transport number = 1 is important for high discharge application since the concentration polarisation in this case will have less effect on the system (e.g. batteries).

Layered nanoparticles other than clays can be also used for intercalation of polymer electrolytes. Layered lithium trivandate (LiV_3O_8), for example, was found successful for the intercalation of poly methyl methacrylate (PMMA) [19]. Ionic conductivity and interfacial stability of a polymer gel electrolyte made of PMMA/LiV_3O_8 saturated with a solution of 1 M $LiClO_4$-PC/DEC (1:1) were superior to the same gel in the absence of LiV_3O_8. X-ray diffraction is a common technique for the characterisation of the intercalation process. For PMMA/LiV_3O_8 (Fig. 9.9), the interlayer distance increased from 6.3 A° ($2\theta = 14°$) for LiV_3O_8 to 12.8 A° ($2\theta = 7°$) for PMMA/LiV_3O_8, proving successful insertion of PMMA chains between the layers pushing them further apart.

Room temperature ionic conductivity of PMMA/LiV_3O_8 gel nanocomposite (1.8×10^{-3} S cm^{-1}) was about four times higher than PMMA gel in the absence of LiV_3O_8 (5.1×10^{-4} S cm^{-1}). This significant increase could be attributed to the fact that the negative $V_3O_8^-$ layers have high dielectric constant ($\varepsilon = 4.5$), which help dissolve more of $LiClO_4$ resulting in higher conductivity. Additional reason for the

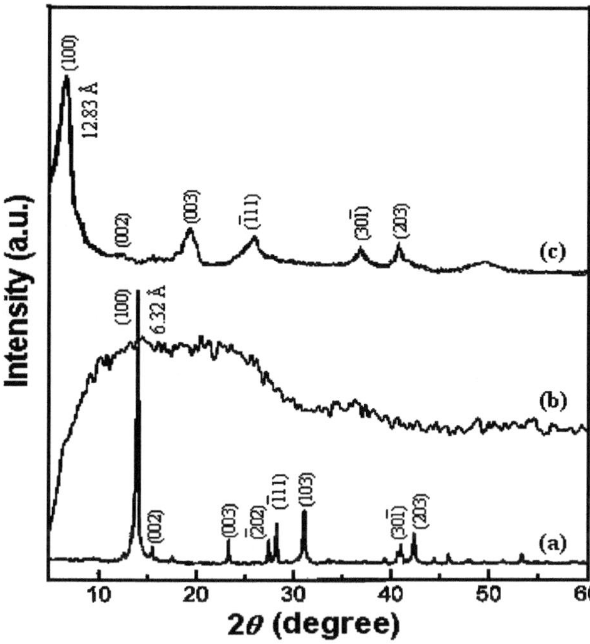

Fig. 9.9 X-ray diffraction of LiV_3O_8 (**a**), PMMA (**b**) and PMMA/LiV_3O_8 (**c**) (Reprinted from [19], Copyright (2011), from Springer)

ionic conductivity enhancement is the possible aggregation of the polymer gel nanocomposite. Scanning electron microscopy (SEM) (Fig. 9.10) showed that PMMA-LiV_3O_8 particles were 2.5 times bigger than LiV_3O_8. This aggregation could be attributed to some polymer chains that are intercalated between layers of two nanoparticles causing LiV_3O_8–LiV_3O_8 cross-linking. Such an aggregation provides a conduction path for Li^+ to transfer from one particle to another.

Polymer gel nanocomposite/electrode interfacial stability was also investigated by monitoring the conductivity over time of a cell made of stainless steel/polymer gel nanocomposite/stainless steel. Figure 9.11 shows that the PMMA gel was stable for 5 days compared to 10 days for PMMA-LiV_3O_8 gel and that the loss of conductivity was much less for the later. The highly porous structure of PMMA gel is expected to be disrupted by the intercalation process which reduces the pathways (pores) for lithium dendrites to grow and thus leading to better electrolyte/electrode interfacial stability in lithium batteries [28].

It is worth mentioning here that inorganic fillers such as SiO_2 and TiO_2 might leach out from nanocomposites during their lifetime in the amorphous state [29]. This could be associated with hazardous effect on humans if these particles are accidently inhaled or digested. They might also have negative effect on the environment and other living species including water, fish, algae and soils. Any interesting future research could focus on finding methods for better fixation of nano-inorganic fillers in the composite, using additives that suppress oxidative damage to polymers

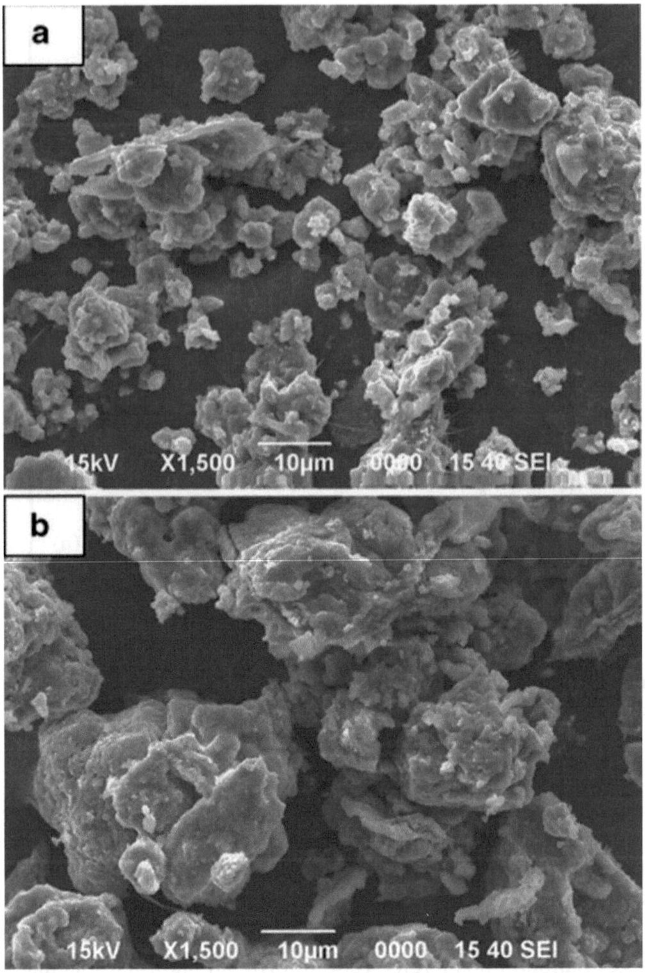

Fig. 9.10 Scanning electron microscopy of LiV_3O_8 (**a**) and PMMA/LiV_3O_8 (**b**) (Reprinted from [19], Copyright (2011), from Springer)

caused by the nanoparticles, modifying nanoparticle surface and improving the design and structure of the composite to release larger (safer) particles.

9.3 Nanocomposites of Liquid Electrolytes 'Soggy Sand'

In this part, we discuss a type of colloidal electrolytes made of insulating solid oxide nanoparticles such as SiO_2, TiO_2 and Al_2O_3 dispersed in conventional lithium liquid electrolytes (e.g. $LiPF_6$-EC/DMC) [6–13] or in ionic liquids [30–32]. The resulting viscous grain ensemble nanocomposite is commonly called 'soggy sand'

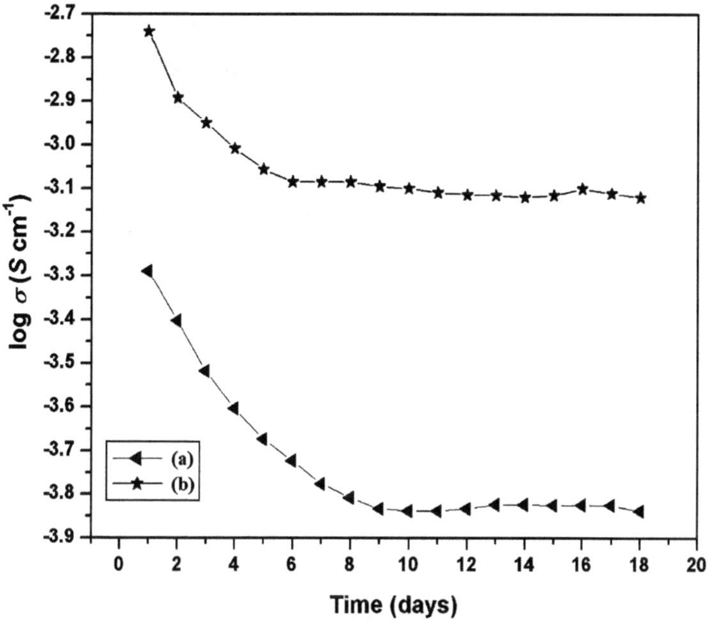

Fig. 9.11 Time dependence of the conductivity of PMMA-1 M $LiClO_4$-PC/DEC (1:1) polymer gel electrolyte (**a**) and PMMA/LiV_3O_8-1 M $LiClO_4$-PC/DEC (1:1) polymer gel electrolyte nanocomposite (**b**) (Reprinted from [19], Copyright (2011), from Springer)

or 'colloidal electrolyte' (Fig. 9.12). The addition of nano-sized particles is expected to enhance ionic conductivities of the original liquid electrolytes [10, 14, 30–32]. For example, silica nanoparticles have increased ionic conductivities of some lithium battery liquid electrolytes by few to ten folds [10]. In addition to conductivity enhancement, the soggy sand texture of this type of materials improves their mechanical properties which widen the range of their applications in electrochemical devices such as batteries [9, 10]. Similar to polymer nanocomposites, the amount of nanoparticles that can be added to liquid electrolytes has an optimum value after which the enhancement in conductivity starts to decline [10].

Maier and Bhattacharyya [8] carried out a simple experiment in which the conductivities of $LiClO_4$/methanol solution doped with different oxides at variable concentrations were monitored (Fig. 9.13). The trend observed was an increase in conductivities with increasing volume fraction of the oxide (φ) until a maximum was reached followed by a decline at higher φ values.

The decline in conductivities at higher oxide fraction is believed to be caused by a path of dry oxide particles blocking the conducting pathways (break down of base properties of the oxide used). Conductivity values of a solution doped with SiO_2, which is highly acidic, were higher compared to solutions doped with Al_2O_3 which is more basic, while TiO_2 resulted in intermediate values for conductivities (Fig. 9.13). These results suggest that the interaction between the oxide and $LiClO_4$

Fig. 9.12 Representation of the synthesis of soggy sand from liquid electrolytes

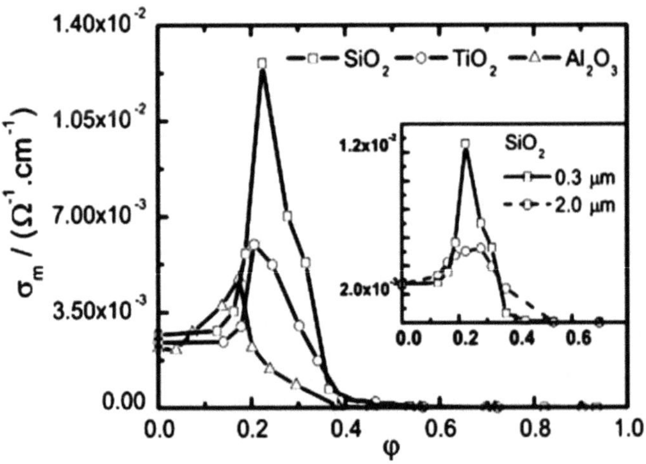

Fig. 9.13 Room temperature ionic conductivity versus volume fraction of TiO_2 (O), SiO_2 (□) and Al_2O3 (Δ) in $LiClO_4$/methanol solution. *Inset*: Ionic conductivity dependence on particle size of SiO_2: 0.3 μm (—) and 2.0 μm (- - -) (Reprinted from [8], Copyright (2011), from John Wiley and Sons)

salt involves adsorption of ClO^- anion by the oxide which enhances the dissociation of the salt and thus increases Li^+ concentration which increases conductivities. This phenomenon was confirmed by zeta-potential measurements which were negative (for negatively charged surfaces) and increased from −18.3 mV for Al_2O_3 to −36.3 mV for SiO_2. Higher values for SiO_2 are attributed to denser surface charge caused by higher ClO^- adsorption. Best conductivity ($\sigma_m = 1.2 \times 10^{-2}\ \Omega^{-1}\ cm^{-1}$) was observed for $LiClO_4$/methanol doped with 0.22 volume fraction SiO_2 which is comparable with electrolytes used in Li batteries. The surface effect of the oxide was also confirmed by a better conductivity enhancement observed for solutions containing silica particles with higher surface area. Figure 9.14 depicts the dissociation of salt enhanced by oxides that happen in both liquid and polymer nanocomposites.

The same research group added SiO_2 particles to a solution that consists of 1 M $LiPF_6$ in EC-DMC (1:1 v/v) and 1 M $LiPF_6$ in DMC [7]. Significant conductivity enhancement was observed which was again attributed to increased salt dissociation due to the adsorption of the anion, PF_6^-, on the surface of the oxide producing

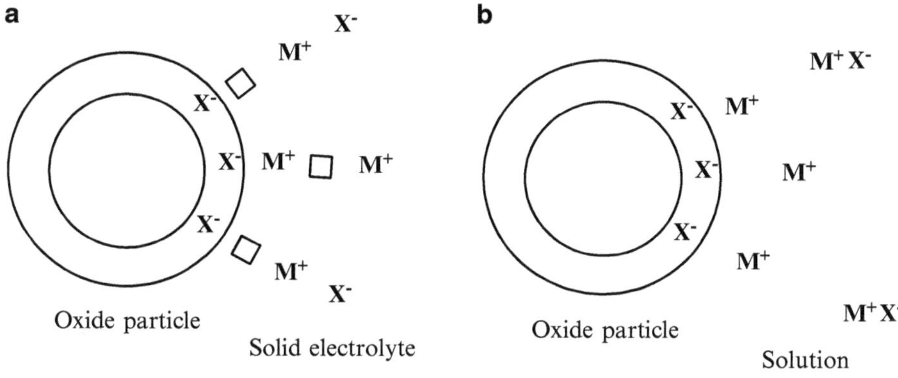

Fig. 9.14 Effect of oxides on salt dissociation in solid electrolytes (**a**) and solution electrolytes (**b**)

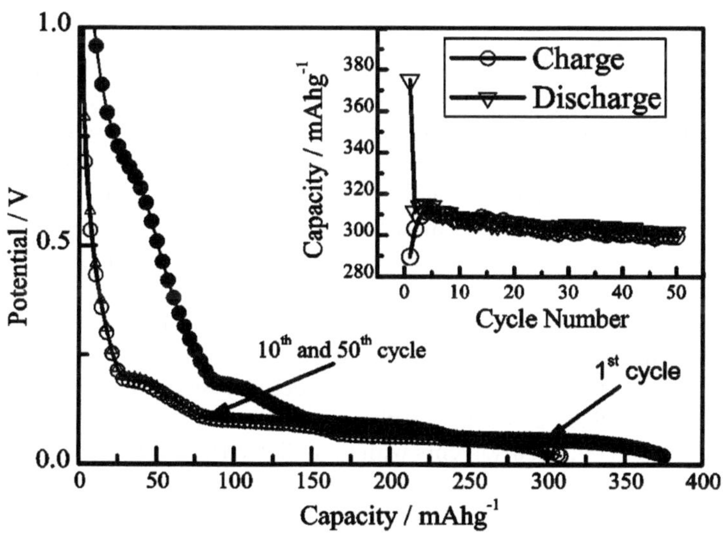

Fig. 9.15 Cycling behaviour of graphite/1 M $LiPF_6$ in EC-DMC (1:1 v/v)-SiO_2 ($\varphi = 0.04$)/Li cell at C/3 rate at 25°C (Reprinted from [7], Copyright (2011), from The Electrochemical Society)

higher concentration of Li^+ ions. The charge–discharge behaviour of a cell made of graphite/1 M $LiPF_6$ in EC-DMC (1:1 v/v)-SiO_2 ($\varphi = 0.04$)/Li (Fig. 9.15) shows that the capacity of the second discharge cycle (310 mAh g^{-1}) remains constant for about 50 cycles with a capacity loss of only 3.2%. These results prove the absence of any detrimental effect that SiO_2 might have towards the electrodes.

The results compare well with the theoretical capacity (370 mAh g^{-1}) and with data obtained for a liquid electrolyte in the absence of silica particles. The formation of solid electrolyte interface (SEI) is clear in the first shoulder of the first discharge cycle. This shoulder is not observed in the following cycles due to the

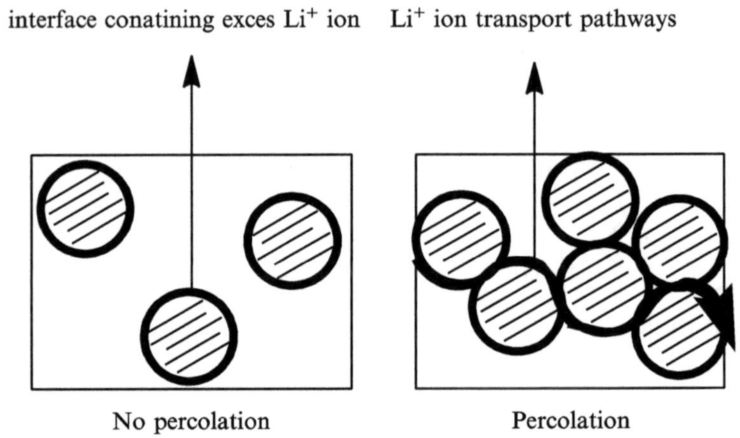

Fig. 9.16 Representation of proposed ionic conductivity mechanism in soggy sand

protection that the SEI layer provides to the electrolyte from the lithiated carbon once it is formed. The charge–discharge behaviour of this soggy sand proved to be similar to the behaviour of liquid electrolytes. Furthermore, the improved mechanical properties provided by the viscous nature of the soggy sand enabled the assembly of the above battery without the need to use a polymer separator.

In addition to enhancing salt dissociation, oxides such as the ones mentioned above are believed to have a percolation effect on ionic conductivity [10]. By this effect, ionic conductivities in soggy sands happen at the interfacial region, a mechanism similar to the one discussed in polymer nanocomposites (Fig. 9.16).

In most of the previously mentioned studies, it was observed that the percolation onset (oxide volume fraction at which conductivity starts to increase abruptly) depends on the dielectric constant of the solvent used to prepare the soggy sands. Soggy sands that contained solvents with high dielectric constant such as methanol ($\varepsilon = 32.6$) had percolation onset that happened at higher oxide volume fraction compared to ones that contained solvents with lower dielectric constant such as THF ($\varepsilon = 7.4$) (Fig. 9.17) [9].

This was attributed to a lower stability of the soggy sand caused by the high polarity of the solvent with high dielectric constant. Based on rheology measurements, soggy sand viscosity decreased with increasing shear rate (Fig. 9.18). These results indicate the formation of percolating network of oxide particles via van der Waals interaction when the thermal events overcome the electrostatic repulsion.

Soggy sand of ionic liquids (ILs) (called ion gels) can also be achieved to enhance their mechanical properties [30–32] for fabrication purposes. ILs recently received considerable attention as a potential replacement for conventional organic electrolytes because of their low volatility, high thermal/electrochemical/chemical stability, non-flammability and high ionic conductivity [10, 14, 30–32]. Despite these advantages, the fluid nature of ILs puts a limit on their

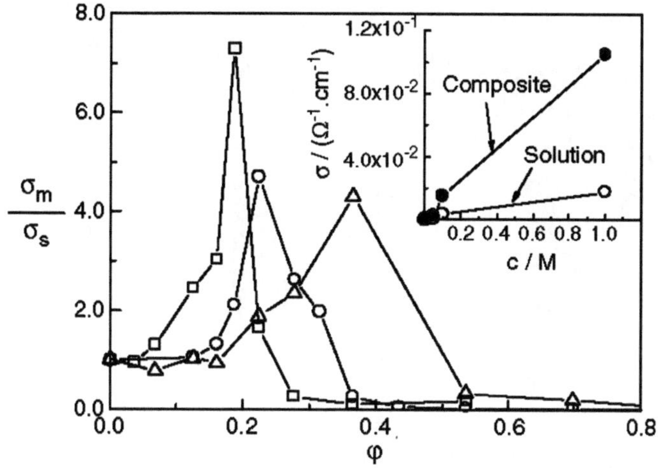

Fig. 9.17 Room temperature ionic conductivity (σ_m) normalised to solution conductivity (σ_s) with volume fraction of SiO_2 in SiO_2/0.1 M $LiClO_4$ solution in: MeOH (O), THF (□) and DMSO (Δ). Inset: Ionic conductivity with $LiClO_4$ concentration in MeOH solution (O) and its SiO_2 composite at loading ($\varphi = 0.28$) (●) (Reprinted from [9], Copyright (2011), from Elsevier)

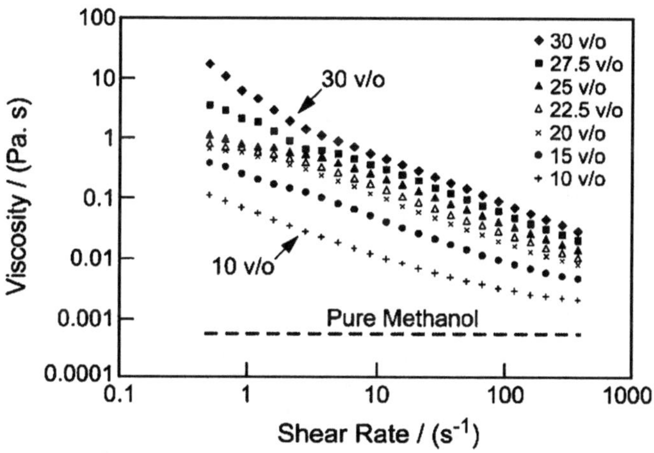

Fig. 9.18 Room temperature viscosity versus shear rate of 0.1 M $LiClO_4$ in MeOH-SiO_2 composites of various SiO_2 contents (Reprinted from [9], Copyright (2011), from Elsevier)

fabrication. Adding inorganic filler to ILs could be a potential solution for this problem since it results in gelation of the ionic liquid and thus better mechanical properties. Figure 9.19 [31] shows the dispersion of different loads of silica nanoparticles in 1-ethyl-3-methyl imidazolium bis(trifluoromethanesulfonyl)amide $[C_2mim][NTf_2]$ with gelation occurring at 3 wt.% and hardness of the gel increasing with silica contents. Rheological measurements of $[C_2mim][NTf_2]$ (Fig. 9.20) indicated a Newtonian fluid behaviour where the viscosity is independent on

Fig. 9.19 Photos of [C_2mim][NTf_2]-SiO_2 dispersions with SiO_2 contents of (a) 0 wt.%, (b) 1 wt.%, (c) 3 wt.%, (d) 5 wt.% and (e) 15 wt.% (Reprinted from [31], Copyright (2011), from American Chemical Society)

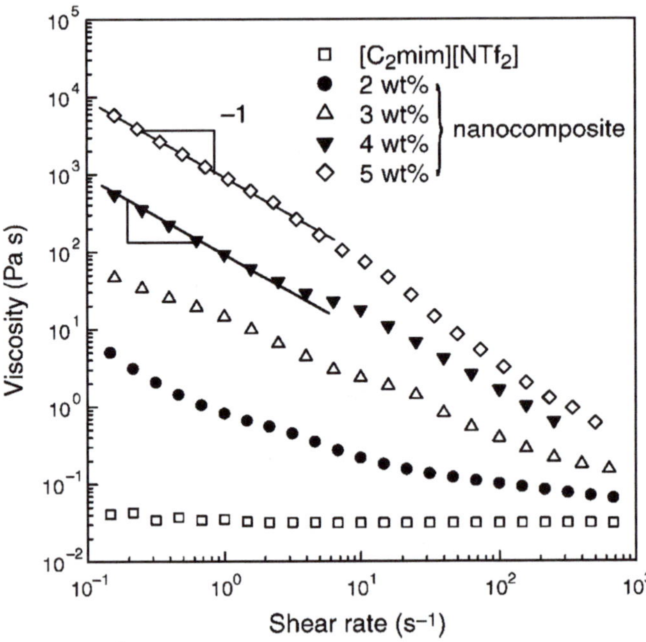

Fig. 9.20 Shear rate versus viscosity for [C_2mim][NTf_2]-SiO_2 with different silica particle content (Reprinted from [31], Copyright (2011), from American Chemical Society)

shear rate. Viscosities of composites with 2–5 silica wt.% on the other hand decreased with increasing shear rate indicating the disruption of the physical bond commonly found between particles in colloidal suspensions. This attraction between particles results in network formation at high silica contents which explain the sharp decrease in viscosity with shear rate for 5 wt.% nanocomposite.

Ionic conductivity of neat $[C_2mim][NTf_2]$ (about 10^{-2} S cm^{-1} at 30°C) was not affected by adding 5 wt.% silica which maintains its application as electrolyte with the extra advantage of enhanced mechanical properties. Gels with higher silica contents exhibited decreased ionic conductivities due to decreased mobility (solidification).

9.4 Nanocomposites of Organic Ionic Plastic Crystals

Organic ionic plastic crystals are a new type of solid state; conducting molecular plastic materials consist entirely of ions (Fig. 9.21). They recently showed significant conductivity at room temperature [5, 33–42]. The rotational and orientational disorders of the ionic molecules in ionic plastic crystals make them possess liquid-like behaviour that induces fast ion motion and thus higher conductivity. The plastic mechanical properties of this type of solid electrolytes combined with their ionic transport ability and the absence of liquid solvents drew the attention of researchers to study them for possible applications in electrochemical devices such as batteries [5, 33–42]. Figure 9.22 lists some ionic plastic crystals that were studied for lithium-

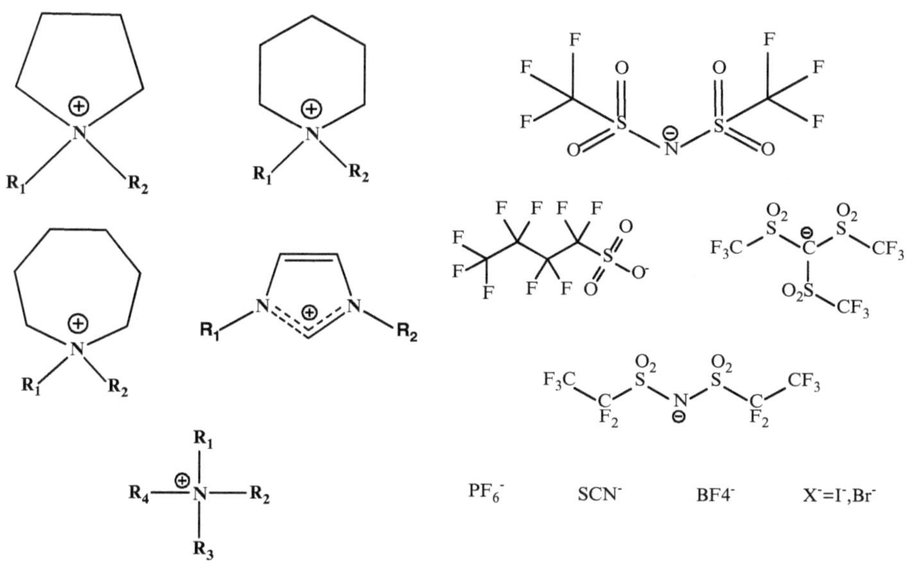

Fig. 9.21 Examples of organic ionic plastic crystals

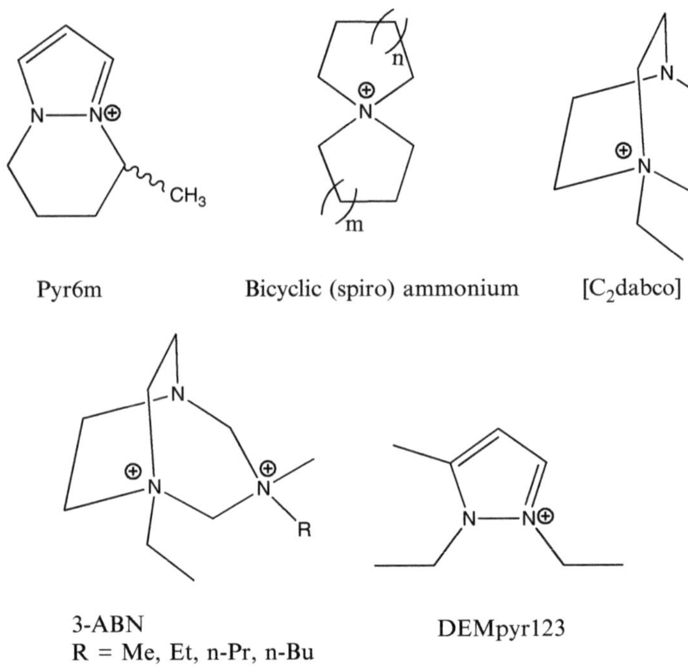

Fig. 9.22 Some ionic plastic crystals studied for lithium-ion batteries

ion batteries [40]. Similar to other types of electrolytes discussed in this chapter, doping with inorganic filler such as lithium salts and oxides (e.g. SiO_2) results in significant increase in conductivity [5, 33–42]. This enhancement in conductivity was attributed to the induced dissociation of ionic aggregates at the interface and increased concentration of charge carriers in a space-charge layer around the filler surface [5, 42]. It was also proposed that the interaction between the ionic crystal molecules and the inorganic additive results in more disorder and hence more mobility of those molecules which enhances ionic transport [35, 40–42].

Shekibi et al. [5] reported two orders of magnitude conductivity enhancement when they doped *N*-methyl-*N*-ethylpyrrolidinium bis(trifluoromethanesulfonyl) amide ([C_2mpyr][NTf_2]) (P12TFSA) with 10 wt.% SiO_2 nanoparticles sized 7, 12, and 16 nm (Fig. 9.23). As one would expect, particles with smallest size (highest surface area) provided maximum enhancement.

Later on, in another study by Shekibi et al. [42], the addition of lithium-functionalised silica nanoparticles to [C_2mpyr][NTf_2] (Fig. 9.24) produced conductivity enhancement consistent with the ones obtained using non-functionalised particles. Lithium-functionalised silica however has the extra advantage of acting as lithium-ion source as well as inorganic filler.

Thermal analysis of the pure [C_2mpyr][NTf_2] (denoted Mx for 'matrix') (Fig. 9.25) shows four phase transitions for the pure ionic crystal in which melting

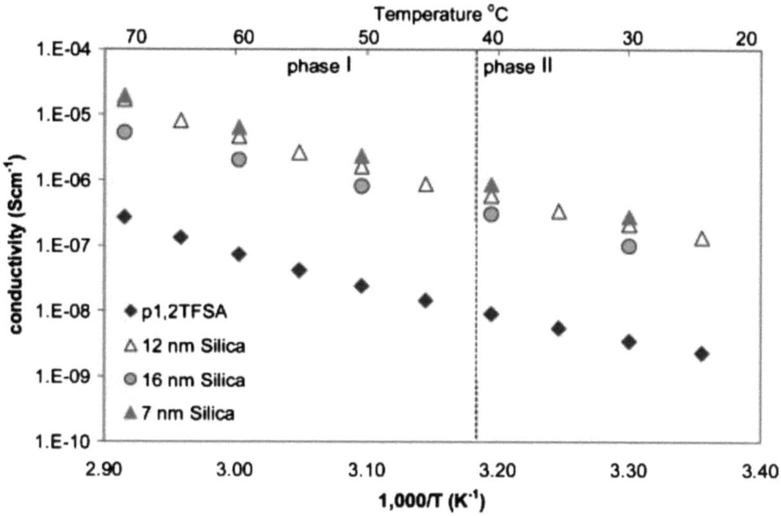

Fig. 9.23 Conductivity versus temperature of pure P12TFSA and P12TFSA combined with 10 wt. % SiO_2 of sizes 7, 12 and 16 nm (Reprinted from [5], Copyright (2011), from American Chemical Society)

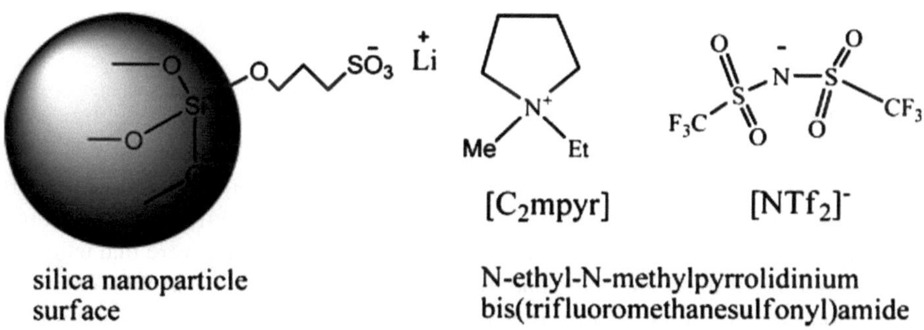

Fig. 9.24 Lithium-functionalised silica nanoparticles added to $[C_2mpyr][NTf_2]$ (Reprinted from [42], Copyright (2011), from The Royal Society of Chemistry)

occurs at ~89°C. Addition of functionalised nanoparticles decreased entropy and resulted in an additional melting stage at ~84°C which became more intense with increasing concentration of Li-SiO_2. The peak at 89°C corresponds to the melting point of pure $[C_2mpyr][NTf_2]$, while the one at 84°C corresponds to the region of $[C_2mpyr][NTf_2]$ in contact with Li-SiO_2. The increase in the peak intensity of the doped material with increasing filler content was attributed to higher interfacial region fraction. The disorder created at the interface causes lower enthalpy of melting which result in entropy decrease.

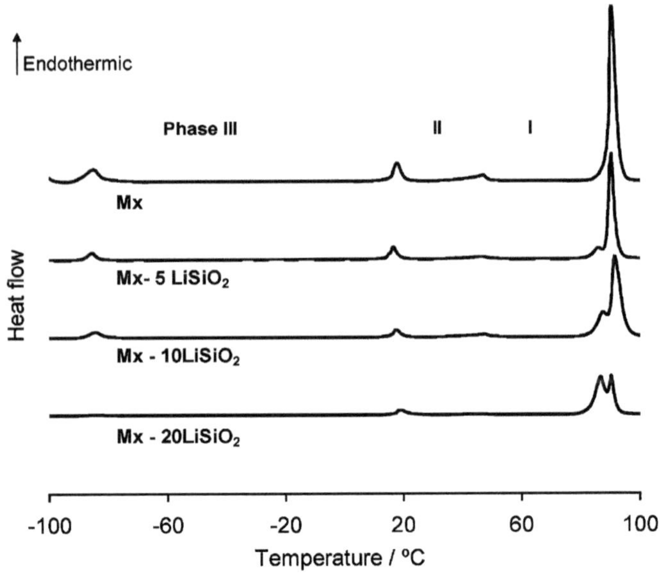

Fig. 9.25 DSC of pure plastic crystal [C_2mpyr][NTf_2] and [C_2mpyr][NTf_2] doped with 5, 10 and 20 wt.% Li-SiO_2 at 10 °C min^{-1} (Reprinted from [42], Copyright (2011), from The Royal Society of Chemistry)

9.5 Summary and Outlook

This chapter outlines the utilisation of various types of inorganic nano-fillers as additives to enhance the performance of electrolytes commonly used in lithium batteries. The main findings of the work presented in this chapter were that using an optimum amount of an inorganic nanoparticle such as ceramics (e.g. SiO_2) or clays (e.g. montmorillonite) with different types of electrolytes produced a significant enhancement in electrochemical properties that are crucial for optimum battery performance; these properties include lithium-ion transport number, ionic conductivity and electrochemical stability. These types of additives were also shown to improve the electrolyte/electrode interfacial stability in some cases, another important factor in battery performance. The degree of the effect of these nano-fillers on the above-mentioned properties depends greatly on the concentration and on the surface properties of the used nanoparticle.

Electrolyte nanocomposites produced this way are highly promising for future application in electrochemical devices. The impact on electrolyte properties, the availability of the materials and the ease of design provide advantages over other materials. Further future research could be devoted to modifying the surface of the nanoparticles to produce systems with more impact on electrolyte properties, testing new materials and optimising the fabrication and design.

References

1. Croce F, D'Epifanio A, Hassoun J, Reale P, Scrosati B (2003) Advanced electrolyte and electrode materials for lithium polymer batteries. J Power Sources 119–121:399–402
2. Blomgren GE (2003) Liquid electrolytes for lithium and lithium-ion batteries. J Power Sources 119–121:326–329
3. Sazhin SV, Harrup MK, Gering KL (2011) Characterization of low-flammability electrolytes for lithium-ion batteries. J Power Sources 196(7):3433–3438
4. Walls HJ, Riley MW, Singhal RR, Spontak RJ, Fedkiw PS, Khan SA (2003) Nanocomposite electrolytes with fumed silica and hectorites clay networks: passive versus active fillers. Adv Funct Mater 13(9):710–717
5. Shekibi Y, Gray-Weale A, MacFarlane DR, Hill AJ, Forsyth M (2007) Nanoparticle enhanced conductivity in organic ionic plastic crystals: space charge versus strain induced defect Mechanism. J Phys Chem C 111(30):11463–11468
6. Balaya P, Bhattacharyya AJ, Jamnik J, Zhukovskii YF, Kotomin EA, Maier J (2006) Nano-ionics in the context of lithium batteries. J Power Sources 159(1 SPEC. ISS):171–178
7. Bhattacharyya AJ, Dollé M, Maier J (2004) Improved Li-battery electrolytes by heterogeneous doping of nonaqueous Li-salt solutions. Electrochem Solid State Lett 7(11):A432–A434
8. Bhattacharyya AJ, Maier J (2004) Second phase effects on the conductivity of non-aqueous salt solutions: "Soggy sand electrolytes". Adv Mater 16(9–10):811–814
9. Bhattacharyya AJ, Maier J, Bock R, Lange FF (2006) New class of soft matter electrolytes obtained via heterogeneous doping: percolation effects in "soggy sand" electrolytes. Solid State Ion 177(26–32 SPEC. ISS.):2565–2568
10. Bhattacharyya AJ, Patel M, Das SK (2009) Soft matter lithium salt electrolytes: ion conduction and application to rechargeable batteries. Monatsh Chem 140(9):1001–1010
11. Das SK, Bhattacharyya AJ (2009) Oxide particle surface chemistry and ion transport in "soggy sand" electrolytes. J Phys Chem C 113(16):6699–6705
12. Das SK, Bhattacharyya AJ (2010) Influence of oxide particle network morphology on ion solvation and transport in "soggy Sand" electrolytes. J Phys Chem B 114(20):6830–6835
13. Edwards WV, Bhattacharyya AJ, Chadwick AV, Maier J (2006) An XAS study of the local environment of ions in soggy sand electrolytes. Electrochem Solid State Lett 9(12):564–567
14. Kumar B, Rodrigues SJ (2004) Ionic conductivity of colloidal electrolytes. Solid State Ion 167 (1–2):91–97
15. Osinska M, Walkowiak M, Zalewska A, Jesionowski T (2009) Study of the role of ceramic filler in composite gel electrolytes based on microporous polymer membranes. J Membr Sci 326(2):582–588
16. Appetecchi GB, Croce F, Persi L, Ronci F, Scrosati B (2000) Transport and interfacial properties of composite polymer electrolytes. Electrochim Acta 45(8–9):1481–1490
17. Forsyth M, MacFarlane DR, Best A, Adebahr J, Jacobsson P, Hill AJ (2002) The effect of nano-particle TiO2 fillers on structure and transport in polymer electrolytes. Solid State Ion 147(3–4):203–211
18. Chen YT, Chuang YC, Su JH, Yu HC, Chen-Yang YW (2011) High discharge capacity solid composite polymer electrolyte lithium battery. J Power Sources 196(5):2802–2809
19. Deka M, Kumar A (2010) Enhanced ionic conductivity in novel nanocomposite gel polymer electrolyte based on intercalation of PMMA into layered LiV_3O_8. J Solid State Electrochem 14(9):1649–1656
20. Deka M, Kumar A (2011) Electrical and electrochemical studies of poly(vinylidene fluoride)-clay nanocomposite gel polymer electrolytes for Li-ion batteries. J Power Sources 196 (3):1358–1364
21. Kim KM, Park N-G, Ryu KS, Chang SH (2002) Characterization of poly(vinylidene fluoride-co-hexafluoropropylene)-based polymer electrolyte filled with TiO2 nanoparticles. Polymer 43(14):3951–3957

22. Sharma AL, Thakur AK (2010) Polymer-ion-clay interaction based model for ion conduction in intercalation-type polymer nanocomposite. Ionics 16(4):339–350
23. Croce F, Settimi L, Scrosati B (2006) Superacid ZrO2-added, composite polymer electrolytes with improved transport properties. Electrochem Commun 8(2):364–368
24. Skaarup S, West K, Julian PM, Thomas DM (1990) Mixed phase solid electrolytes with nonconducting polymer binder. Solid State Ion 40–41(Part 2):1021–1024
25. Skaarup S, West K, Zachau-Christiansen B (1988) Mixed phase solid electrolytes. Solid State Ion 28–30(Part 2):975–978
26. Croce F, Sacchetti S, Scrosati B (2006) Advanced, lithium batteries based on high-performance composite polymer electrolytes. J Power Sources 162(1):685–689
27. Adebahr J, Best AS, Byrne N, Jacobsson P, MacFarlane DR, Forsyth M (eds) (2003) Ion transport in polymer electrolytes containing nanoparticulate TiO2: the influence of polymer morphology. Phys Chem Chem Phys 5:720–725, The Royal Society of Chemistry
28. Song JY, Wang YY, Wan CC (1999) Review of gel-type polymer electrolytes for lithium-ion batteries. J Power Sources 77(2):183–197
29. Reijnders L (2009) The release of TiO2 and SiO2 nanoparticles from nanocomposites. Polym Degrad Stab 94(5):873–876
30. Moganty SS, Jayaprakash N, Nugent JL, Shen J, Archer LA (2010) Ionic-liquid-tethered nanoparticles: hybrid electrolytes. Angew Chem Int Ed 49(48):9158–9161
31. Ueno K, Hata K, Katakabe T, Kondoh M, Watanabe M (2008) Nanocomposite ion gels based on silica nanoparticles and an ionic liquid: ionic transport, viscoelastic properties, and microstructure. J Phys Chem B 112(30):9013–9019
32. Zhang J, Zhang Q, Li X, Liu S, Ma Y, Shi F, Deng Y (2010) Nanocomposites of ionic liquids confined in mesoporous silica gels: preparation, characterization and performance. Phys Chem Chem Phys 12(8):1971–1981
33. Abu-Lebdeh Y, Alarco P-J, Armand M (2003) Conductive organic plastic crystals based on pyrazolium imides. Angew Chem Int Ed 42(37):4499–4501
34. Abu-Lebdeh Y, Austin E, Davidson IJ (2009) Spiro-ammonium imide salts as electrolytes for lithium batteries. Chem Lett 38(8):782–783
35. Adebahr J, Ciccosillo N, Shekibi Y, MacFarlane DR, Hill AJ, Forsyth M (2006) The "filler-effect" in organic ionic plastic crystals: enhanced conductivity by the addition of nano-sized TiO2. Solid State Ion 177(9–10):827–831
36. Cooper EI, Angell CA (1986) Ambient temperature plastic crystal fast ion conductors (PLICFICS). Solid State Ion 18–19(Part 1):570–576
37. Fukada S-I, Yamamoto H, Ikeda R, Nakamura D (1987) Hydrogen-1 nuclear magnetic resonance, differential thermal analysis, X-ray powder diffraction and electrical conductivity studies on the motion of cations, including self-diffusion in crystals of propylammonium chloride and bromide as well as their n-deuterated analogues. J Chem Soc 10:3207–3222
38. MacFarlane DR, Forsyth M (2001) Plastic crystal electrolyte materials: new perspectives on solid state. Adv Mater 13(12–13):957–966
39. MacFarlane DR, Meakin P, Amini N, Forsyth M (2001) Structural studies of ambient temperature plastic crystal ion conductors. J Phys Condens Matter 13(36):8257–8267
40. Pringle JM, Howlett PC, MacFarlane DR, Forsyth M (2010) Organic ionic plastic crystals: recent advances. J Mater Chem 20(11):2056–2062
41. Pringle JM, Shekibi Y, MacFarlane DR, Forsyth M (2010) The influence of different nanoparticles on a range of organic ionic plastic crystals. Electrochim Acta 55(28):8847–8854
42. Shekibi Y, Pringle JM, Sun J, Pas SJ, Rocher NM, Clare BR, Hill AJ, MacFarlane DR, Forsyth M (2010) Lithium-functionalised silica nanoparticles for enhanced ionic conductivity in an organic ionic plastic crystal. J Mater Chem 20(2):338–344

Chapter 10
Micro-scaled Three-Dimensional Architectures for Battery Applications

Matthew Roberts, Phil Johns, and John Owen

Abstract The concept of 3D microbatteries provides an approach which could result in a step change in the energy and power per footprint of surface-mountable rechargeable batteries for microelectromechanical systems (MEMS) and other small electronic devices. The proposed structure has a high aspect ratio microstructured current collector coated in the three battery active layers (cathode, anode and electrolyte), each layer being a few microns in thickness; this reduces the length of the diffusion path through the layers, maximising the power capability. The high aspect ratio of these batteries also allows for significant increases in the energy storage per footprint area. This chapter outlines the design principles for 3D microbatteries and estimates the geometrical and physical requirements of the materials. Relevant examples of microbattery half-cells and full cells are presented to illustrate the key fabrication methods. Moreover, the same basic concepts and techniques presented could be used in the future to fabricate batteries at the nanoscale.

10.1 Introduction

10.1.1 Why Do We Need 3D Structures for Microbatteries?

Small devices such as medical implants, microsensors, self-powered integrated circuits or microelectromechanical systems (MEMS)[1, 2] need packaged rechargeable batteries with dimensions on the scale of 1–10 mm^3 and a high energy storage

M. Roberts • P. Johns • J. Owen (✉)
School of Chemistry, University of Southampton, Southampton SO172LU, Hampshire, United Kingdom
e-mail: jro@soton.ac.uk

Y. Abu-Lebdeh and I. Davidson (eds.), *Nanotechnology for Lithium-Ion Batteries*, Nanostructure Science and Technology, DOI 10.1007/978-1-4614-4605-7_10,

density. Therefore, a lithium-ion system should be preferable as it will provide the highest energy density of available technologies suitable for the application. The recent surge in development of MEMS is a particular driving force for development of a reliable and versatile lithium-ion microbattery. MEMS devices have been used in applications such as inkjet printer cartridges, accelerometers, miniature robots, microengines, locks, inertial sensors, microtransmissions, micromirrors, microactuators, optical scanners, fluid pumps, transducers and chemical, pressure and flow sensors [3]. This field will play an ever-increasing role in modern life as these products begin to filter down into everyday items. The incorporation of localised power in the form of batteries would be a big improvement in many of these applications. One difficulty is the concept of recharging these tiny devices. However, microdocking stations for charging could be envisaged. Rechargeable lithium-ion systems could also be extremely useful if these microdevices had a micro solar cell also attached.

Over the last 15 years or so, thin-film lithium-ion microbatteries [4, 5] have emerged as surface-mountable devices up to about 10 μm thick. The provision of increased power levels to modern MEMS devices has become an increasing challenge because of the limited energy and power available per area of footprint on the substrate. Increasing the thickness does not solve this problem in a thin-film cell because this also increases the current path length, leading to a reduction in power density. Conventional routes to solving this problem in the battery world would be to wind the thin film up, including a large surface area in a small volume. However, this is not suitable for most thin-film systems as the components tend to be brittle and winding the electrodes will cause fractures, breaks and short circuits. This limits these designs to planar systems which need large footprint areas for large capacities. This leads to the concept of *capacity per footprint area* (μA. h cm^{-2}), which is a key consideration for the construction of microbatteries. Battery technologists typically characterise charge storage in terms of gravimetric (units mA h g^{-1}) and volumetric capacities (units mA h cm^{-3}). However, in the case of microbattery applications where the limitation is the area available, the relevant specification is capacity per footprint area.

One definition [6] of the term '3D battery' reads 'cells comprising anodes and cathodes which have active surface areas exposed in three dimensions'. Although this could include the composite electrodes used in the thick-film (powder–binder composite) cells, it is normally reserved for cells assembled using micro-architectured or microfabricated porous electrodes. The term 'semi-3D' can be applied to the combination of a microfabricated electrode (i.e. with three-dimensionally exposed active area) connected to another conventional electrode *via* a planar separator. A more advanced concept, herein called 3D (or full 3D), is a design which folds the complete thin-film cell structure from the planar geometry into a thick laminate or network placed on a small footprint, so that the overall current path remains small. The following chapter will address both of these concepts and show how they can improve device performance specifications. It will then discuss examples of cell designs and fabrication methods, with particular reference to materials deposition techniques.

Fig. 10.1 A thin-film cell (Reprinted with permission from The Royal Society of Chemistry)

10.1.2 Planar (2D) Cells

10.1.2.1 Thin-Film Cells

'Thin film' means a planar semiconductor device that is made by physical or chemical vapour deposition, and the materials are solid ceramics or glasses. Thin-film microbatteries (Fig. 10.1) are designed for small-scale applications where high storage capacities are not required. Starting with a thin current collector, the cell is built by depositing layers of the lower electrode, electrolyte, upper electrode and a second current collector to form the battery. The thickness is limited to a few micrometres by the maximum thickness each layer can have before mechanical stresses cause fracture.

A major power limitation for the thin-film cell is due to the ohmic drop in the electrolyte/separator layer, which increases with the separator thickness, causing the maximum power to decrease. In the absence of other limitations, we can estimate the maximum power available *per footprint (area)*, P_A, from the resistance x area product as follows. The maximum power is delivered at half the short-circuit current, I_{SC}, where the ohmic loss is half the open-circuit voltage, V_{oc}.

$$R \times A = \frac{L_S}{\sigma} \tag{10.1}$$

and

$$P_A \sim \frac{V_{oc}}{2} \times \frac{V_{oc}}{2RA} = \frac{\sigma V_{oc}{}^2}{4L_S} \tag{10.2}$$

where σ = conductivity and L_S = separator thickness.

Alternatively, energy (E)/power ratio can be expressed as a discharge time constant, τ,

$$\tau = \frac{E_A}{P_A} \sim \frac{V_{oc}Q_A}{2} \times \frac{4L_S}{\sigma V_{oc}} \sim Q_V \times \frac{2L_S L_E}{\sigma V_{oc}} \tag{10.3}$$

where E_A is the specific energy per footprint.

Q_A and Q_V are specific capacity (charge) of the electrode per footprint and volume and L_E is the electrode thickness.

Thin solid electrolytes like lithium phosphorus oxynitride ('LiPON') and lithium borophosphate ('LiBP') [7] have been used, and despite their rather low conductivities (e.g. LiPON ~10^{-5} to 10^{-6} S cm^{-1}), they can support a modest current density due to their low thickness (<5 μm).

The energy available per footprint increases with the thickness of the cathode and anode layers. Increasing the electrode thickness will at some point lead to power limitations due to slow diffusion in the electrode rather than the low conductivity of the electrolyte. In that case, we can estimate the maximum power using Fick's laws according to the diffusion coefficient for lithium in the solid materials. The rate of diffusion determines the shortest discharge time for 50% discharge, $\tau_{0.5}$, and the corresponding maximum power density through the electrode thickness as follows:

$$\tau_{0.5} \sim \frac{{L_E}^2}{3D_{Li}} \tag{10.4}$$

where L_E is the electrode thickness and D_{Li} is the lithium diffusion coefficient.

$$P_A \sim \frac{0.5E_A}{\tau} \sim \frac{1.5E_A \times D_{Li}}{{L_E}^2} \tag{10.5}$$

The result is that for a given energy per footprint, the power per footprint and the maximum rate (1/τ), for half discharge will both be inversely proportional to the square of the thickness. The low diffusion coefficients (<10^{-9} cm^2s^{-1}) for lithium ions in solids will ensure that the thin-film construction can never deliver high energy and power simultaneously.

Bates et al.[4] reported 50% DoD at rates of over 50 C (~1 min charge or discharge) using a cell of total thickness 15 μm with a $LiCoO_2$, LiPON electrolyte and Li counter electrode. Although this level of performance is impressive, the total cell capacity is only 170 μA h (taken at a rate of 1 C). This means that although the cell can be charged and discharged very efficiently at high rates, only small amount of charge and energy can be stored and therefore only small devices can be powered.

Table 10.1 compares the result with the predictions of Eqs. 10.3 and 10.4 with typical parameter values reported in the literature. The table makes the point that micro dimensions enable fast discharge even if the materials have very low conductivities and diffusion coefficients.

10.1.2.2 Composite Electrodes and Thick-Film Cells

Deposition of composite materials in layers from solvent dispersions, e.g. using doctor blade, laser coating or inkjet equipment, is the preferred technology for making thick-film cells from, say, 20 μm to 1 mm overall thickness. The materials

Table 10.1 Comparisons between experiment and theory for a Li/LiPON/$LiCoO_2$ thin-film cell

LiPON solid electrolyte		$LiCoO_2$ cathode material	
Σ/Scm^{-1}	1×10^{-6}	D/cm^2s^{-1}	1×10^{-10}
L_S/cm	3×10^{-4}	L_E/cm	2.5×10^{-4}
L_E/cm	2.5×10^{-4}		
$Q_v/C\ cm^{-3}$	500		
V_{oc}/V	4		
Time constant t/s	40	*Time constant* $t_{0.5}/s$	200
Experimental t/s	70	*Experimental t/s*	70

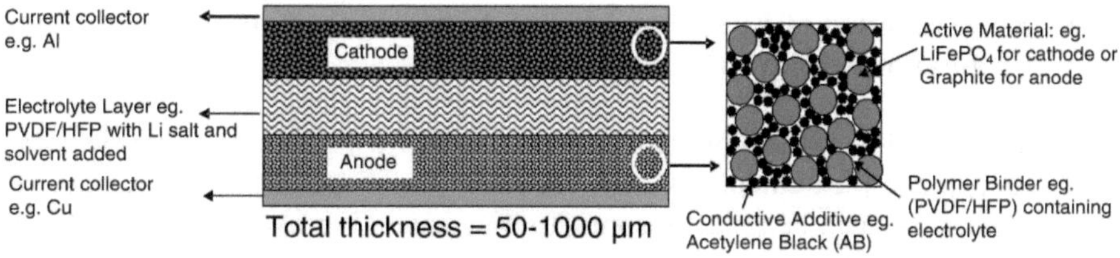

Fig. 10.2 A thick-film cell (Reprinted with permission from The Royal Society of Chemistry)

are usually ground to a small particle size and fabricated into composite porous electrode structures with a polymer binder to give the film mechanical strength. A liquid electrolyte contained in the pores provides ionic pathways, and a conductive additive, typically acetylene black, provides electronic pathways to the surfaces of the active material particles where the redox reaction occurs. (The polymer can also be chosen to have a dual function as the binder and the electrolyte, e.g. polyethylene oxide (*PEO*) containing a lithium salt ($LiPF_6$) [8].) Much thicker layers can be used in this case because the effective conductivity and diffusion coefficient for lithium are enhanced by the ionic conductivity of the infused electrolyte. Conventional lithium-ion batteries found in applications such as mobile phones or laptops are typically formed from five flexible films 20–100 μm in thickness as shown schematically in Fig. 10.2.

A thick-film cell has been described by Kim et al. [9] who used a laser printer to deposit thick films of porous battery materials on to metallic current collectors, separated by a gel polymer electrolyte to make microbatteries. Cathode and anode inks ($LiCoO_2$ and mesoporous carbon microbead (MCMB), respectively) were deposited onto their respective current collectors using the described laser printing process. The gel polymer electrolyte used was PVdF–HFP/1 M $LiPF_6$ in propylene carbonate (PC)/ethylene carbonate (EC)/dimethyl carbonate (DMC) (1:1:3). The rate performance of the cell was shown to be independent of the electrode thickness, in contrast to thin-film sputtered cells where the rates of discharge decreased rapidly with electrode thickness. The authors briefly compared the capacity (μA h cm^{-2}) of the laser-printed thick-film microbattery to that of the thin-film sputtered microbattery and reported low-rate capacities that were an order of magnitude greater for the laser-printed system.

In the above case, the improvement was probably due to the higher conductivity of the gel polymer electrolyte compared with that of glass or ceramic. A similar situation exists when a liquid electrolyte is contained in porous polymer – the current path is essentially perpendicular to the plane of the separator on the microscopic scale. Provided that the microscopic path is relatively non-tortuous, we can use Eq. 10.2 as a reasonable prediction of the power limitation due to the electrolyte resistance.

10.1.2.3 Estimation of Energy and Power Densities per Footprint for Thick-Film Cells

Accurate estimation of the power limitation due to diffusion in composite electrode materials is a complex calculation that is outside the scope of this discussion, but useful estimations can be made in some limiting cases. For example, if *solid-state* diffusion is the limiting factor, we now have a discharge time that depends on the particle *radius* R rather than the electrode thickness:

$$\tau_D \approx \frac{R^2}{D_{Li}} \tag{10.6}$$

Reduction of the particle size of the electrode material can, in principle, alleviate the problem of solid-state diffusion, so that the power density should only be limited by ion and electron transport in the *composite*. For short pulse discharge, we can use the De Levie expression [10] to obtain:

$$D_{comp} = \frac{(\sigma_i.\sigma_e)}{C_v(\sigma_i + \sigma_e)} \approx \frac{\sigma_i}{C_v} for\ \sigma_e << \sigma_i \tag{10.7}$$

where σ_e is the electronic conductivity, etc.

In the latter case, the power per footprint area decreases with the electrode thickness as in Eq. 10.4, and we again obtain a limitation in the electrode thickness which limits the storage capacity as before because of the need to retain a specified power per footprint area.

Generally, the low diffusivity of solid electrodes is bypassed by the liquid or polymer electrolyte of the composite provided that the electrode particles are small enough and the electrolyte is sufficiently conductive.

Another limiting factor can become dominant in the *continuous* discharge of composite electrodes, which is the limited rate of *salt diffusion* in the composite [11]. The following approximation was suggested to estimate this effect:

$$\tau \sim \frac{L^2}{D_{salt}} \times \frac{(1 - T_+)[Li]}{[salt]} \tag{10.8}$$

where $T_+ = Li^+$ transference number, or the number of moles of Li crossing per Faraday of charge passed, *[Li]* = change in lithium concentration in the electrode during discharge and *[salt]* = salt concentration in the separator. Obviously, if the transference number for cations is close to one, as it is believed to be for many solid and glassy electrolytes, this limitation will not apply.

Again, we have a simple expression that approximately quantifies the common result that for a given time constant for discharge, the ionic conductivity required of the electrolyte varies as the square of the electrode thickness, L^2. Therefore, we can enable the use of relatively low-conductivity electrolytes, e.g. dry polymers or glasses; by reducing the thickness by just one order of magnitude, we can compensate for a reduction in conductivity by two orders of magnitude. Such a reduction in thickness, however, will reduce the energy density per footprint as in the thin-film cell. Therefore, for applications that require high power and energy per footprint or for devices constructed with a poorly conducting, liquid-free electrolyte, we seek a method of minimising the ionic current path between the two electrodes. This will be found in the principles and descriptions of the 3D cells described below.

10.2 Semi-3D Microbatteries

The following section examines some microbattery architectures that illustrate the development of thin-film technology towards the full 3D configurations.

10.2.1 Nanoarchitectured Current Collectors and 'Semi-3D' Cells

Several types of nanoarchitectured electrodes have been described as alternatives to the composite electrode described above. They may be defined here as electrodes that are carefully fabricated to optimise the ionic and electronic current paths, e.g. as shown in Fig. 10.3, by depositing a thin layer of active material on a nanoarchitectured current collector array. The design will ensure a small tortuosity factor,

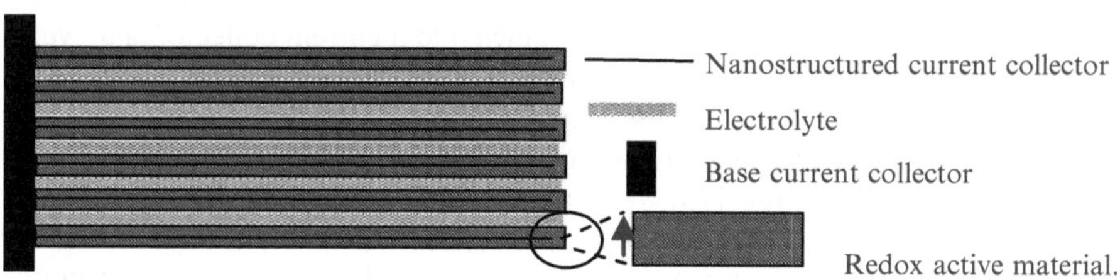

Fig. 10.3 Schematic diagram of a 3D nanostructured current collector coated in redox-active material. This constitutes a 'semi-3D' when coupled with a second electrode via a planar separator (Reprinted with permission from The Royal Society of Chemistry)

Fig. 10.4 (**a**) Low- and (**b**) high-magnification SEM images of Ni nanorod-supported $LiCoO_2$ deposits (Reprinted from [14], Copyright (2011), from John Wiley and Sons)

leading to a higher effective diffusion coefficient than that obtainable from a random composite electrode. The theory of the composite electrode presented above may be applied most easily to the example of Fig. 10.3 because the *effective* ionic conductivity due to the electrolyte within the channels is precisely the bulk conductivity value multiplied by the ratio of the cross-sectional area of the ionic current path to the total area of the base current collector. These electrodes can then be cycled against a standard 2D electrode such as lithium or graphite.

10.2.2 Semi-3D Cathodes

This section will look at some of the most relevant examples of 3D cathodes which have been developed. These examples typically involve a 3D current collector which a thin film of cathode material has then been applied. This process will significantly enhance the amount of active material per footprint area (as in thick-film cells) while maintaining short diffusion distances (as in thin-film cells). These cells are cycled versus a planar negative or anode material with a conventional separator and electrolyte.

V_2O_5 aerogels have been suggested for use as a 3D network cathode [12]. In this report, powders of V_2O_5 aerogels were contacted to a current collector and cycled versus Li. Large double-layer capacitance is observed as well as some redox chemistry. It is conceived that these materials could be developed into a 3D microbattery if electrolyte and anode layers were interpenetrated into the porous 3D structure. Other aerogels [13] would be appropriate for use in the same way.

A nanostructured $LiCoO_2$ has been developed as an active material for 3D microbattery positive electrodes [14]. A conformal layer of nanostructured $LiCoO_2$ was applied to Ni and Al nanorods prepared by templated electrodeposition. The nanostructured $LiCoO_2$ was synthesised by thermal decomposition of spray-coated sol–gel precursors; examples are shown in Fig. 10.4.

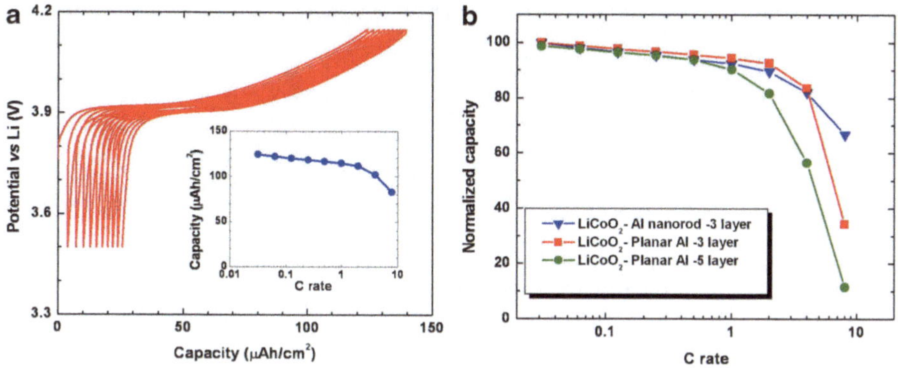

Fig. 10.5 (**a**)Charge–discharge galvanostatic curves for aluminium nanorod-supported $LiCoO_2$ deposits cycled at a rate of C/10 versus Li and using a charge cut-off voltage of 4.15 V. *inset*: rate capability plot for the same electrode. (**b**) Normalised capacity and rate capability plots for Al nanorod-supported $LiCoO_2$ electrode. $LiCoO_2$ film deposited on planar Al foil with different thickness (three and five layers of spray coating) (Reprinted from [14], Copyright (2011), from John Wiley and Sons)

Electrochemical characterisations of the nanorod-supported $LiCoO_2$ 3D positive electrode were then made. The electrodes were fabricated with uniform thickness of $LiCoO_2$ deposit (three layers of spray coating), and the geometrical area was fixed to be 1.3 cm^2. The Al nanorod-supported $LiCoO_2$ electrodes were characterised for their performance as cathodes in Li half-cells which were cycled in a galvanostatic mode at a rate of C/10 and with a charge cut-off voltage of 4.15 V (Fig. 10.5a). A well-defined plateau around 3.9 V was observed corresponding to the first-order phase transition between two hexagonal phases (during Li de-insertion and insertion). The cycling shows negligible hysteresis, and the electrodes were found to exhibit excellent capacity retention. In Fig. 10.5b, normalised capacity is plotted versus rate to highlight the high-rate performance of 3D positive electrode. Excellent rate capability is observed for the Al nanorod-supported $LiCoO_2$ electrode compared to their planar counterparts and is shown to recover ~70% of its total capacity at a high rate of 8C.

Golodnitsky et al. electrodeposited nanosized particles of copper sulphide cathodes on the 3D perforated silicon substrates (Fig. 10.6) [15]. The morphology and composition of the cathodes have been controlled by varying the operating parameters, such as current density, pH and temperature, of an electrolyte. The addition of a polymer to the electrolyte bath enables the formation of sulphur-rich thick porous layers. This is not possible without the additive, which serves to decrease the internal stresses in the bulk of the deposit.

When the semi-3D cells with Ni or Au current collectors were electrochemically characterised, capacities between 1.0 and 2.5 mAh cm^{-2} with CuS cathodes were seen, depending on the morphology and composition of the cathode; the cells ran for >400 reversible cycles showing low degradation (Fig. 10.7). This is in good agreement with the geometrical area gain (AG) factor of 9 for the perforated

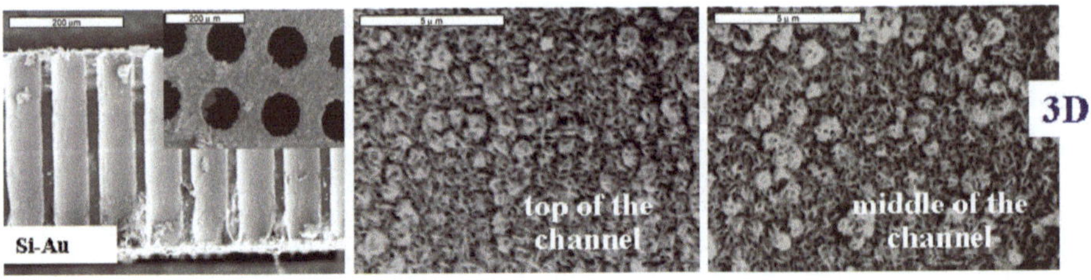

Fig. 10.6 CuS electrodeposited on perforated (3D) silicon substrates (Reprinted from [15], Copyright (2011), from The Electrochemical Society)

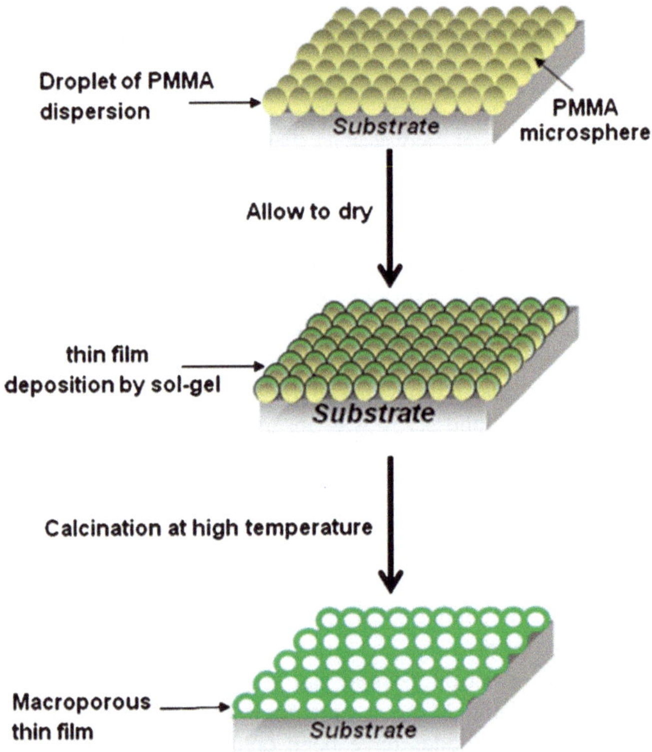

Fig. 10.7 Schematic diagram of the fabrication process of 3D $LiMn_2O_4$ thin-film electrodes presented by Park et al. (Reprinted from [17], Copyright (2011), from Springer)

substrate. At constant charge/discharge current, the semi-3D microbatteries with modified copper sulphide cathode retain approximately 80% of the initial capacity when the discharge rate increases from 120 μA cm^{-2} to 2 mA cm^{-2}. The semi-3D cell with sub-micron-thick modified CuS cathode can withstand 100C rate and can be charged in 0.6 min. Under these conditions, however, the capacity of battery is only 30% of its initial value.

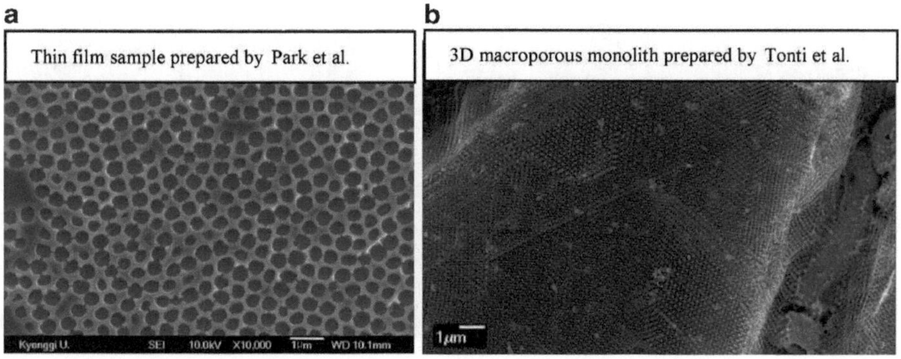

Fig. 10.8 SEM images of 3D macroporous $LiMn_2O_4$ ((**a**) Reprinted from [17], Copyright (2011), from Springer. (**b**) Reprinted from [16], Copyright (2011), from American Chemical Society)

An approach has been presented by Tonti et al. [16] and Park et al. [17] involving the concept of using three-dimensionally macroporous $LiMn_2O_4$ as cathodes for 3D microbatteries. These preparations typically start with the fabrication of an opal template using polystyrene spheres. This template is then filled with a sol–gel preparation mixture for $LiMn_2O_4$. The composite is then heated in air to form $LiMn_2O_4$, and the polystyrene beads are removed via combustion. A scheme presented by Park et al. for this synthesis is shown in Fig. 10.7.

The samples prepared using this approach are shown in Fig. 10.8 A high degree of order can be seen with a hole size of ~500 μm. These inverse opal structures give a large area gain per layer of spheres; with each added layer, an increase in surface area of pi is given.

Park et al. have shown that there is a strong dependence on capacity as a result of the use of different preparation conditions. The best electrochemical performance was noted for the films prepared with a synthesis temperature of 800°C. They also showed that the adhesion of the film to the substrate was extremely good even after cycling, and little or no damage to the film was observed on cycling. Tonti et al. tested their films in an aqueous electrolyte and observed extremely high rates of discharge (50% DoD observed at 1,000C). The capacity retention on cycling was also reported to be amongst the highest seen in the literature.

10.2.3 Semi-3D Anodes

The following section will now look at relevant examples of 3D lithium battery anodes. These are fabricated using similar approaches to the 3D cathodes and are again cycled with a conventional lithium battery electrolyte versus a planar negative electrode.

Teixidor et al. [18] presented the fabrication and characterisation of carbon pillars as electrodes for lithium-ion microbatteries. The authors used lithographic

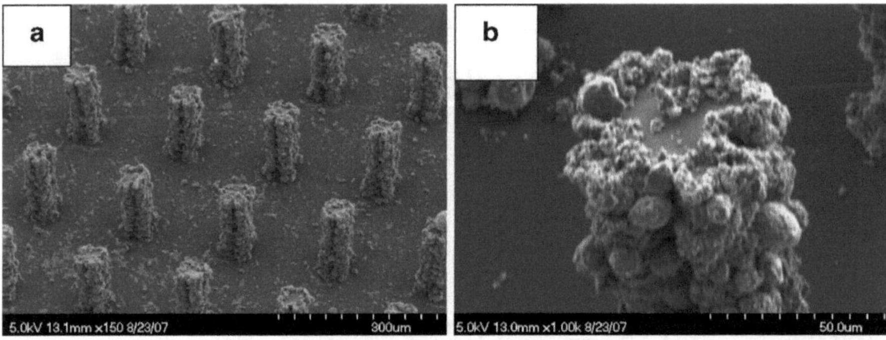

Fig. 10.9 Hybrid carbon-MEMS/MCMB electrodes (Reprinted from [19], Copyright (2011), from Elsevier)

patterning and subsequent pyrolysis of the cross-linked photoresist to produce a variety of different-shaped carbon pillar electrodes similar to that of the work by Min et al. [19] The mechanical 'robustness' of various different pillar geometries was compared. MCMBs were spin coated onto the carbon pillars, either dispersed in a solvent or dispersed in photoresist. By using the photoresist as a solvent, it was possible to enhance the adhesion of the MCMB particles onto the polymer microstructures by curing the dispersion with UV light. Figure 10.9 contains SEM images of the MCMB-coated carbon pillars.

The effect of the fabrication conditions, i.e. the pyrolysis temperature, heating rate and mass loading of MCMBs on the carbon pillars, was studied in detail. The carbon pillar electrodes that showed the greatest capacities (mAh cm^{-2}) were, not surprisingly, for the pillars with the densest coating of MCMBs. However, when the authors compared the gravimetric capacity for electrodes coated in MCMBs to those not coated with MCMBs, the not coated samples showed the highest capacities. The loss in gravimetric capacity for the MCMB-coated samples was attributed to a thickness effect limiting the discharge of the MCMB coatings.

Fleischauer et al. [20] have used a high-vacuum physical vapour deposition technique to deposit porous thin films of high aspect ratio silicon posts. Figure 10.10 shows the film of silicon posts roughly 500 nm in height deposited on a single crystal silicon wafer (the deposited silicon was shown to be amorphous by XRD).

Electrochemical characterisation was performed by galvanostatic cycling of the silicon-based half-cells; stainless steel discs were used as the silicon current collector along with a Celgard 2302 microporous polypropylene separator, 1 M $LiPF_6$ in EC:DMC (1:2 vol:vol) electrolyte and lithium foil as the counter electrode.

The half-cells were cycled at C rates of roughly C/8, C/4 and C/2 for 10 cycles at each C rate before repeating the pattern. Good capacity retention was reported (after the initial insertion) at all three rates, and most capacity loss associated with higher rate cycling was recovered as the C rate was reduced; the rate capability was attributed to the porosity of the silicon films. These results are summarised in Fig. 10.11.

Fig. 10.10 SEM of 500-nm-thick Si films deposited on a Si (100) wafer (Reprinted from [20], Copyright (2011), from The Electrochemical Society)

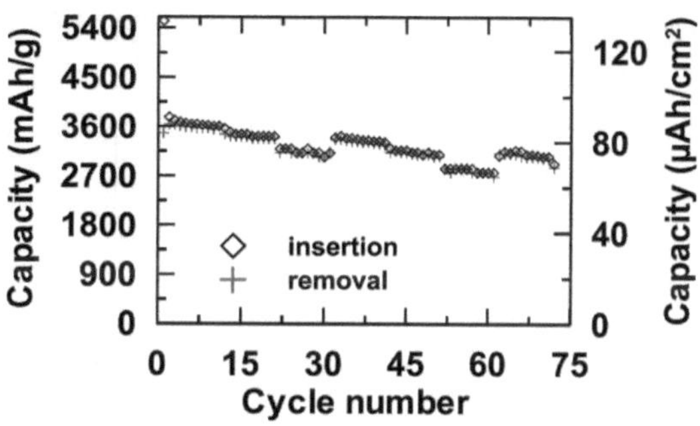

Fig. 10.11 Capacity as a function of cycle number in mAh g^{-1} (*left axis*) and µAh cm^{-2} (*right axis*) for half-cells of high aspect ratio silicon posts (Reprinted from [20], Copyright (2011), from The Electrochemical Society)

The authors finish by pointing out the porous thin films of silicon pillars had a capacity of over 90 µA h cm^{-2} and envisage higher capacities simply by increasing the thickness of the silicon film.

In the work by Simon et al., atomic layer deposition was used to deposit thin (~20 nm) and conformal layer of TiO_2 onto the nanorods [21]. The aluminium nanorod TiO_2 half-cells were cycled versus Li in a 1 M $LiClO_4$:PC electrolyte and showed a capacity (mA h cm^{-2}) roughly 10 times greater than the same 2D half-cell (TiO_2 deposited on aluminium plate). The system also showed good high-rate performance, showing 40% and 35% of slow cycling capacity (C/5) at 10 and 20C (6- and 3-min charges and discharges), respectively.

The same research group also investigated the deposition of CuSb electrodes onto the Cu nanorod substrates discussed earlier [22]. To prepare nanostructured Cu_2Sb active material, they electrodeposited Sb and alloyed this with the Cu from nanorod current collector. To promote the diffusion of the electroactive species within the 3D structure and thus to obtain a uniform coverage of the complex 3D surface of the Cu nanorod current collectors, the electrodeposition was performed using pulsed-current

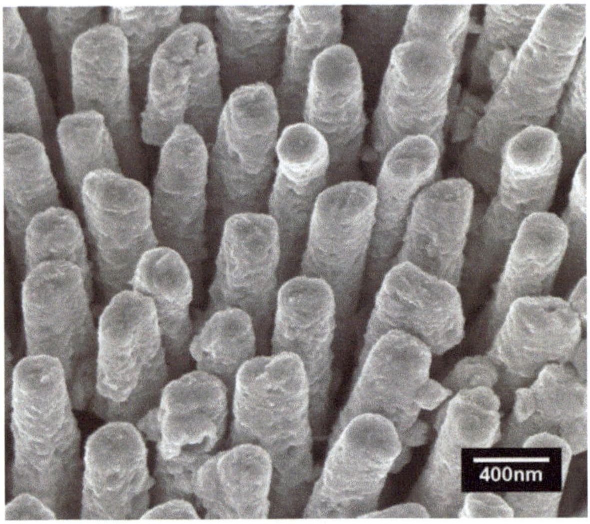

Fig. 10.12 SEM images of Cu nanorod current collectors coated with Sb under optimised current pulse conditions (Reprinted from [22], Copyright (2011), from Cambridge University Press)

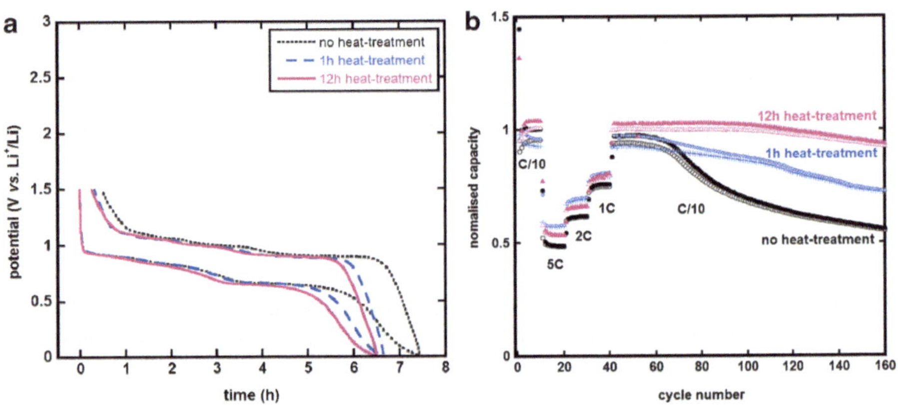

Fig. 10.13 (**a**) Cycling profile (fifth cycle) and (**b**) capacity retention upon cycling of 3D Cu_2Sb electrodes annealed at 120 °C for 1 and 12 h. The non-annealed electrodes are also compared (Reprinted from [22], Copyright (2011), from Cambridge University Press)

steps rather than a simple galvanostatic technique. Homogeneous and conformal Sb deposits (Fig. 10.12) were obtained under the optimised conditions.

The best performance was observed when an annealing step was used. After the electrodeposition, a thermal annealing at 120°C in vacuum was used to promote the full alloying of Sb with the Cu current collectors. Independent of the amount of elemental Cu or Sb, only the Cu_2Sb phase is expected to form. The voltage profile and capacity stability of 3D electrodes annealed for 1 and 12 h are shown and compared with non-annealed 3D Cu_2Sb electrode (Fig. 10.13). The typical plateaus observed during lithiation and delithiation of Cu_2Sb were visible, and the plateau observed during charge suggests that Cu_2Sb is formed back at the end of the charge. The capacity retention upon cycling is greatly improved (at least doubled) by the annealing step and with the annealing time. The complete formation of the alloy Cu_2Sb which presents lower volume expansion percentage than pure Sb and probably the extended availability of Cu to be re-inserted in the structure are most likely the reasons for the observed increase of the 3D electrode cycling life. The capacities observed were around 300 μA h cm^{-2}.

10.2.4 Semi-3D Cathode and 3D Anodes Combined

The logical progression after the development of 3D cathodes and anodes for lithium batteries is to try and combine these two components. This section looks at a few examples where this has been attempted. The 3D cathodes and anodes are combined using a conventional separator and electrolyte, and there is no interdigitation.

Chamran et al. [23] have proposed a microbattery system based on vanadium oxide nanoroll (VONR) and mesoporous carbon microbead (MCMB) electrode arrays.

The electrode arrays were formed from colloidal suspensions of the active material, a PVDF binder and solvent. The suspension was forced into a silicone template (prepared by 'photo-assisted anodic etching') using a combination of overpressure, provided by syringing the electrode material from one side of the template, and negative pressure, from application of a vacuum at the other (see Fig. 10.14).

Silver epoxy was spread on the side of the template, acting as a current collector and also providing mechanical support for the electrode pillars. The Si template was removed via etching in XeF_2, leaving the pillars of electrode material. The pillars of VONR and MCMB were deposited in separate moulds on separate substrates. The two were then combined with the addition of a separator similar to a conventional battery. The area capacity (mAh cm^{-2}) gain for the 3D VONR half-cells is quoted as being five times that of the 2D system.

Another approach used by Kim et al. [9] was markedly different to that of most other papers on the fabrication of the lithium microbatteries. Here, a laser was used to deposit thick films of porous battery materials on to metallic current collectors, which were separated by a gel polymer electrolyte. Cathode and anode inks ($LiCoO_2$ and mesoporous carbon microbead (MCMB), respectively) were deposited onto their respective current collectors using the described laser printing process. The gel polymer electrolyte used was PVdF–HFP/1 M $LiPF_6$ in propylene carbonate (PC)/ethylene carbonate (EC)/dimethyl carbonate (DMC) (1:1:3).

The resulting microbattery structure is shown in Fig. 10.15.

The electrochemical characterisation of the microbattery showed gravimetric capacities as expected for the electrode materials within the inks and that the capacity per footprint area was dependent upon the electrode thickness, Fig. 10.16.

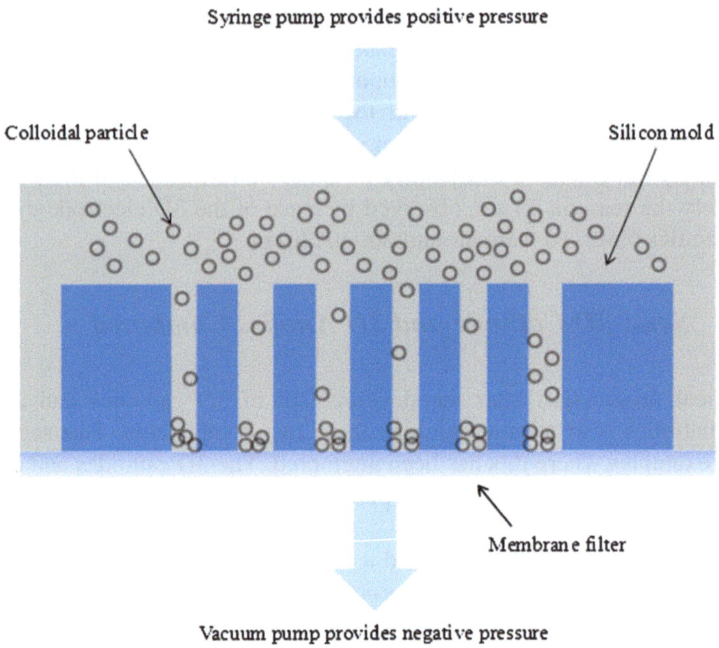

Fig. 10.14 Schematic of colloidal process using forced infiltration of the suspension in the silicon mould [24]

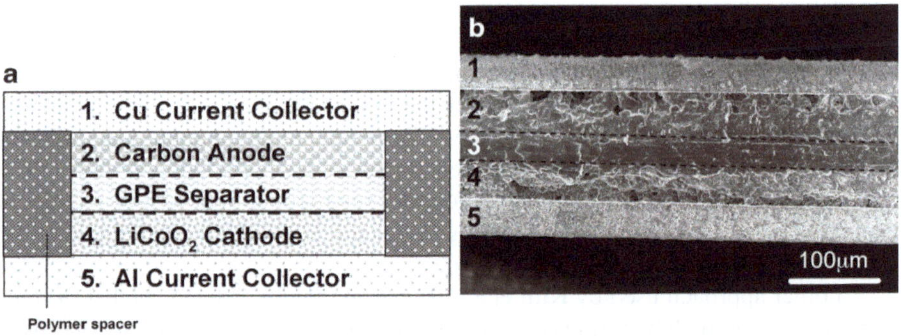

Fig. 10.15 (**a**) Cross-sectional schematic diagram (not to scale) of a typical lithium-ion microbattery used for this work. (**b**) Cross-sectional SEM micrograph of a packaged thick-film lithium-ion microbattery. The layers visible (from top to bottom) are *(1)* Cu current collector, *(2)* carbon anode, *(3)* GPE soaked separator, *(4)* $LiCoO_2$ cathode and *(5)* Al current collector. The GPE soaked separator is marked by *black dashed lines* (Reprinted from [9], Copyright (2011), from Elsevier)

The rate performance of the material was shown to be independent of the electrode thickness, Fig. 10.16. This independence of electrode thickness is in contrast to thin-film sputtered cells where the resistance of the electrodes limits rates of discharge. The authors briefly compare the capacity ($\mu A\ h\ cm^{-2}$) of the

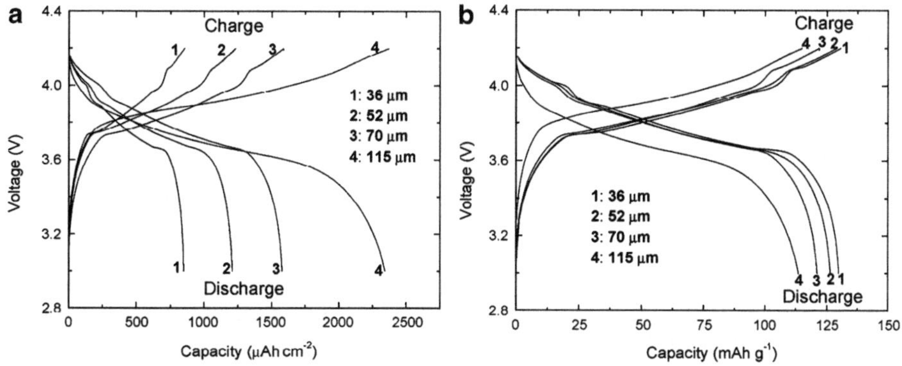

Fig. 10.16 Charge/discharge curves (fifth cycle) as a function of capacity per (**a**) active electrode area and (**b**) cathode mass for packaged lithium-ion microbatteries (Al/$LiCoO_2$/GPE/MCMB/Cu) with $LiCoO_2$ cathodes of different thicknesses (36, 52, 70 and 115 μm). Microbatteries are charged at a constant current (100_Acm^{-2}) between 4.2 and 3 V. The active electrode area is 0.49 cm^2 (Reprinted from [9], Copyright (2011), from Elsevier)

laser-printed thick-film microbattery to that of the thin-film sputtered microbattery and report for the laser-printed system the low-rate capacities are an order of magnitude greater.

10.3 The Full Interdigitated 3D Microbattery

The following section has shown that 3D architectured battery materials can be produced. Enhanced capacities per footprint area and rate improvements have also been shown. The final stage of development is to move to what can be referred to as a full interdigitated 3D microbattery system. In this category, the microscale cathode and anode components of a battery are fully interdigitated such that the diffusion distance of Li from the cathode to the anode is significantly reduced.

10.3.1 Three-Dimensional Microbattery Designs and Fabrication Methods

Several structures have been proposed as possibilities for this design shown in Fig. 10.17 [1, 2, 25]. They are all based on the five-layer concept of Fig. 10.1, in which the current collectors form two closely spaced interpenetrating networks and the electrode/electrolyte/electrode sandwich forms the interface separating the two current collectors shown in black and red. The power density may again be estimated from the Eqs. 10.1, 10.2, 10.3, 10.4, and 10.5 above, while recognising that neither the thickness of the electrode L_E nor the separator, L_S, can exceed the spacing between the positive and negative current collectors. Two main topologies can be distinguished as interdigitated or interlocked as shown in Fig. 10.17a and b.

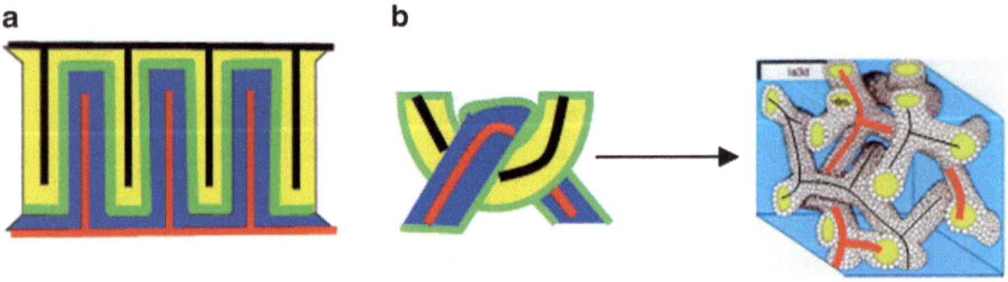

Fig. 10.17 Two topologies for the 3D microbattery. (**a**) Interdigitated or interlaced topology. (**b**) Interlocked or sponge topology where two electrode networks fill each other's pores (Reprinted with permission from The Royal Society of Chemistry)

Further differences arise from the detailed geometric arrangements and fabrication methods for the interdigitated or interlaced topology. The easiest concept to visualise is that of Fig. 10.6a where the two-dimensional diagram can represent either a cross section through an array of interpenetrating trench structures as in a thick version of the 2D cell described above or two arrays of interpenetrating columns as current collectors. In either case, the first two layers, an active material (cathode or anode) and an electrolyte/separator, are deposited conformally, leaving enough space for a current collector of the opposite polarity to the base. One problem here is to optimise the fabrication of the final two layers – in particular, how to ensure that the space left after deposition of the second active material is sufficient and precise to ensure continuity of the final current collector if it is required to compensate for a poorly conductive active material. Specific example structures that have been proposed as possibilities for these kinds of structures are shown in Fig. 10.18 [1, 2, 25].

Concentric – Fig. 10.18 shows a 3D ordered array of pillars used as a substrate; this is coated with layers of cathode, electrolyte and anode. A further current collector is added, and the battery stack is formed. Carbon microelectromechanical systems [26] and perforated silicon [27] substrates have been used extensively for this particular configuration.

Interdigitated cylindrical – This design consists of cathode and anode pillars arranged in an interdigitated format as shown in Fig. 10.18. The space between the pillars is then filled with electrolyte. Methods to fabricate microbatteries this way are typically attempted using carbon microelectromechanical systems [19].

Interdigitated plate – Normally, a trench-like structure (Fig. 10.18) is made in this design. Using this trench substrate, layers of cathode, electrolyte and anode are then coated on followed by a final current collector layer. This has typically been attempted in silicon substrates.

Aperiodic – An aperiodic structure is shown in Fig. 10.18; this is where there is no systematic ordering of the components. However, the three layers are microscopically mixed with the cathode and anode being separated by an electrolyte. This could be from the concentric coating of a metal or carbon foam.

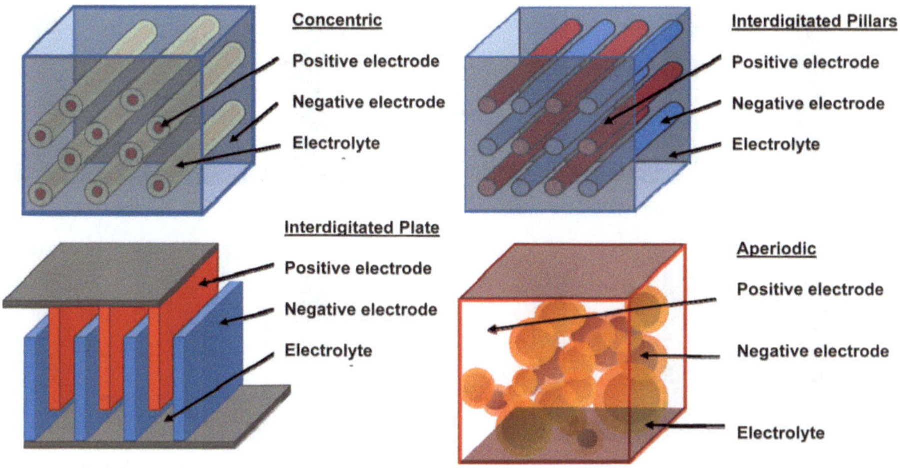

Fig. 10.18 Schematics of some of the most common 3D battery (Reprinted from [1], Copyright (2011), from The American Chemical Society)

All of the configurations shown in Fig. 10.18 possess the short diffusion path between the cathode, anode and electrolyte, allowing for high DoDs at short discharge times as predicted by the Eq. 10.8 and a much improved capacity per footprint area over thin-film cells.

Although there are several possibilities for fabrication of the microbattery, one possible route is shown in Fig. 10.19. The following discussion will critically examine the fabrication method outlining many of the relevant considerations. The first step is to design a microstructured high aspect ratio current collector. The literature is replete with examples of such structures [19, 28–31]. However, suitable candidates must meet a number of criteria:

- Enough room must exist in the structure to accommodate all the layers that will be deposited to make the 3D microbattery.
- The aspect ratio must be high enough such that a significant area gain is achieved.
- It should be electrochemically inert over the voltage range of operation (typically 2–5 V vs. Li for a cathode and 0–2 V vs. Li for an anode).
- The material must also be electronically conducting. (However, in some cases, an insulating high aspect ratio substrate coated in a conducting layer could be used.)

Once a suitable candidate for this substrate has been established, the first layer is deposited. Numerous techniques could be used for this including electrochemical deposition [15], vapour deposition techniques [31] and slurry deposition [32]. The deposit should again meet a number of crucial criteria:

- Electrochemically active over an appropriate potential range (typically 2–5 V vs. Li for a cathode and 0–2 V vs. Li for an anode).

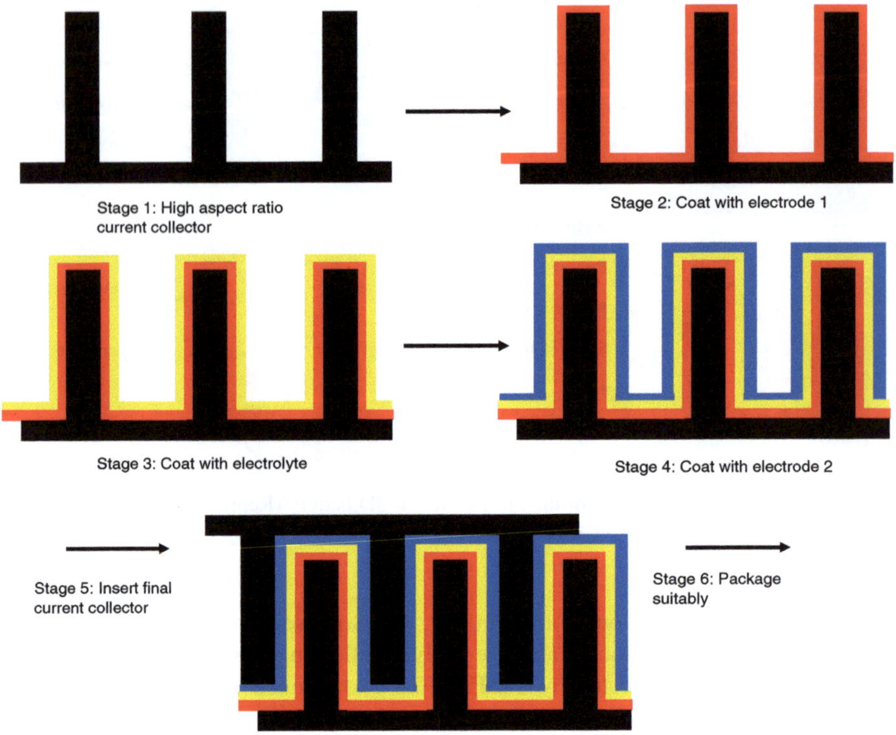

Fig. 10.19 Schematic of the general fabrication process for full 3D microbatteries

- Ideally conformal over the high aspect ratio substrate such that a suitable amount of room is still available for the remaining components.
- Compatibility with the remaining layers must also be considered.

Probably, the most critical phase in the fabrication process comes next, the application of an electrolyte layer [23, 32, 34]. The battery will only operate successfully if the following considerations are met:

- Conformal coating with no defects or pinholes
- Reasonably high ionic conductivities depending on the thickness of the layer ($10^{-3}$to 10^{-7} S cm^{-2})
- Electronically insulating
- Electrochemically stable over the operation potential window (between 0 and 5 V)
- Chemically compatible with the other battery components

The second electrode either the cathode or anode is coated onto the electrolyte layer, using one of the aforementioned techniques. The qualities that this electrode must possess are similar to the first electrode; however, if a cathode was deposited first, an anode must be second and vice versa.

The addition of a second current collector completes the fabrication of the stack. This could be done using methods such as vapour deposition technique, slurry deposition or electroless deposition. The material must also meet criteria c and d of the first current collector.

A final packaging step is required to complete the cell as a product. This is a crucial step as micro-sized hermetically sealed packaging must be developed. This packaging must also be chemically and electrochemically stable.

10.3.2 Examples of Full 3D Microbatteries

This section will look at a few examples of the 3D lithium microbattery systems that have been or are under development which were not part of the FP7 Superlion. A review of some of the most relevant papers regarding full 3D microbattery systems is followed by a review of 3D microbattery cathodes, anodes and modelling of these systems.

10.3.2.1 3D Lithium-Ion Microbattery Systems

Dokku et al. [35] created a system based on microarrays of gold current collectors, coated in $LiMn_2O_4/Li_{4/3}Ti_{5/3}O_4$ with a gel polymer electrolyte. Photolithography was used to pattern a SiO_2 substrate with microarrays of gold current collectors. Sol–gel precursors of $LiMn_2O_4$ and $Li_{4/3}Ti_{5/3}O_4$ were then deposited onto the current collectors, using a microinjection system, before the precursors were calcined to form the electrode materials. Thermal polymerisation of methyl methacrylate in the presence of 1 M $LiClO_4$ EC:DMC (1:1) electrolyte gave a sheet of gel polymer electrolyte based on poly(methyl methacrylate) (PMMA) and $LiClO_4$; this was placed onto the microarray of $LiMn_2O_4$ and $Li_{4/3}Ti_{5/3}O_4$, and a lithium foil was placed on top (Fig. 10.20). This fabrication technique differs somewhat from that proposed in Fig. 10.19 as two-dimensional interdigitated current collector tracks are laid down first. Then the cathode, anode and electrolyte are applied. Although the strategy may be slightly different, many of the considerations are similar.

The microelectrode arrays of $LiMn_2O_4$ and $Li_{4/3}Ti_{5/3}O_4$ were characterised individually, using the lithium foil as the counter electrode. The lithium insertion/extraction behaviour is shown in Fig. 10.21; the reversibility of the CVs was used to suggest good cyclability of the electrodes.

Dokku et al. then proceeded to test the microelectrode arrays against each other, using $LiMn_2O_4$ as the cathode and $Li_{4/3}Ti_{5/3}O_4$ as the anode. The cell showed charge and discharge plateaus of 2.55 and 2.4 V, respectively; the rate performance of the cell is shown in Fig. 10.22. It is worth noting here that a >50% DoD is seen at 50C (~1 min charge discharges).

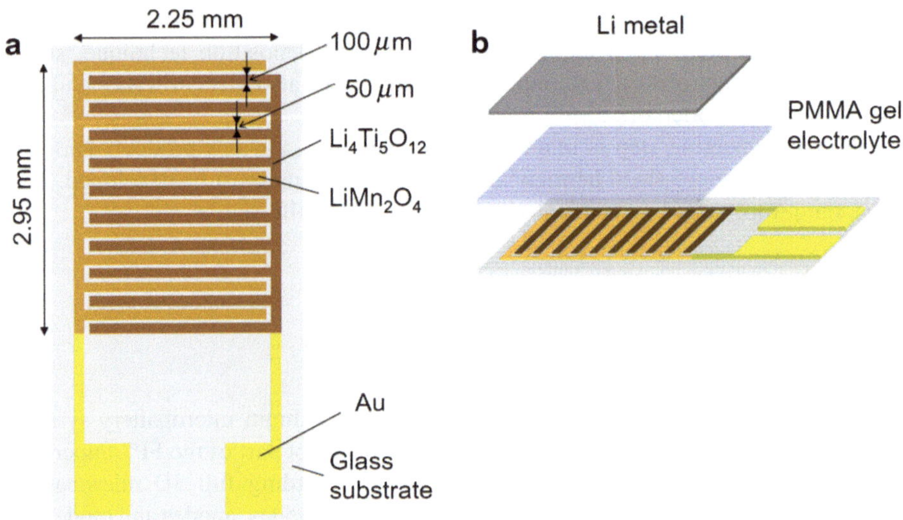

Fig. 10.20 Schematic illustrations of microarray electrodes of $LiMn_2O_4$ and $Li_{4/3}Ti_{5/3}O_4$ (**a**) and assembly of electrochemical cell (**b**) (Reprinted from [35], Copyright (2011), from Elsevier)

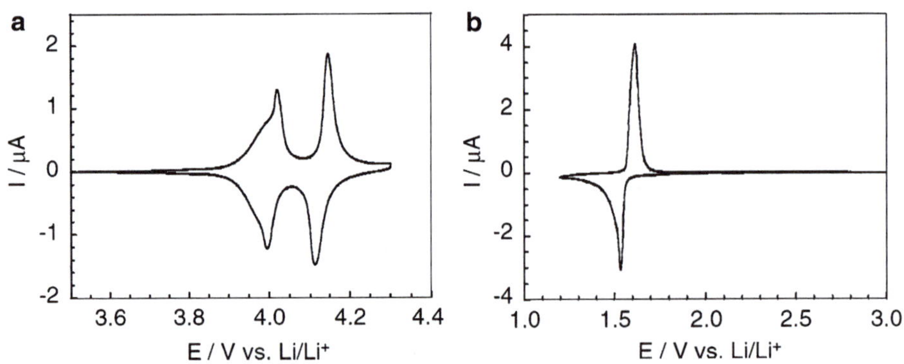

Fig. 10.21 Cyclic voltammograms of microarray electrodes of $LiMn_2O_4$ (**a**) and $Li_{4/3}Ti_{5/3}O_4$ (**b**) (Reprinted from [35], Copyright (2011), from Elsevier)

The authors noted that although the system had good rate performance, which they attributed to the ionic conductivity of the polymer electrolyte and the short diffusion path of the lithium ions, the energy density compared unfavourably to thin-film sputtered systems. Shortening the distance between the microelectrode arrays and increasing the thickness of the electrodes were suggested as possible methods of improving the energy density.

Work carried out by Nathan et al. [32] focused on the deposition of conformal battery materials on glass or silicon 'microchannel plates' (MCP, essentially silicon or glass wafers perforated by a regular array of microchannels). This work was

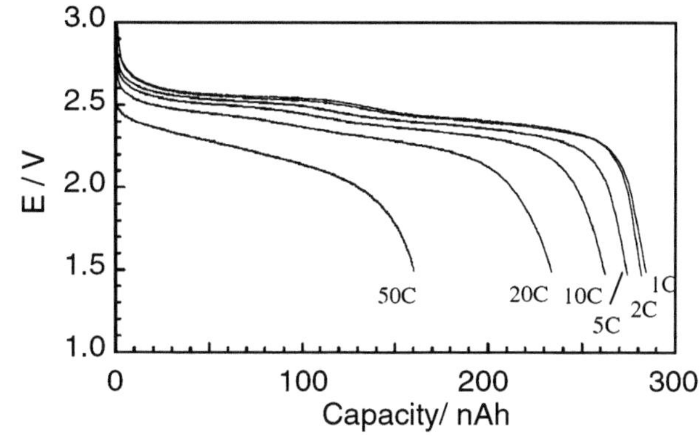

Fig. 10.22 Discharge capacity (nAh) plotted at varying C rate for the $LiMn_2O_4$/PMMA gel/$Li_{4/3}Ti_{5/3}O_4$ microbattery; the cell was charged at 1C for all rates (Reprinted from [35], Copyright (2011), from Elsevier)

probably the first literature example of a 'functioning full 3D' lithium-ion microbattery.

The structure of the 3D microbattery was based on high aspect ratio channels onto which thin films of cathode and polymer electrolyte were deposited; the remaining volume in the channel was filled with a slurry containing the anode. Fabrication began by depositing a current collector onto the silicon substrate, in this case Ni, onto which an electrodeposited cathode (molybdenum sulphide) was plated. The polymer electrolyte was based on PVDF and was deposited onto the molybdenum sulphide through the depth of the microchannel using what was described as 'sequential spin-coating and vacuum-pulling steps'. The anode was a slurry containing mesoporous microbeads (MCMBs), polymer binder and solvent (also deposited into the microchannel using sequential spin-coating and vacuum-pulling steps). Once constructed, the whole assembly was soaked in 1 M $LiPF_6$ 1EC:1DEC or 1 M $LiBF_4$ 1 EC:9 DEC under vacuum for 10 h. Lithium foil placed on top of the structure provided lithium intercalation into the anode. One question that has arisen with this 3D microbattery structure is how to make electronic connection both anode and cathode independently; in this case, the anode and polymer electrolyte were back polished to reveal the cathode/Ni current collector. Figure 10.23 shows the schematic 3D microbattery construction. This design exactly follows the preparation scheme described in Fig. 10.19.

The 3D microchannel plate (molybdenum sulphide/polymer electrolyte/MCMB soaked in 1 M $LiPF_6$ 1EC:1DEC or 1M $LiBF_4$ 1 EC:9 DEC) and the equivalent 2D planar microbatteries were tested in parallel by galvanostatic cycling, between 2.2 and 1.3 V versus lithium, in order to compare the available capacity (mAh/cm^{-2}). The 3D system had much greater capacity (>1 mA h cm^2) than that of the 2D (~0.015 mA h cm^2), due to the area gain (the capacity gain directly corresponds to the increase in surface area). Although there was some indication of the rate performance of the 3D system, the paper does not show results comparing the capacity at high rate of the 3D MCP and 2D systems. It is worth noting the authors have published several papers based on similar microchannel plate systems [26, 27, 32].

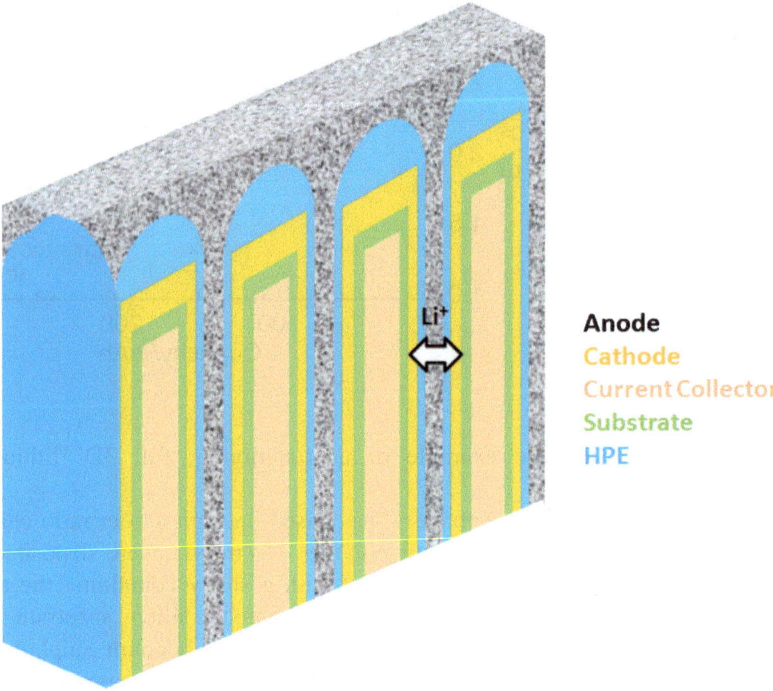

Fig. 10.23 Schematic view of a 3D microbattery showing the substrate (perforate silicon), current collector (Au of Ni), cathode (CuS_x, etc.), polymer electrolyte (HPE, hybrid polymer electrolyte) and anode

A paper reported by Notten et al. [31] focused on all-solid-state 3D microbatteries based on silicon. The difference between this and other proposed systems based on silicon was in the use of a solid-state electrolyte. The paper demonstrated the advantage of using the solid electrolyte LiPON over conventional liquid electrolytes for thin-film silicon anodes. Thin-film planar half-cells based on a silicon substrate, a barrier layer (TiN or TaN, to protect the silicon substrate from lithium insertion) and a silicon anode (50 nm) were cycled versus lithium using conventional organic lithium-ion and solid-state LiPON electrolytes. Evidence that the LiPON electrolyte suppressed SEI formation on the silicon was shown in the form of SEM images; this led to increased cycling stability, when compared to the organic lithium-ion electrolyte.

By using single crystalline silicon as highly ordered high-surface-area substrate/current collectors containing high aspect ratio pores (reactive ion etching), the authors suggested an all-solid-state 3D microbattery based on the coherent deposition of the various battery layers (barrier layer, solid-state electrolyte, cathode and second current collector) as visualised in Fig. 10.24. Although the 3D structure is suggested, the silicon cycling results shown in the paper referred only to the 2D thin-film cells.

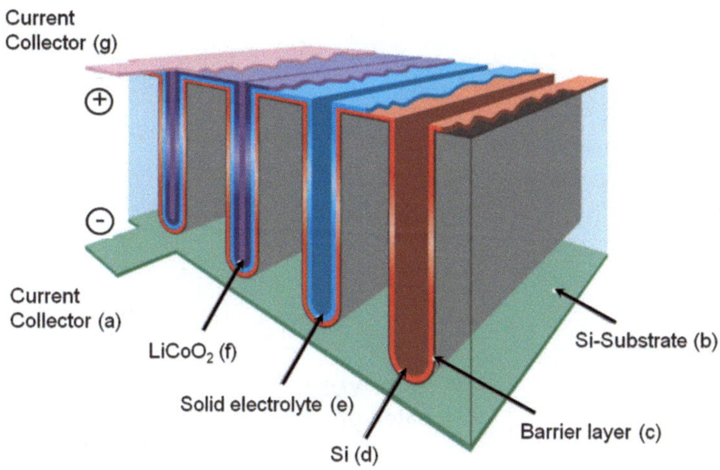

Fig. 10.24 3D integrated all-solid-state lithium-ion battery for which surface enlargement has been accomplished by electrochemical or reactive ion etching (RIE) of a silicon substrate (a) (Reprinted from [31], Copyright (2011), from John Wiley and Sons)

The carbon/polypyrrole [19] three-dimensional battery proposed by Min et al. was based on the high aspect ratio electrode array system proposed in previous papers by the same authors [36] and others [37]. Here, however, the authors proposed a slightly modified design based on polypyrrole as the cathode and carbon as the anode. The cathode and anode current collectors were both carbon and produced by photolithographic patterning of silicon wafers. The carbon tracks and high aspect ratio pillars were produced from pyrolysis of cross-linked polymer-based photoresists. Dodecylbenzenesulfonate-doped polypyrrole (PPYDBS) was electrodeposited onto one of current collector arrays to form a cathode; the second array of carbon pillars was used as the anode, and 1 M $LiClO_4$ in 1:1 EC-DMC electrolyte completed the battery.

Figure 10.25 shows the completed arrays of carbon and PPYDBS electrodes on their individual current collectors.

The authors compared the gravimetric capacity of half-cells of the 2D and 3D PPYDBS electrodes and found the 3D configuration to have slightly better performance (37.9 mAh g^{-1} at 1.15C for the 3D and 23.4 mAh g^{-1} at 0.9C for the 2D configuration). The increase in performance of the 3D PPYDBS electrodes was attributed to the larger active surface area and the effect of the electrolyte penetration into the entire electrode as compared to the planar front that the electrolyte makes with the 2D PPYDBS electrode.

The authors also presented results of a prototype full 3D microbattery system where the PPYDBS cathode was cycled against the carbon pillar anode. However, alongside a lower-than-expected capacity, the system was found to have problems with electronic short circuits, leading to self-discharge and limited discharge capacities (Fig. 10.26), and with large internal resistances attributed to the relatively high resistance of the carbon current collector arrays. Although the

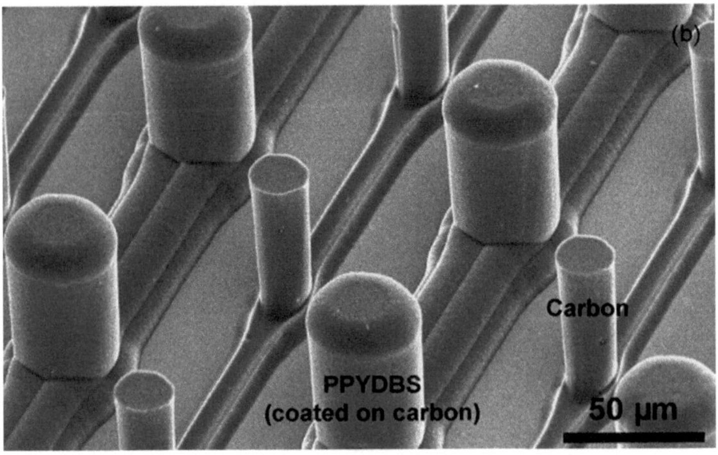

Fig. 10.25 Carbon and PPYDBS electrodes on individual current collector arrays (Reprinted from [19], Copyright (2011), from Elsevier)

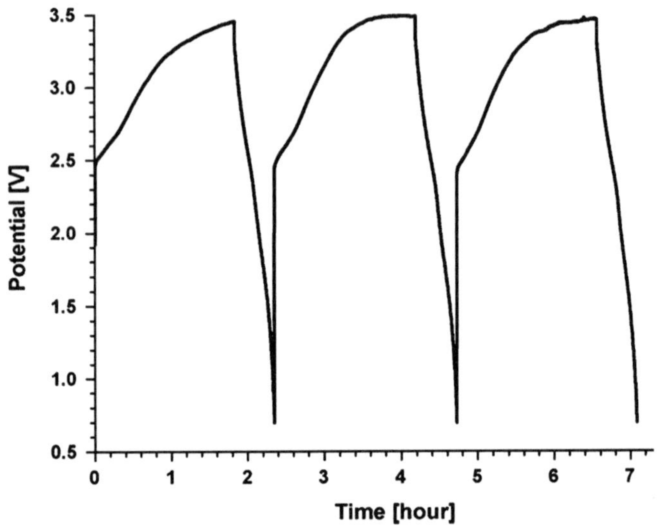

Fig. 10.26 Charge/discharge characteristic for the 3D C/PPYDBS interdigitated microbattery; the discharge current was 20 μAcm^{-2} (0.46C) and the charge current 90 μAcm^{-2} (2.1C) (Reprinted from [19], Copyright (2011), from Elsevier)

microbattery demonstrated showed some serious shortcomings, the work raised some interesting questions regarding the need for a separator in this type of microbattery configuration to exclude the possibility of shorts.

The lithium microbattery system proposed by Kotobuki et al. [38] was based around a 'honeycomb'-structured $Li_{0.35}La_{0.55}TiO_3$ (LLT) solid electrolyte. The schematic 'honeycomb'-type configuration is illustrated in Fig. 10.27.

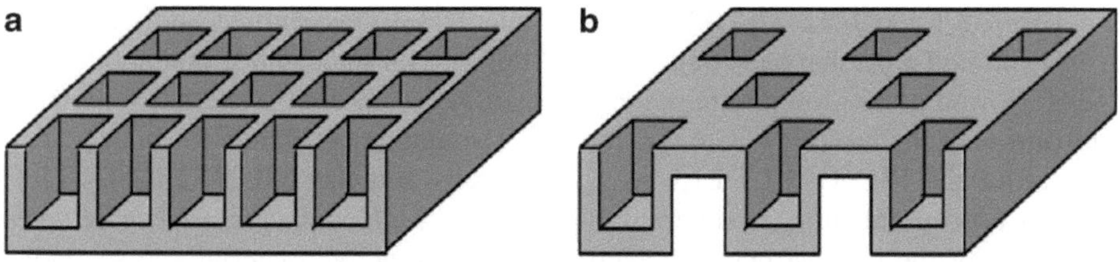

Fig. 10.27 Illustration of LLT honeycomb structures, (**a**) half honeycomb structure with 400 holes on one side of LLT membrane and (**b**) full honeycomb structure with 200 holes on each side of LLT membrane. The hole size is 180 × 180 × 1180 μm (Reprinted from [38], Copyright (2011), from The Electrochemical Society)

Fig. 10.28 Cross section of $LiCoO_2$/LLT/$Li_4Mn_5O_{12}$ cell (Reprinted from [38], Copyright (2011), from The Electrochemical Society)

Sol–gel precursors of the cathode and anode materials $LiCoO_2$ and $Li_4Mn_5O_{12}$ were injected (vacuum impregnation) into opposing sides of the microstructured electrolyte and subsequently calcined to form the full 3D microbattery (Fig. 10.28).

Cyclic voltammetry of the solid electrolyte/lithium half-cell showed the electrolyte to be stable in the region 2.5–5 V versus lithium. However, the CV showed some redox activity below 2.5 V versus Li, attributed to reduction of titanium in the $Li_{0.35}La_{0.55}TiO_3$. To avoid this problem, any breakdown of this electrolyte $Li_4Mn_5O_{12}$ was used as the anode (Li inserts into $Li_4Mn_5O_{12}$ 2.8 V vs. Li).

The sol–gel derived $LiCoO_2$ and $Li_4Mn_5O_{12}$ was cycled versus lithium and characterised according to redox potential and theoretical capacity (mA.h g^{-1}); the authors reported that both cathode and anode material performed according to expectations when tested in a normal cell configuration. When tested, the $LiCoO_2$/LLT/$Li_4Mn_5O_{12}$ microstructured half-cells performed notably worse than expected, 0.22 and 0.99 mA.h g^{-1} for $LiCoO_2$ and $Li_4Mn_5O_{12}$, respectively.

The full $LiCoO_2/LLT/Li_4Mn_5O_{12}$ was successfully assembled and tested; the cell exhibited a discharge voltage of ~1 V but as with the cathode and anode half-cells showed a very low discharge capacity of 7.3 μA.h cm^{-2} stated as only 0.1% utilisation of the limiting $LiCoO_2$ electrode. The authors attributed the poor performance of the $LiCoO_2/LLT/Li_4Mn_5O_{12}$ half-cells and the full 3D $LiCoO_2/LLT/Li_4Mn_5O_{12}$ microbattery to several factors; firstly, high contact resistance between the walls of the microstructured electrolyte and the active material and secondly, size of the 'honeycomb' electrolyte. The depth of the pores in the electrolyte was 180 μm, meaning a large diffusion distance of the lithium ion from the centre of the pore to the electrolyte. The authors noted that a reduction in the size of the electrolyte pore should improve the available capacity of the system.

10.3.2.2 Non-Li 3D Microbattery Systems

The work of Chamran et al. has demonstrated high performance from a 3D zinc air primary microbattery [39]. Initially, an array of Zn pillars was fabricated by electroplating Zn into a porous silicon template. This was then connected as a half-cell to a carbon and MnO_2 air cathode. This battery is reported to show enhanced rate capability (still utilising >90% of the available energy density at ~5C) over conventional systems (which are reported to show <10% the available energy density at ~0.1C). The use of such a device was also demonstrated by powering LEDs.

A 3D nickel–zinc battery was proposed by Chamran et al. [40] which had an overall electrochemical reaction shown in Eq. 10.9. This is a secondary non-lithium system:

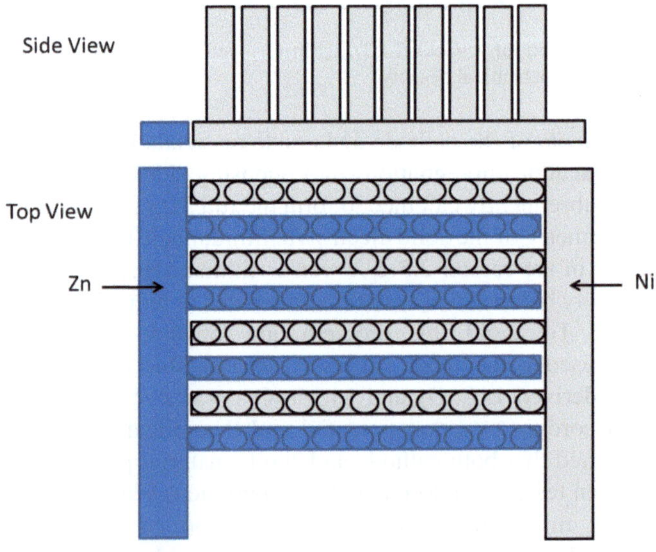

Fig. 10.29 Schematic of the nickel–zinc system

$$Zn + 2\ NiOOH + H_2O \leftrightarrow ZnO + 2Ni(OH)_2 \tag{10.9}$$

This system employed pillars of Ni and Zn grown on a 2D interdigitated current collector as shown in Fig. 10.29 This design was fabricated using a photoresist template base with Ni and Zn pillars grown through a silicon mask. This method produced batteries with a discharge voltage of between 1.75 and 1.65 V. The principle has been demonstrated over several cycles; however, significant corrosion of the Zn led to failure of the microbattery

10.4 Summary and Outlook

Many possible configurations and designs for 3D batteries, semi-3D batteries and 2D batteries have been introduced in this chapter. The principles of the design necessary to achieve high rates of discharge have also been described. The progression in designs from the thin- and thick-film conventional approaches through the semi-3D cathodes and anodes and finally to the full 3D microbattery are shown. The examples of the fabrication of these batteries have highlighted the rich variety of chemistries and techniques that are required. This field has rapidly grown in the last few years and will likely dramatically increase in the future. This introduction is written while these devices are in there infancy; however, the basic concepts and techniques will remain relevant and useful.

Acknowledgement The authors would like to thank the EU FP7 project Superlion for continued support.

References

1. Long JW, Dunn B, Rolison DR, White HS (2004) Three-dimensional battery architectures. Chem Rev 104(10):4463–4492
2. Rolison DR, Long JW, Lytle JC, Fischer AE, Rhodes CP, McEvoy TM, Bourg ME, Lubers AM (2009) Multifunctional 3D nanoarchitectures for energy storage and conversion. Chem Soc Rev 38(1):226–252
3. Vittorio SA (2001) http://www.csa.com/discoveryguides. [Online]. Available: http://www.csa.com/discoveryguides/mems/overview.php
4. Bates J, Dudney NJ, Neudecker B, Ueda A, Evans CD (2000) Thin-film lithium and lithium-ion batteries. Solid State Ionics 135(1–4):33–45
5. Souquet J, Duclot M (2002) Thin film lithium batteries. Solid State Ionics 148(3–4):375–379
6. Hart RW, White HS, Dunn B, Rolison DR (2003) 3-D microbatteries. Electrochem Commun 5(2):120–123
7. Jones S, Akridge J (1995) Development and performance of a rechargeable thin-film solid-state microbattery. J Power Sources 54(1):63–67
8. Xu K (2004) Nonaqueous liquid electrolytes for lithium-based re-chargeable batteries. Chem Rev 104(10):4303–4417

9. Kim H, Auyeung R, Pique A (2007) Laser-printed thick-film electrodes for solid-state rechargeable Li-ion microbatteries. J Power Sources 165(1):413–419
10. Owen JR (1997) Rechargeable lithium batteries. Chem Soc Rev 26:259–267
11. Johns P, Roberts M, Wakizaka Y, Sanders JH, Ow-en J (2009) How the electrolyte limits fast discharge in nanostructured batteries and supercapacitors. Electrochem Commun 11(11):2089–2092
12. Dong W, Rolison DR, Dunn B (1999) Electrochemical properties of high surface area vanadium oxide aerogels. Electrochem Solid-State Lett 3(10):457
13. Long JW, Swider-Lyons KE, Stroud RM, Rolison DR (2000) Design of pore and matter architectures in manganese oxide charge-storage materials. Electrochem Solid-State Lett 3(10):453
14. Shaijumon MM, Perre E, Daffos B, Taberna P, Tarascon J-M, Simon P (2010) Nanoarchitectured 3D cathodes for Li-ion microbatteries. Adv Mater 22:4978–4981
15. Mazor H, Golodnitsky D, Burstein L, Peled E (2009) High power copper sulfide cathodes for thin-film microbatteries. Electrochem Solid-State Lett 12(12):232–235
16. Tonti D, Torralvo MJ, Enciso E, Sobrados I, Sanz J (2008) Three-dimensionally ordered macroporous lithium manganese oxide for rechargeable lithium batteries. Chem Mater 20(14):4783–4790
17. Park BG, Kim S, Kim I-D, Park YJ (2010) Structural and electrochemical performance of three-dimensional LiMn2O4 thin film. J Mater Sci 45(14):3947–3953
18. Teixidor G, Zaouk R, Park B, Madou M (2008) Fabrication and characterization of three-dimensional carbon electrodes for lithium-ion batteries. J Power Sources 183(2):730–740
19. Min H-S, Park BY, Taherabadi L, Wang C, Yeh Y, Zaouk R, Madou MJ, Dunn B (2008) Fabrication and properties of a carbon/polypyrrole three-dimensional microbattery. J Power Sources 178(2):795–800
20. Fleischauer MD, Li J, Brett MJ (2009) Columnar thin films for three-dimensional microbatteries. J Electrochem Soc 156(1):A33
21. Cheah SK, Perre E, Rooth M, Fondell M, Hårsta A, Nyholm L, Boman M, Lu J, Simon P, Edstro K (2009) Self-supported three-dimensional nanoelectrodes for microbattery applications. Nano Lett 9(9):3230–3233
22. Perre E, Taberna PL, Mazouzi D, Poizot P, Gustafsson T, Edström K, Simon P (2010) Electrodeposited Cu2Sb as anode material for 3-dimensional Li-ion microbatteries. J Mater Res 25(8):1485–1491
23. Chamran F, Yeh Y, Min H-S, Dunn B, Kim C-J (2007) Fabrication of high-aspect-ratio electrode arrays for three-dimensional microbatteries. J Microelectromech Syst 16(4):844–852
24. Zadin V, Kasemägi H, Aabloo A, Brandell D (2010) Modelling electrode material utilization in the trench model 3D-microbattery by finite element analysis. J Power Sources 195(18):6218–6224
25. Dunn B, Kim CJ, Tolbert S (2010) Three-dimensional microbatteries for MEMS/NEMS technology. In: Proceedings of the 2010 IEEE 23rd international conference on micro electro mechanical systems (MEMS), Hong Kong, pp 164–167
26. Golodnitsky D, Nathan M, Yufit V, Strauss E, Freed-man K, Burstein L, Gladkich A, Peled E (2006) Progress in three-dimensional (3D) Li-ion microbatteries. Solid State Ionics 177(26–32):2811–2819
27. Golodnitsky D, Yufit V, Nathan M, Shechtman I, Ripenbein T, Strauss E, Menkin S, Peled E (2006) Advanced materials for the 3D microbattery. J Power Sources 153(2):281–287
28. Attard GS, Bartlett PN, Coleman NRB, Elliott JM, Wang JH (1997) Mesoporous platinum films from lyotropic liquid crystalline phases. Science 278(5339):838–840
29. Elliott JM, Attard GS, Bartlett PN, Coleman NRB, Merckel DAS, Owen JR (1999) Nanostructured platinum (H I -ePt) films: effects of electrodeposition conditions on film properties. Chem Mater 11(7):3602–3609
30. Ripenbein T, Golodnitsky D, Nathan M, Peled E (2009) Electroless nickel current collector for 3D-microbatteries. J Appl Electrochem 40(2):435–444

31. Notten PHL, Roozeboom F, Niessen RAH, Bag-getto L (2007) 3-D integrated all-solid-state rechargeable batteries. Adv Mater 19(24):4564–4567
32. Nathan M, Golodnitsky D, Yufit V, Strauss E, Ripenbein T, Shechtman I, Menkin S, Peled E (2005) Three-dimensional thin-film Li-ion microbatteries for autonomous MEMS. J Microelectromech Syst 14(5):879–885
33. Long JW, Rhodes CP, Young AL, Rolison DR (2003) Ultrathin, protective coatings of poly (o-phenylenediamine) as electrochemical proton gates: making mesoporous MnO 2 nanoarchitectures stable in acid electrolytes. Nano Lett 3(8):1155–1161
34. El-Enany G, Lacey MJ, Johns PA, Owen JR (2009) In situ growth of polymer electrolytes on lithium ion electrode surfaces. Electrochem Commun 11(12):2320–2323
35. Dokko K, Sugaya J, Nakano H, Yasukawa T, Matsue T, Kanamura K (2007) Sol–gel fabrication of lithium-ion micro-array battery. Electrochem Commun 9(5):857–862
36. Wang C, Taherabadi L, Jia G, Madou M, Yeh Y, Dunn B (2004) C-MEMS for the manufacture of 3D microbatteries. Electrochem Solid-State Lett 7(11):A435
37. Kinoshita K, Song X, Kim J, Inaba M (1999) Development of a carbon-based lithium microbattery. J Power Sources 81–82(1–2):170–175
38. Kotobuki M, Suzuki Y, Munakata H, Kanamura K, Sato Y, Yamamoto K, Yoshida T (2010) Fabrication of three-dimensional battery using ceramic electrolyte with honey-comb Structure by sol–gel process. J Electrochem Soc 157(4):A493
39. Chamran F, Min H, Dunn B, Kim CCJ (2007) Zinc-air microbattery with electrode array of zinc microposts, Mechanical and Aerospace Engineering Dept., 2 Materials Science and Engineering Dept. In: Proceedings of the Micro Electro Mechanical Systems, MEMS. IEEE 20th international conference, Kobe, pp 871–874
40. Chamran F, Min H, Dunn B (2006) Three-dimensional nickel-zinc microbatteries. In: Proceedings of the Micro Electro Mechanical Systems, MEMS 2006, Istanbul. 19th IEEE international conference, vol 2, pp 950–953

Index

Y. Abu-Lebdeh and I. Davidson (eds.), *Nanotechnology for Lithium-Ion Batteries*, Nanostructure Science and Technology, DOI 10.1007/978-1-4614-4605-7,

MIX
Papier aus verantwortungsvollen Quellen
Paper from responsible sources
FSC® C105338

If you have any concerns about our products,
you can contact us on
ProductSafety@springernature.com

In case Publisher is established outside the EU,
the EU authorized representative is:
Springer Nature Customer Service Center GmbH
Europaplatz 3, 69115 Heidelberg, Germany

Printed by Libri Plureos GmbH
in Hamburg, Germany